Certified Wireless Security Professional (CWSP)
Study and Reference Guide
(CWSP-206)

Copyright © 2019 by CertiTrek Publishing and CWNP, LLC. All rights reserved. Printed in the United States of America. Except as permitted under the United States Copyright Act of 1976, no part of this publication may be reproduced or distributed in any form or by any means or stored in a database or retrieval system, without the prior written permission of the publisher.

All trademarks or copyrights mentioned herein are the possession of their respective owners and CertiTrek Publishing makes no claim of ownership by the mention of products that contain these marks.

Errata, when available, for this study guide, can be found at: www.cwnp.com/errata/

First printing: September 2019, version 1.0

ISBN: 9780997629040

In addition to the authors of this book, listed in the About the Authors section of the Introduction, CWNP would like to say a special thanks to all those involved in the development of materials for CWSP-206 from the Job Task Analysis (JTA) through to materials review and feedback. These individuals include Ryan Adzima, Robert Bartz, Tom Carpenter, Brett Hill, Scott Lester, Manon Lessard, James Palmer, and Heather Williams. If we have left out your name it is only because so many helped and not because you were not appreciated. Many thanks to all of you.

Table of Contents

Table of Contents	iii
Introduction	iv
Extended Table of Contents	xii
Chapter 1: Security Fundamentals	1
Chapter 2: Wireless Security Challenges	45
Chapter 3: Security Policy	101
Chapter 4: Authentication	131
Chapter 5: Authentication and Key Management	193
Chapter 6: Encryption	221
Chapter 7: Security Design Scenarios	247
Chapter 8: Secure Roaming	279
Chapter 9: Network Monitoring	313
Chapter 10: WPA3 and OWE	367
Chapter 11: Penetration Testing	403
Appendix A: Up and Running with Kali Linux	441
Glossary: A CWNP Universal Glossary	443

Introduction

The Certified Wireless Security Professional (CWSP) is a WLAN subject matter expert (SME) who can assist in the creation and implementation of an organization's enforceable security policy by following applicable regulations, standards, and accepted best practices. This SME can identify and mitigate threats to a network. A CWSP effectively uses appropriate tools and procedures to ensure the ongoing security of the network.

The CWSP-206 exam consists of 60 multiple choice, single correct answer questions and is delivered through Pearson VUE. The candidate can register for the exam at the Pearson VUE website (https://home.pearsonvue.com/). The candidate will have 90 minutes to take the exam and must achieve a score of 70% or greater to earn the CWSP certification. If the candidate desires to become a Certified Wireless Network Trainer (CWNT) the passing score must be 80% or greater. A CWNT is authorized to teach official CWNP courses for certifications in which they hold the CWNT credential.

Book Features

The CWSP Certified Wireless Security Professional Study and Reference Guide includes the following features:

- End of chapter review questions. The review questions at the end of each chapter, when appropriate, are intended to help you ensure proper reading of the chapter. They are not intended to simulate exam questions and should not be assumed to be questions that will be presented on the exam.

- Notes with special indicators. The notes throughout the book fall into one of three categories, as outlined in Table i.1.

- CWNP official glossary. A glossary of terms provided at the end of the book that helps you as a reference while reading.

- Complete coverage of the CWSP-206 objectives. Every objective is covered in the book, and each chapter lists the major objective category covered within.

Icon	Description
	Note: A general note related to the current topic.
	Defined Note: A note providing a concise definition of a term or concept.
	Exam Note: A note providing tips for exam preparation.

Table i.1: Book Note Icons

Authors

The following individuals wrote one or more chapters in this book:

- Lee Badman – Primary Author
- Robert Bartz – Primary Author
- Tom Carpenter – Author and General Editor
- Brett Hill – WPA3
- Phil Morgan - Pentesting

All of these authors are CWNEs and experts in the field with more than fifty years of experience working with various wireless technologies among them.

At CWNP, we would like to thank all of these authors for their hard work and dedication in making this book a reality. You have brought significant value to thousands.

The authors would like to dedicate this book to future CWSPs: you and your passion for wireless are what makes this all worthwhile.

CWSP-206 Objectives

The CWSP-206 exam tests your knowledge against four knowledge domains as documented in Table i.2. The CWSP candidate should understand these domains before taking the exam. The CWSP-206 objectives follow.

Knowledge Domain	Percentage
Security Policy	10%
Vulnerabilities, Threats, and Attacks	30%
WLAN Security Design and Architecture	45%
Security Lifecycle Management	15%

Table i.2: CWSP-206 Exam Knowledge Domains with Percentage of Questions in Each Domain

1.0 Security Policy – 10%

1.1 Define WLAN security Requirements

- Evaluate and incorporate business, technical, and applicable regulatory policies (for example, PCI-DSS, HIPAA, GDPR, etc.)
- Involve appropriate stakeholders
- Review client devices and applications
- Review WLAN infrastructure devices

1.2 Develop WLAN security policies

- Translate security requirements to high-level policy statements
- Write policies conforming to common practices including definitions of enforcement and constraint specification
- Ensure appropriate approval and support for all policies
- Implement security policy lifecycle management

1.3 Ensure proper training is administered for all stakeholders related to security policies and ongoing security awareness

2.0 Vulnerabilities, Threats, and Attacks – 30%

2.1 Identify potential vulnerabilities and threats to determine the impact on the WLAN and supporting systems and verify, mitigate, and remediate them

- Use information sources to identify the latest vulnerabilities related to a WLAN including online repositories containing CVEs
- Determine the risk and impact of identified vulnerabilities
- Select appropriate actions to mitigate threats exposed by vulnerabilities
    - Review and adjust device configurations to ensure conformance with security policy
    - Implement appropriate code modifications, patches and upgrades
    - Quarantine unrepaired/compromised systems
    - Examine logs and network traffic where applicable
- Describe and detect possible, common WLAN attacks including eavesdropping, man-in-the-middle, cracking, phishing, and social engineering attacks
- Implement penetration testing procedures to identify weaknesses in the WLAN
    - Use appropriate penetration testing processes including scope definition, information gathering, scanning, attack, and documentation procedures
    - Select and use penetration testing tools including project documentation, scanners, hardware tools, Kali Linux, protocol analyzers, WLAN auditing tools (software and hardware)
- Implement network monitoring to identify attacks and potential vulnerabilities
    - Use appropriate tools for network monitoring including centralized monitoring, distributed monitoring, and Security Information Event Management (SIEM) systems
    - Implement mobile (temporary), integrated and overlay WIPS solutions to monitor security events

2.2 Describe and perform risk analysis and risk mitigation procedures

- Asset management
- Risk Ratings
- Loss expectancy calculations
- Develop risk management plans for WLANs

3.0 WLAN Security Design and Architecture – 45%

3.1 Select the appropriate security solution for a given implementation and ensure it is installed and configured according to policy requirements

- Select and implement appropriate authentication solutions
    - WPA/WPA2-Personal (Pre-Shared Key)
    - WPA/WPA2-Enterprise
    - WPA3-SAE and 192-Bit enterprise security
    - 802.1X/EAP
    - Understand the capabilities of EAP methods including EAP-TLS, EAP-TTLS, PEAP, EAP-FAST, EAP-SIM, and EAP-GTC
    - Guest access authentication
- Select and implement appropriate encryption solutions
    - Encryption methods and concepts
    - TKIP/RC4
    - CCMP/AES
    - SAE and 192-bit security
    - OWE
    - Virtual Private Network (VPN)
- Select and implement wireless monitoring solutions
    - Wireless Intrusion Prevention System (WIPS) - overlay and integrated
    - Laptop-based monitoring with protocol and spectrum analyzers
- Understand and explain 802.11 Authentication and Key Management (AKM) components and processes
    - Encryption keys and key hierarchies
    - Handshakes and exchanges (4-way, SAE, OWE)
    - Pre-shared keys
    - Pre-RSNA security (WEP and 802.11 Shared Key authentication)
    - TSN security
    - RSN security
    - WPA, WPA2, and WPA3

3.2 Implement or recommend appropriate wired security configurations to support the WLAN

- Physical port security in Ethernet switches

- Network segmentation, VLANs, and layered security solutions
- Tunneling protocols and connections
- Access Control Lists (ACLs)
- Firewalls

3.3 Implement authentication and security services

- Role-Based Access Control (RBAC)
- Certificate Authorities (CAs)
- AAA Servers
- Client onboarding
- Network Access Control (NAC)
- BYOD and MDM

3.4 Implement secure transitioning (roaming) solutions

- 802.11r Fast BSS Transition (FT)
- Opportunistic Key Caching (OKC)
- Pre-Shared Key (PSK) - standard and per-user

3.5 Secure public access and/or open networks

- Guest access
- Peer-to-peer connectivity
- Captive portals
- Hotspot 2.0/Passpoint

3.6 Implement preventative measures required for common vulnerabilities associated with wireless infrastructure devices and avoid weak security solutions

- Weak/default passwords
- Misconfiguration
- Firmware/software updates
- HTTP-based administration interface access
- Telnet-based administration interface access
- Older SNMP protocols such as SNMPv1 and SNMPv2

4.0 Security Lifecycle Management – 15%

4.1 Understand and implement management within the security lifecycle of identify, assess, protect, and monitor

- Identify technologies being introduced to the WLAN
- Assess security requirements for new technologies
- Implement appropriate protective measures for new technologies and validate the security of the measures
- Monitor and audit the new technologies for security compliance (Security Information Event Management (SIEM), portable audits, infrastructure-based audits, WIPS/WIDS)

4.2 Use effective change management procedures including documentation, approval, and notifications

4.3 Use information from monitoring solutions for load observation and forecasting of future requirements to comply with security policy

4.4 Implement appropriate maintenance procedures including license management, software/code upgrades, and configuration management

4.5 Implement effective auditing procedures to perform audits, analyze results, and generate reports

- User interviews
- Vulnerability scans
- Reviewing access controls
- Penetration testing
- System log analysis
- Report findings to management and support professionals as appropriate

Each chapter of the CWSP Study and Reference Guide lists the major objectives (x.x level objectives) covered in that chapter on the chapter title page.

Extended Table of Contents

Table of Contents --- iii

Introduction --- iv

    Book Features --- v

    Authors --- vi

    CWSP-206 Objectives --- vii

Extended Table of Contents --- xii

Chapter 1: Security Fundamentals --- 1

    Objectives Covered --- 1

    A Brief History of Wireless Security --- 2

    RF Does not Respect Boundaries --- 5

    Usage Threat Assessment --- 7

    Network Extension --- 8

    CWNA Security Review --- 10

    Industry Organizations --- 15

    Terminology Review --- 19

    Home Office Security --- 21

    Small Business Security --- 22

    Large Enterprise Security --- 22

    Public Network Security --- 24

    Remote Access Security --- 25

    Security and the OSI Model --- 26

    Security Analysis Basics --- 29

    Chapter Review --- 39

    Review Questions --- 40

    Review Answers --- 43

Chapter 2: Wireless Security Challenges — 45

- Objective Covered — 45
- Passive WLAN Discovery — 46
- Active WLAN Discovery — 48
- Discovery Hardware — 50
- Discovery Software — 53
- Weakest Link — 54
- SSID Hiding — 57
- MAC Address Filtering — 60
- Open System Authentication — 64
- Wired Equivalent Privacy (WEP) — 65
- Shared Key Authentication — 68
- Extensible Authentication Protocol (EAP) — 69
- Eavesdropping — 72
- Social Engineering — 76
- RF DoS — 79
- Layer 2 (MAC) DoS — 81
- Peer-to-Peer — 82
- Man in the Middle (MITM) — 83
- Management Interface Exploits — 86
- Authentication Cracking — 87
- Encryption Cracking — 88
- Other Common Concerns — 90
- General Recommended Practices — 92
- Chapter Summary — 94
- Review Questions — 95

Review Answers —99
Chapter 3: Security Policy —101
    Objectives Covered —101
    Security Policy Defined —102
    Regulations —104
    Legal Considerations —105
    Policy Importance —106
    Risk Assessment —109
    Document and Define —111
    Buy-In and Training —111
    Response —112
    Enforcement —113
    Monitor and Audit —115
    Review and Revise —116
    Password Policy —117
    Additional Policies —118
    Security Baselines —119
    Device Management —120
    BYOD Policy —121
    Social Networking Policy —124
    Chapter Summary —125
    Review Questions —126
    Review Answers —130
Chapter 4: Authentication —131
    Objectives Covered —131
    Authentication —132

    Choosing the Right Credentials — 136
    Passphrase-based Security — 140
    WPA or WPA2 Personal — 144
    AAA — 153
    Mutual Authentication — 155
    Authorization — 157
    Accounting — 160
    Network Access Control (NAC) — 161
    WPA & WPA2 Enterprise — 162
    Single Building or Campus — 170
    Branch Option #1 — 170
    Branch Option #2 — 171
    IEEE 802.1X/EAP Framework — 174
    Chapter Summary — 186
    Review Questions — 188
    Review Answers — 192

Chapter 5: Authentication and Key Management — 193
    Objectives Covered — 193
    Terminology — 194
    Pre-Robust Security Networks — 196
    Robust Security Networks (RSN) — 197
    802.11 Association — 202
    Key Hierarchy — 203
    4-Way Handshake Frames — 212
    Group Key Handshake — 213
    Chapter Summary — 214

Review Questions	215
Review Answers	219

Chapter 6: Encryption — 221
Objectives Covered	221
Terminology	222
Symmetric Key Encryption	223
Asymmetric Key Encryption	224
Stream Ciphers	226
Block Ciphers	227
Frame Encryption	228
Encryption and Decryption	229
Encryption Algorithms	230
WEP (Pre-RSNA)	231
TKIP (WPA)	236
CCMP (WPA2)	239
Chapter Summary	242
Review Questions	243
Review Answers	246

Chapter 7: Security Design Scenarios — 247
Objectives Covered	247
Virtual Private Networking Basics	248
Common VPN Protocols	250
VPN Functionality	252
Common Wireless VPN Uses	253
Tunneling and Split Tunneling	258
Public Access Networks	260

Captive Portals --- 263
Network Segmentation --- 266
BYOD --- 270
MDM --- 271
Client Management Strategies --- 272
Chapter Summary --- 273
Review Questions --- 274
Review Answers --- 278

Chapter 8: Secure Roaming --- 279
Objectives Covered --- 279
IEEE 802.11 Roaming Basics --- 280
PSK Roaming (WPA/WPA2) --- 283
Basic Roaming Review --- 288
Wi-Fi Voice-Personal Certification --- 290
Wi-Fi Voice-Enterprise Certification --- 292
Preauthentication --- 293
PMK Caching --- 295
Opportunistic Key Caching (OKC) --- 298
802.11-2016 (802.11r) Fast Transition (FT) --- 299
Single Channel Architecture --- 306
Chapter Summary --- 308
Review Questions --- 309
Review Answers --- 312

Chapter 9: Network Monitoring --- 313
Objectives Covered --- 313
Secure Management Protocols --- 314

xvii

WLAN Monitoring	316
Rogue AP and Client Detection	319
WIPS - Features	325
Enforcing Functional Policy	326
Security Monitoring	326
Reporting and Auditing	327
PCI Compliance as an Example	328
HIPAA Compliance as an Example	332
Auditing and Forensics	333
Audit Methods	334
Audit Tools	334
Enterprise WIPS Topology	336
Integrated Spectrum Analysis	346
New 802.11 Challenges	346
Monitoring in the Cloud	347
WNMS Security Features	348
WLAN Controllers	349
Distributed Protocol Analysis as a Monitoring Solution	349
IEEE 802.11 Frames vs. 802.3 Frames	351
Frame Exchanges	353
Spectrum Analysis	354
Physical Layer Defenses	355
Laptop-Based Intrusion Analysis	356
Specialty Analysis Devices	359
Chapter Summary	360
Review Questions	361

Review Answers --- 365
Chapter 10: WPA3 and OWE --- 367
Objectives Covered --- 367
The Need for WPA3 --- 369
WPA3 Core Components Explained --- 374
WPA3-Personal --- 377
WPA3-Enterprise --- 381
Wi-Fi Easy Connect --- 385
OWE (Enhanced Open) --- 387
Review Questions --- 395
Review Answers --- 401
Chapter 11: Penetration Testing --- 403
Objectives Covered --- 403
What About Your Hat? --- 404
Hacking Process Phases --- 405
Wireless Penetration Testing --- 407
Common Vulnerabilities and Exposures --- 411
National Vulnerability Database --- 412
CISA and US-CERT --- 414
Risk --- 416
Impact --- 417
White Box vs. Black Box --- 419
Hardware and Software Selection --- 420
Discovery --- 421
Promiscuous and Monitor Mode --- 421
Discovery Tools --- 422

Kali Linux Discovery Tools --- 423
Linux Wi-Fi Tools --- 424
Attacking --- 430
Chapter Summary --- 435
Review Questions --- 437
Review Answers --- 440
Appendix A: Up and Running with Kali Linux --- 441
Glossary: A CWNP Universal Glossary --- 443

Chapter 1: Security Fundamentals

Objectives Covered

3.6   Implement preventative measures required for common vulnerabilities associated with wireless infrastructure devices and avoid weak security solutions

4.2   Use effective change management procedures including documentation, approval, and notifications

There is no single correct approach to wireless security. How you approach real-world WLAN security as a wireless professional will be based on the particulars of each situation and will ultimately depend on the policies that guide operational goals for each implementation. It's important for those embarking on the CWSP journey to understand that we're never really done learning about security. The material you will learn in this course is comprehensive, but it has a shelf life. In a few years, some of the knowledge that you acquire here will "age out" and become obsolete in practice, while new methods and security standards will emerge. It has been this way since the inception of early 802.11. Furthermore, how WLAN-specific security fits into overall organizational security postures, also continuously evolves as organizations modernize their policies. It's important to realize that CWSP mastery will provide you with a formidable body of security knowledge, but also know that wireless security is just part of the bigger networking security story.

In this chapter, you will examine the foundational CWSP material at a high level, adding depth to it in subsequent chapters. Many topics in this book are purposefully repetitive in that they are introduced early in the text and then expanded on as we get deeper into the various concepts that use those topics. As with the CWNA material, few CWSP topics stand-alone, suitable to discuss once and then put aside. The interdependencies of wireless security require us to revisit the same fundamental principles often as we continue to tie them all together throughout the text to build the full breadth of modern Wireless Local Area Network (WLAN) security concerns and options.

In the process of studying CWSP, you will explore the importance of policy in shaping wireless security solutions, security basics, and you will briefly review the security knowledge gained from CWNA to refresh that material. You will examine organizations that shape the WLAN landscape, important WLAN security terminology, and some of the more common vulnerabilities associated with 802.11 networks. Our first chapter introduces a range of topics and serves as a preview of the rest of the book.

A Brief History of Wireless Security

We are currently operating in an exciting time for wireless, with a rising IoT tide, 802.11ax and WPA3 taking root, and ever more client devices finding their way to the wireless network. Almost any modern wireless vendor pitch will eventually touch on

Artificial Intelligence (AI) and Software-Defined functionality, but we can't really embrace the security ramifications of today's Wi-Fi without reviewing where the industry started and how it has developed from the security perspective.

Any WLAN implementation should be designed with a secure foundation that provides Confidentiality, Integrity, and Availability. You can easily remember these functional pieces from the acronym CIA (CIA may not be applied in public open network scenarios, but understanding the trade-offs of forgoing CIA for ease-of-use is equally important). Maintaining secure network communications is as important in wireless networking as it is anywhere else in the IT realm. In the early days of standards-based wireless networking, the notion of what constituted good wireless security was flawed, and the only option to secure WLAN communications was Wired Equivalent Privacy (WEP). A 40-bit key was used to protect the wireless network from casual eavesdropping. In addition to the key, WEP also used a 24-bit Initialization Vector (IV) as part of the encryption and decryption process. This 24-bit IV was relatively short, cryptographically speaking, allowing the IV to be reused with the same key and therefore causing WEP to be vulnerable to intrusion if enough frames with unique IVs were captured. You'll often see the 40-bit key and the 24-bit initialization vector together referred to as the 64-bit WEP key. Though optional, some manufacturers allowed for the use of a 104-bit key. Again, the 24-bit IV was used, and it was common to see the two components referred to as the 128-bit WEP key. The developers of the early standard meant well but could not predict that hacking wireless networks would fast become a blood sport that would show just how weak WEP was.

Several high-profile hacks of WEP fueled the evolution of standards-based WLAN security, and eventually lead to the adoption of the 802.11i amendment to the standard. 802.11i has since been rolled into 802.11-2012 and carried into 802.11-2016. This amendment provided the still-current concept of the robust security network association (RSNA). An RSNA is defined as an association between a pair of stations (STAs), which includes a 4-way handshake between the STAs. As per the 802.11 standard, a STA is defined as a "logical entity that is a singly addressable instance of a medium access control (MAC) and physical layer (PHY) interface to the wireless medium (WM)." The STA designator includes stations that are either APs or client devices, which is an important point to remember. An RSNA does not allow the use of legacy 802.11 Shared Key Authentication and only allows devices to connect to the network using 802.11 Open System Authentication. We'll talk about both of these concepts and what else is needed for RSNA later in the book.

Eventually, the 802.11i security amendment also introduced a new term: Pre-RSNA. It's important to note that the Pre-RSNA networks allow use of the legacy WEP cipher suite (using the Rivest Cipher 4 (RC4) algorithm) for data confidentiality, 802.11 Open System or Shared Key authentication methods and a single, weak Integrity Check Value (ICV) algorithm. Again, we've already established that WEP was weak; therefore, Pre-RSNA is also weak.

With the advent of the 802.11-2012 standard (which was superseded by 802.11-2016), two classes of security algorithms used with standard-based 802.11 wireless networking were defined:

1. Robust Security Networks (RSNs) – which will allow only RSNAs and do not allow WEP

2. Pre-RSNA Networks – which do allow WEP

You might wonder why WEP was carried forward to Pre-RSNA and why Pre-RSNA was kept around as an option when RSNA was introduced. Recall from your CWNA studies that the notion of backward compatibility is prevalent in 802.11. This reality is very much the case when it comes to wireless security as well and is a double-edged sword pitting convenience versus security. Although WEP has long-since been broken, it's still a selectable (though ill-advised) option in many of the newest WLAN systems and this point is often lamented by wireless professionals.

 The 802.11-2016 standard (which supersedes 802.11-2012) allows STAs to operate simultaneously with pre-RSNA and RSNA algorithms, but RSNA forbids the use of Shared Key 802.11 authentication, which Pre-RSNA allows. For RSNA, only the Open System Authentication mechanism can be used.

The 802.11-2016 standard RSNA defines several security features in addition to those of pre-RSNA networks, including:

4

- Enhanced authentication mechanisms for STAs
- Key management (generation and distribution) algorithms
- Strong cryptographic key establishment
- Enhanced cipher suite solution in Counter Mode with Cipher-Block Chaining Message Authentication Code Protocol (CCMP) with the use of Advanced Encryption Standard (AES)
- An optional, transitional cipher suite, Temporal Key Integrity Protocol (TKIP) with the use of RC4
- Fast basic service set (BSS) transition (FT) mechanisms
- Enhanced cryptographic encapsulation mechanisms for robust management frames

Transitional security networks (TSNs) allows for both RSNA-level security and Pre-RSNA-level security. A TSN is identified by the indication in the Robust Security Network information element (RSN-IE or RSNE) of Beacon frames, in which the group cipher suite in use is Wired Equivalent Privacy (WEP). Yes, this is a lot to grasp, but don't get discouraged. All of this information will be explored in more depth as you continue reading in later chapters.

RF Does not Respect Boundaries

Wireless networks use radio frequency (RF) for communications, which makes for unique security challenges when compared to wired network technologies like Ethernet. Like other types of radio-based communications, 802.11 wireless signals can easily pass through many types of obstacles and various construction materials. As Wi-Fi becomes ever more pervasive, outdoor deployments in open spaces are commonplace. Too often, signals propagate beyond their intended target spaces, allowing eavesdroppers and intruders to monitor the air for valuable information from afar. WLAN designers, administrators, and users all must take their own special precautions to ensure that transmitted data remains private because physical RF security is not easily achieved. The wireless network truly presents a target for hackers of various skill levels that is vastly more accessible than is the case for the wired LAN.

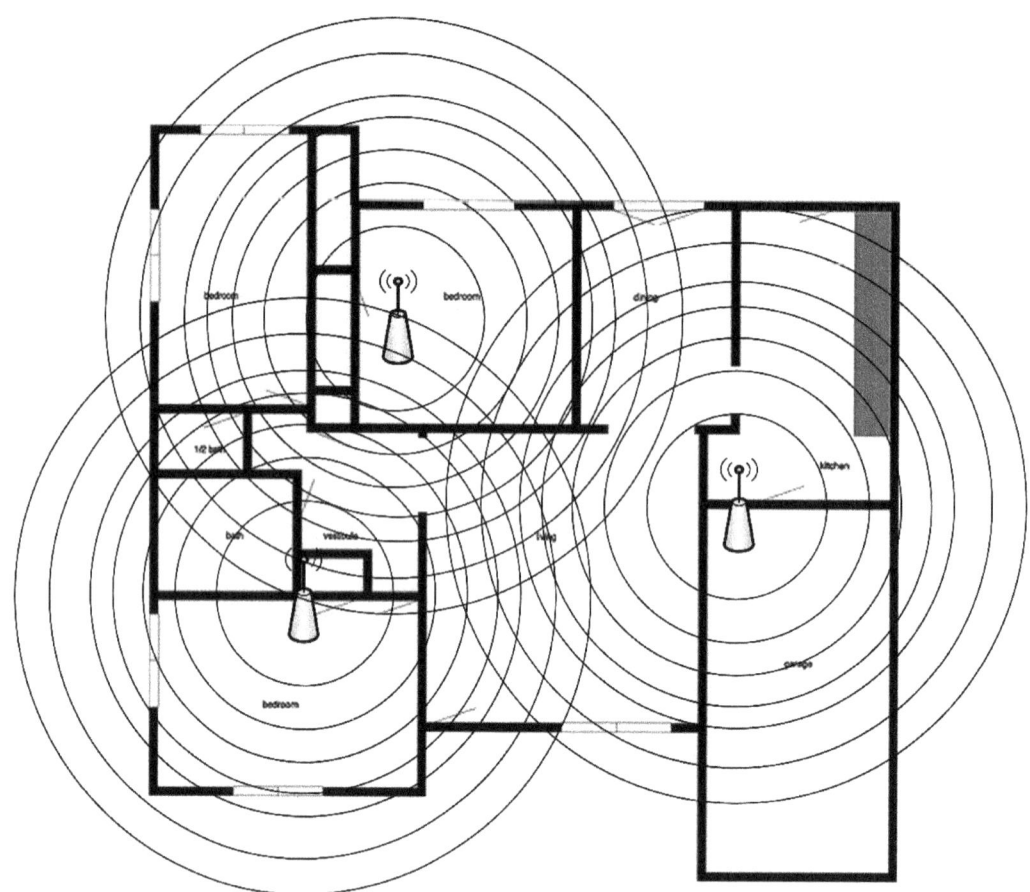

Unbounded Signals are Easily Seen by Attackers

Wireless professionals must deal with the reality that someone monitoring the air with easily obtainable software utilities may never be seen as he or she eavesdrops on wireless networks. Keep in mind that, for WLAN communications to be successful, devices must be able to "hear" or receive each other's low-power transmissions. This implies a certain degree of proximity as a requirement. However, with the proper free or low-cost tools and equipment, an intruder may be able to monitor RF communications from a surprising distance. While likely going unnoticed, she may well be able to leverage the information harvested from the unbounded medium. Just because the wireless network isn't logically aware of the intruder, does not mean that they will not be able to gather and use the information that is exchanged between legitimate wireless network devices. The software utilities needed for this type of eavesdropping are generally available online (and often include training videos). Even someone with a limited skill set will be able to easily and clandestinely gather valuable information from a wireless network that is lacking an adequate security solution. The advent of wireless networking was unfortunately paralleled by the advent of the notion of listening to wireless networks by both genuine hackers and those who aspire to become proficient in hacking. CWSP is all about protecting the target.

Usage Threat Assessment

Relying on legacy 802.11 and weak security mechanisms carries significant risk and can be equated with simply putting up a low fence to keep intruders off of your property. There can be no expectation of real security, as these mechanisms at best present a hurdle to network entry and not a proper barrier. Legacy solutions are those that are deprecated over time by the 802.11 standard because they were never very good security solutions in the first place. Yes, many of these options are still available because of the strong emphasis on backward compatibility that pervades the 802.11 framework, but the security-minded WLAN administrator recognizes that using legacy security methods is not a good strategy and works to educate the customer on the topic.

 The current 802.11 standard, as amended, has deprecated both WEP and TKIP as security solutions. Newly installed systems should no longer use these options. TKIP is roughly equivalent to WPA.

It is also important to understand the various wireless vulnerabilities that exist for the different types of WLAN deployments, and that new vulnerabilities will come along periodically. Whether the WLAN is a home, small office, or enterprise installation, all WLANs have their share of weaknesses for personal, and business uses. Some vulnerabilities are common across all scenarios (like malicious jamming), while others are more specific to individual types of environments. We will address personal and small business wireless environments first.

Considerations for Personal Network Usage Threat Assessment

- Anonymous intruders may perform illegal computer activities through open wireless networks

- Intruders may compromise a home user's privacy
- Intruders may learn financial, medical and personal information for use in identity theft
- Intruders may tamper with home user's files and information
- Intruders may insert malware, viruses, rootkits, or backdoors onto the home user's network
- Intruders could camp on the personal network and use so much bandwidth that the rightful owner's performance suffers

Considerations for Home/Small Business Usage Threat Assessment

- Intruders may gather financial details of home-based businesses
- Intruders may eavesdrop on business communications of employees working from home
- Intruders may hijack logins to corporate accounts
- Intruders may insert malware, viruses, rootkits, or backdoors onto the corporate network through remote connections
- Intruders might leverage the WLAN to compromise or disrupt wired network devices

Network Extension

Now let's consider larger environments. Due to the diverse set of network topologies, each with their own unique security concerns and requirements, it is important that a wireless security designer consider the entire scope of the wireless deployment and how it fits into the overall network paradigm. Keep in mind that early wireless networks were more of an extension to an existing wired network infrastructure with a limited number of APs and a small number of users, whereas today Wi-Fi may be the dominant access method for many networked environments.

Early wireless networks allowed users to access network resources but were usually limited in privilege. With the growing reliance on wireless, the need for stronger, more robust security policies and solutions became necessary. The days of Wi-Fi being merely a convenient, limited accessory to wired networking are long past.

Since most wireless installations occur at the edge of a wired LAN infrastructure, any weaknesses in the wireless segments can lead to the exploitation of vulnerabilities on the wired segments as well. WLANs have become a tightly integrated network access resource, often with the same privileges as given to users of the wired LAN. The implication here is that a breach of the wireless network may well provide an ingress point to critical business resources that typically reside on the wired LAN. We'll learn later about the importance of wireless penetration testing as a defense against the use of Wi-Fi as an attack vector on the wired network.

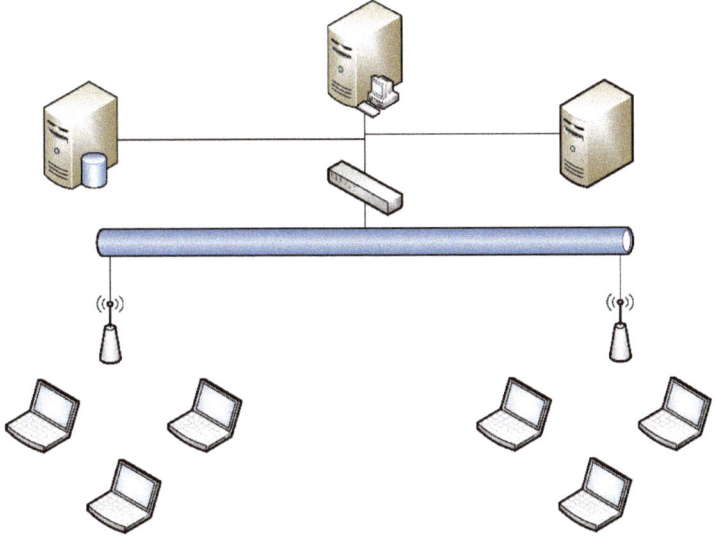

Access WLANs are Client Access Networks

Network targets for intruders may include:

- Databases
- Application Servers
- Management Devices
- File Servers

- Client devices
- Network infrastructure devices
- Unpermitted network access

CWNA Security Review

The following is a review of WLAN security concepts that you may have learned in your CWNA studies, from equivalent wireless training or from personal experience.

 When I teach any of the P-level certification classes (CWSP, CWAP, or CWDP), I always like to remind students that CWNA knowledge is fair game on the exam. What I mean by this is simple: an exam item on a P-level certification can, with complete fairness, ask a question that requires both existing CWNA knowledge and new CWSP knowledge. Because the P-level certifications stipulate the CWNA as a prerequisite, all of that knowledge is testable as well. However, I do have just a bit of insight into this and I assure you, arcane knowledge is not used in this way. -Tom

IEEE 802.11 Open System Authentication

Open System Authentication is a required component for 802.11 devices to connect to a WLAN and is considered a null authentication algorithm. Open System authentication consists of two 802.11 management frames. These frames are not a request and response but merely identified as Authentication. For the most part, this authentication will always be successful from the device's perspective. Simply put, a device says, "I'd like to communicate with you," and the other device says, "and I would like to talk with you as well." Though the mechanism is called Authentication, it's also more or less guaranteed, and there is no credential checking or other vetting used.

Without any other additional authentication mechanisms, Open System Authentication will allow all information sent across the air to be in clear or plain text and, therefore, remains vulnerable to eavesdropping. Most wireless hotspots will use only Open System Authentication (to provide ease of connection), and users will have to supply additional authentication methods or encryption solutions such as virtual private networks (VPNs) to secure their wireless transmissions. Where other authentication and encryption mechanisms are used, like with 802.1X, it can be confusing that "Open" System Authentication is still in play. It's a fundamental 802.11 construct that stays with us even when the wireless network it's used on is not "open."

WEP

As mentioned earlier in the chapter, WEP was intended as an early way to protect information on a wireless network from casual eavesdropping using a 40-bit key and a 24-bit initialization vector (IV). When wireless networking was just gaining traction, robust security was not really considered as a concept that would be needed. It was unfortunately discovered early on that WEP was functionally weak and was soon broken as a result of the way the IV was used in conjunction with the 40-bit encryption key. Although it does not provide adequate wireless security by any measure and should not be employed, as a rule, WEP may still be found within a dwindling number of wireless networks due to the continued use of legacy (usually specialized) equipment. Rather than continuing to use WEP because older or poorly built client devices require this dated option, it is highly recommended to upgrade to newer devices that will support a more secure solution such as CCMP/AES (roughly equivalent to WPA2).

From the perspective of "best practices," WEP, TKIP, and Shared Key authentication are all mechanisms deemed to be outdated and ineffective for securing 802.11 networks. They are deprecated as of the latest 802.11-2016 standard. When something is "deprecated" in networking or programming, it means that you are strongly encouraged to find other ways of accomplishing what the deprecated feature does because it has been proven to be faulty in some significant way or will soon be replaced and removed. Network equipment vendors may decline to provide support for deprecated features.

Organizations have largely moved to using CCMP/AES while abandoning deprecated wireless security solutions. Every life-cycle-refresh helps in this evolution, but the rare corner case where outdated security measures must stay in place, for whatever reason, will be with us into the future. Where they are still present, you generally can't remove the actual options of using WEP and TKIP from wireless hardware. Regardless, you should avoid their use and select stronger options every time you have a technically viable choice.

802.11 Shared Key Authentication

This wireless authentication method is NOT WPA/WPA-2 PSK, or pre-shared key usage, despite what its nomenclature may imply. Shared Key Authentication was defined in the original 802.11 standard in 1997 as a way to provide both 802.11 authentication and data encryption, which was accomplished through the use of WEP. Because WEP is highly vulnerable to eavesdropping and is not an adequate security solution, this authentication method also adds no value to 802.11 WLAN technology, despite perhaps seeming more robust than Open System Authentication to the uninitiated. Unlike Open System authentication, which uses two 802.11 management frames, Shared Key authentication requires four 802.11 management frames. With a challenge string sent in clear text in the second frame, Shared Key authentication can easily be exploited allowing for unauthorized users to authenticate to the wireless network and view user data that should actually be secured.

Wi-Fi Protected Access (WPA) and WPA2

Given that the original 802.11 standard-based wireless security methods turned out to be devastatingly weak, the Wi-Fi Alliance created a pre-802.11i stop-gap certification known as Wi-Fi Protected Access (WPA). While 802.11i was being developed, WPA was introduced as a temporary security method. It can take quite a while for amendments to the 802.11 standard to be written and adopted, and the Wi-Fi Alliance did well bridging the gap between poor security and the more robust mechanisms that would eventually follow when it created the WPA certification. This certification provided wireless device manufacturers with the operational framework to either build new equipment or to provide firmware updates to existing hardware that provided legitimate security options.

The original WPA interoperability certification was based on the fact that Temporal Key Integrity Protocol (TKIP) provided an enhancement to WEP on pre-RSNA equipment and allowed for the protection of 802.11 data frames. Equipment that supported legacy WEP and was capable of TKIP (usually determined by what specific components with which it was built) could be upgraded through firmware. The 802.11i amendment was ratified in 2004. Given the success of the WPA certification in bringing a new level of security awareness, the Wi-Fi Alliance created a post-802.11i certification known as WPA2. Based on the 802.11i amendment to the standard, the WPA2 certification requires support for CCMP/AES and optionally allows TKIP/RC4 for backward compatibility for legacy clients. WPA2 was pivotal in providing a level of security that enabled Wi-Fi to explode as a trusted access method. In 2018, WLAN security took another leap forward with WPA3. We will cover the specifics of WPA3 in this text, but at the time of publication, WPA3 has not yet made it into the market with any significant adoption.

Wi-Fi Protected Access (WPA) Personal Mode

WPA personal mode (WPA-Personal) was created to provide individual users with an easy, but stronger, way to secure their 802.11 wireless networks versus past methodologies. Greater security was accomplished by entering a passphrase (used to create a pre-shared key) on all wireless devices that would be part of the same BSS. A passphrase can be a maximum of 63 ASCII characters in length. From the passphrase that is entered into the device, an algorithm is used to create a 256-bit pre-shared key (PSK). Although this key is secure, using a weak passphrase can make the wireless network vulnerable to intrusion. It is very common for users to pick a short and easily guessed passphrase, of which we will learn more about the dangers later in the book. A WPA network will use TKIP/RC4 as the cipher suite and encryption method. WPA-Personal mode is also called WPA Pre-Shared Key (WPA-PSK) based on the use of a single key for all 4-way handshakes (of which you will learn detail later in the book). WPA was better than WEP, but more could be done to secure individual users' WLAN connectivity.

Wi-Fi Protected Access 2 (WPA2) Personal Mode

WPA2-Personal was created based on the ratification of the 802.11i amendment. This new amendment provided the capability of using an even stronger method to secure 802.11 wireless networks. A WPA2 network can use CCMP/AES as the cipher suite and encryption method for securing wireless communications but allows for optional TKIP/RC4 support for backward compatibility for older devices. A WPA2 passphrase uses the same concepts as WPA but allows for stronger security. As noted earlier, 802.11 associations for devices that are capable of CCMP/AES will be classified as RSNAs. WPA2 Personal Mode is currently the general standard mechanism for securing home and many small-business wireless network environments, as well as many high-roaming scenarios with mobile devices not supporting fast roaming methods.

Wi-Fi Protected Access 3 (WPA3) Personal Mode

WPA3-Personal is the latest in the series of frameworks that seek to secure individual client device security in the absence of a centrally managed methodology. Although WPA3 has significant advantages over WPA2, the Wi-Fi Alliance made sure that the overall feel of WPA3 to end-users is no different than it was with WPA2. However, PSK has been replaced with SAE, which stands for Simultaneous Authentication of Equals. The user still enters a passphrase, but the SAE cryptography characteristics provide a higher degree of protection against dictionary attacks – even offline attacks. It also allows users to enter less sophisticated, easier-to-remember passwords and protects data even if a password was compromised through a technique called Forward Secrecy.

Wi-Fi Protected Access (WPA) and WPA2 Enterprise Mode
WPA and WPA2 enterprise modes (WPA- and WPA2-Enterprise) are far more robust methods of securing enterprise wireless networks. Compared to Personal Mode, these modes use a much more sophisticated process to secure 802.11 wireless communications. Enterprise Mode relies on another IEEE standard, 802.1X, which provides port-based access control and uses Extensible Authentication Protocol (EAP), which is an Internet Engineering Task Force (IETF) standard. It's important to note that 802.1X is not a wireless-specific standard and it has great usefulness in the security of many network types. WPA/WPA2 Enterprise Mode provides user-based access control and a much better authentication process for large wireless networks. The same cipher suites and encryption methods, TKIP/RC4 and CCMP/AES, are used as in Personal Mode; however, the enhanced key generation and implementation process are what makes the two modes different.

WPA3-Enterprise Mode
Recognizing the need to enhance the underpinning cryptography that enables wireless security as well as addressing the unique needs of Internet of Things (IoT devices, often "headless" with no direct user-accessible configuration interface), the Wi-Fi Alliance developed WPA3. It's a major security upgrade for the WLAN industry and is expected to propagate in parallel with the new 802.11ax standard (expected to ratify in 2020). Highlights of WPA3 include a new suite of security features:

- Mandatory use of Protected Management Frames (PFM)
- Disallowance of legacy security protocols
- Minimum of 192-bit encryption

- Authenticated encryption based on GCMP-256 (256-bit Galois/Counter Mode Protocol)
- Key derivation and confirmation using 384-bit HMAC (Hashed Message Authentication Mode) with Secure Hash Algorithm (HMAC-SHA384)
- Key establishment and authentication: Elliptic Curve Diffie-Hellman (ECDH) exchange and Elliptic Curve Digital Signature Algorithm (ECDSA) using a 384-bit elliptic curve
- Robust management frame protection built on BIP-GMAC-256: (256-bit Broadcast/Multicast Integrity Protocol Galois Message Authentication Code)

> You will learn much more about WPA, WPA2, and WPA3 in later chapters. Remember, this chapter is giving you a broad overview of security issues and technologies used in WLANs.

Industry Organizations

Standardization and certification are as important to network security as they are to the basic operations of wireless networks. While proprietary solutions may provide some security advantages due to their secrecy, standardized security mechanisms are central to modern WLANs and the development of widely deployed wireless products.

Several different organizations play a role in standards-based WLAN technology, with each contributing various aspects to wireless technology and security. Their individual influences are combined and codified, with broad industry dissemination with the goal of compatibility. This facilitates the ability for manufacturers to design and build equipment that will operate together in a mixed environment regardless of which company produced the devices. A non-proprietary approach helps to grow the wireless technology and therefore makes it more affordable and supportable for homes, small offices, and enterprise companies. Given that modern client devices may be used in all three environments on the same day, wide-scale adoption of standards is that much more important. The three main industry organizations responsible for the standards-based approach that has become the hallmark of the wireless industry are discussed here: IEEE, Wi-Fi Alliance, and IETF.

Institute of Electrical and Electronics Engineers (IEEE)

The IEEE is a nonprofit organization responsible for generating a variety of technology standards, including those related to information technology. The IEEE is the world's largest technical professional society. Since 1997 the IEEE has released a series of standards related to WLANs. Most important of these to the CWSP is the 802.11 standard. The IEEE has also given us the 802.3 Ethernet standard and the 802.1X Network Access Control standard. Both are also very important to the wireless professional.

Internet Engineering Task Force (IETF)

The IETF is responsible for creating Internet standards and promoting Internet technology and usage through the adoption of Request for Comment (RFC) documents. An RFC is a document created by engineers and scientists and designed to define innovation and technology that works with the Internet. If an RFC is approved by the IETF, it will eventually become an Internet standard. The IETF has provided several important RFC's that aid in securing wireless networks. These RFC's include Remote Authentication Dial-In User Service (RADIUS), EAP, and Internet Protocol Security (IPSec). As with the IEEE, the IETF has interests far beyond wireless, but many of their initiatives end up being widely used in LAN, WLAN, and WAN applications.

Wi-Fi Alliance

The Wi-Fi Alliance was created to both promote wireless networking technology and to provide interoperability testing of WLAN equipment. The Wi-Fi Alliance is responsible for many WLAN interoperability certifications and has been instrumental in the growth and mass adoption of wireless as a network access method. The WPA and WPA2 certifications helped to move the industry forward by providing secure WLAN communications through interoperability testing and removing a major barrier to business use of Wi-Fi. WPA3 is expected to further solidify Wi-Fi as the predominant access technology in a growing number of networks.

Wi-Fi Alliance Compliance

Recall that the IEEE creates standards which manufacturers use to design the WLAN equipment that we use in our wireless networks. The IEEE does not perform any compliance or interoperability testing and leaves that up to the individual manufacturers. To promote interoperability across WLAN devices, the Wi-Fi Alliance maintains many certification programs to verify device compliance with generally specified interoperability parameters. This testing, though not always comprehensive, provides some basic assurance that equipment from different manufacturers will work together when used within the same environment. It is important to consider the role of compatibility testing when selecting products and security solutions.

The security certifications of the Wi-Fi Alliance include WPA-Personal, WPA-Enterprise, WPA2-Personal, WPA2-Enterprise, WPA3-Personal, WPA3-Enterprise, Wi-Fi Protected Setup (WPS), and many different EAP types. In addition to these, the Voice-Enterprise certification addresses fast, secure transition and is intended for larger wireless networks that support fast transitions between APs. This certification defines the requirements for voice quality, mobility, power save mechanisms (which will help to prolong battery life) and of course, wireless security. Given that the wireless industry continues to evolve, we should expect future certifications from the Wi-Fi Alliance as needs arise.

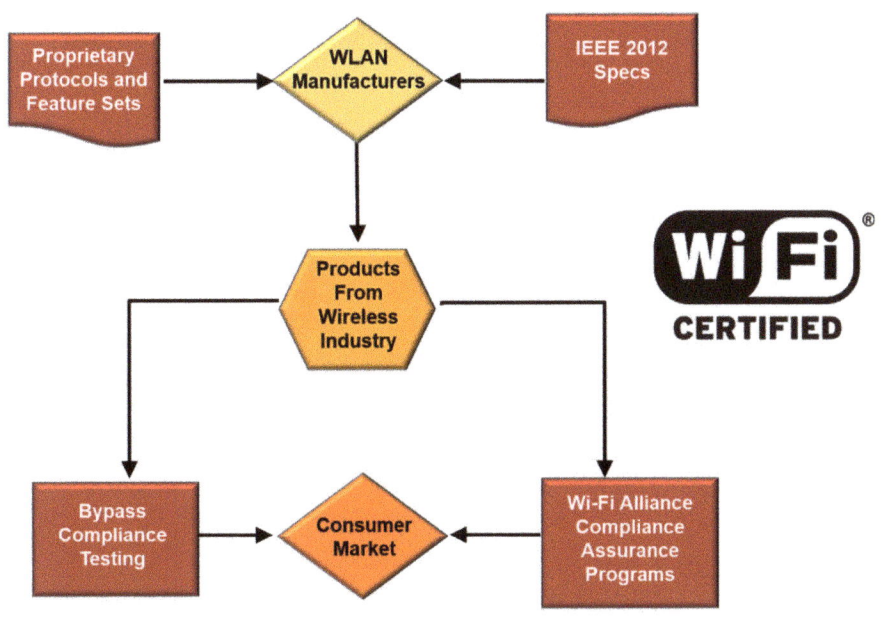

Wi-Fi Alliance Certification Process

Product Certificates

Product certificates provide a quick and easy reference to determine which security certifications a device has received from the Wi-Fi Alliance. Unless a proprietary solution is intentionally selected for added security (not uncommon in government applications), it is always recommended to use equipment that is Wi-Fi Certified. Using devices that are certified by the Wi-Fi Alliance will help ensure interoperability between manufacturers and provide a higher quality user experience along with easier support.

To search for Wi-Fi certified devices, enter the following link into your web browser: https://www.wi-fi.org/product-finder. From this web page, you can search by certificate ID, device model number, keyword, company, category, and other criteria. The Wi-Fi Alliance currently includes testing for several different EAP types (we'll discuss how EAP is used with 802.1X later).

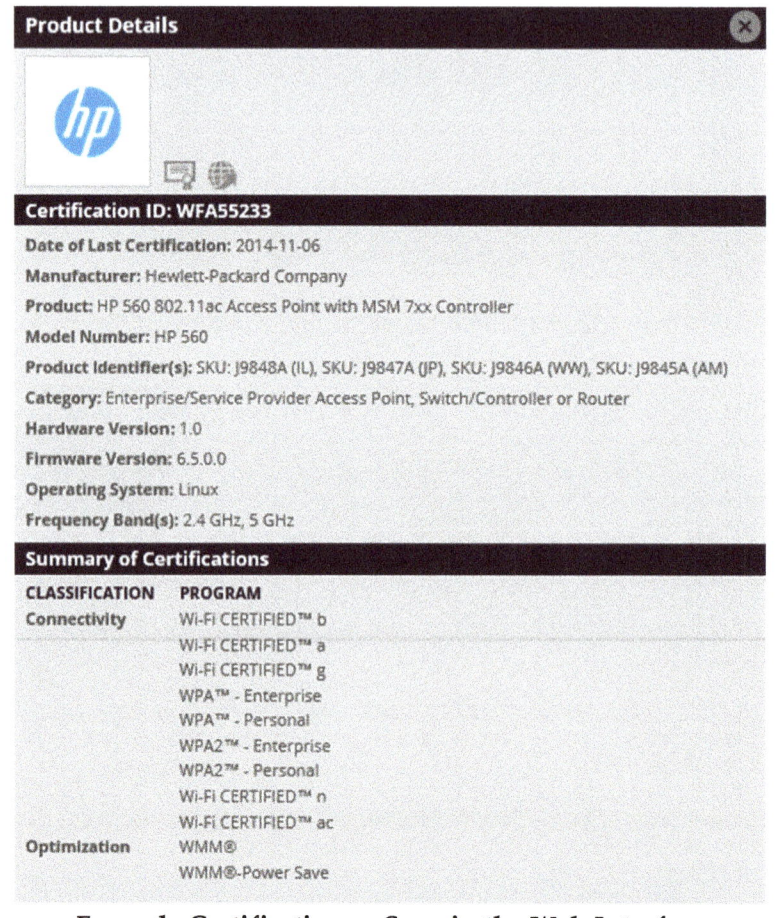

Product Details

Certification ID: WFA55233
Date of Last Certification: 2014-11-06
Manufacturer: Hewlett-Packard Company
Product: HP 560 802.11ac Access Point with MSM 7xx Controller
Model Number: HP 560
Product Identifier(s): SKU: J9848A (IL), SKU: J9847A (JP), SKU: J9846A (WW), SKU: J9845A (AM)
Category: Enterprise/Service Provider Access Point, Switch/Controller or Router
Hardware Version: 1.0
Firmware Version: 6.5.0.0
Operating System: Linux
Frequency Band(s): 2.4 GHz, 5 GHz

Summary of Certifications

CLASSIFICATION	PROGRAM
Connectivity	Wi-Fi CERTIFIED™ b
	Wi-Fi CERTIFIED™ a
	Wi-Fi CERTIFIED™ g
	WPA™ - Enterprise
	WPA™ - Personal
	WPA2™ - Enterprise
	WPA2™ - Personal
	Wi-Fi CERTIFIED™ n
	Wi-Fi CERTIFIED™ ac
Optimization	WMM®
	WMM®-Power Save

Example Certification as Seen in the Web Interface

You can search the certificates by following this procedure:

1) Open your Web Browser.

2) Navigate to https://www.wi-fi.org/product-finder

3) Search for *IAP-224*.

4) On the results page, click the PDF icon in the lower-left corner of the Aruba Instant Access Point IAP-224 result.

5) Open the PDF in your PDF viewer application.

6) On the first page, note the certifications awarded:

    a. Wi-Fi CERTIFIED™ a, b, g, n, ac

    b. WPA™ – Enterprise, Personal

    c. WPA2™ – Enterprise, Personal

    d. WMM®

    e. WMM®-Power Save

7) On the second page, note that all eight current, testable EAP types are certified with this AP.

8) Further, on the second page, note that you can see the number of tested spatial streams. For this device, it supports 3 spatial streams in both 2.4 and 5 GHz.

Terminology Review

Understanding the basic definitions for the following terms will ease the learning process throughout this book:

AAA - Authentication, Authorization, and Accounting (AAA) is a set of separate security functions performed on WLANs to identify and validate a user identity (Authentication), apply specific policies and privileges to his/her network access (Authorization), and monitor the actions performed while this user is associated to the network (Accounting).

Access Control - The prevention of unauthorized use of resources. Access Control is a generic networking term referring to the mechanisms by which access to network resources is controlled.

Authentication - The service that identifies a STA as a member of a group of STAs authorized to join or associate with another STA. Authentication validates user identity to determine permission.

CIA- Confidentiality, Integrity, and Availability

Cipher Suite - A set of one or more algorithms designed to provide data confidentiality, data authenticity or integrity, and/or replay protection.

Encryption - To alter a data stream using a secret code or algorithm so as to be unintelligible to unauthorized parties.

RADIUS - Remote Authentication Dial-In User Service (RADIUS) is an authentication protocol used to provide centralized AAA services for a network.

RSN – A Robust Security Network (RSN) is a network that allows only robust security network associations (RSNAs) by the exclusion of WEP.

SAE- Simultaneous Authentication of Equals. Introduced in WPA3-Personal, replaces PSK with new key-handling methodology derived to prevent KRACK attack and similar.

802.1X/EAP - An enterprise authentication mechanism in which port-based access control (802.1X) is employed with a form of the Extensible Authentication Protocol (EAP) to authenticate STAs.

VPN- Virtual Private Network- (simplified) an encrypted secure, virtual extension of the corporate network to a remote client

WPA-Personal - Security certification specified by the Wi-Fi Alliance in which passphrase-based authentication is paired with the TKIP cipher suite for encryption.

WPA-Enterprise - Enterprise security certification specified by the Wi-Fi Alliance in which 802.1X/EAP authentication is paired with the TKIP cipher suite for encryption.

WPA2-Personal - Security certification specified by the Wi-Fi Alliance in which passphrase-based authentication is paired with the AES-CCMP cipher suite for encryption, with optional TKIP support.

WPA2-Enterprise - Enterprise security certification specified by the Wi-Fi Alliance in which 802.1X/EAP authentication is paired with the AES-CCMP cipher suite for encryption, with optional TKIP support.

WPA3- Personal- Security certification specified by the Wi-Fi Alliance in which Simultaneous Authentication of Equals (SAE) replaces PSK functionality. Optional at the time of this writing.

WPA3-Enterprise- Enterprise security certification specified by the Wi-Fi Alliance, which requires Management Frame Protection (MFP), disallows legacy security types and uses GCMP-256 encryption. Is optional at time of this writing.

Home Office Security

The wireless security solution that you ultimately choose will depend on several factors which include the number of APs, number, and type of client devices, and the intended use of the wireless network. Home and home office installations typically consist of one wireless AP and a limited number of devices that will associate to the network. For modest wireless networks like this, WPA2 passphrase (or WPA3 SAE) will be adequate. Indeed, this is currently far and away the most common way to secure home networks.

Using a strong passphrase and following general wireless security best practices will usually suffice for this type of network. Manufacturers of home-based WLAN equipment will sometimes try to ease the process of securing wireless home routers by providing default security mechanisms including pre-supplied passphrases. Depending on how it is implemented, this could be a security risk. As a best practice, reconfigure the WLAN AP or router to use a strong passphrase with a mix of more than fifteen characters. In addition to this best practice, consider the following important list:

- Change all default settings including the SSID, passphrase and device login credentials
- Do not use WEP

- Use only client devices that will support WPA2, minimally
- Use only CCMP/AES
- Always use strong passphrases and change them often
- Change your passphrase if you lose a client device
- Be mindful of how family members share your passphrase with visitors
- Disable Wi-Fi Protected Setup (WPS) features as many implementations introduce vulnerabilities
- Periodically upgrade devices to the latest firmware versions that are available

Small Business Security

Small business wireless may require more than one AP. Depending on the number of APs and connected wireless devices, the same security best practices as applied to home office security may be applicable here. Small business Wi-Fi might be controller-based or cloud-based, which could provide the opportunity to use stronger security mechanisms such as 802.1X/EAP.

In addition to the home office security best practices, small business security should only use WPA2 for CCMP/AES. Most client devices used in these environments will support WPA2. If they don't, then they should be scrutinized for replacement with something more robust.

Many small businesses cannot justify a dedicated information technology professional, so using enterprise-level security solutions such as 802.1X/EAP may be a challenge as configuring advanced security is usually beyond the skillset of non-IT employees. In these cases, it may be necessary for an employee to get the proper training to support advanced security requirements if policy requires more advanced security methods. Alternatively, outside consultants can be hired to assist with advanced security solutions based on specific business needs.

Large Enterprise Security

Large enterprise wireless networks require careful planning in order to ensure successful deployment, and wireless security plays a major role in that planning. At this scale, security needs are much more policy-driven and granular than in small networks, and the CWSP must be intimately acquainted with the particulars of enterprise wireless security, which are covered in great detail throughout this book. As a CWSP working in large, complex environments, you may find yourself working with multiple security paradigms at a single site.

802.1X addresses port-based access control and helps to provide a secure, scalable, and manageable security solution for enterprise wireless networks. To the uninitiated, the notion that port-based security is relevant in wireless where there are no obvious physical ports for clients can be confusing. We'll talk more about what port-based really means in the context of wireless security later in the book, but for now, know that virtual ports are used within the AP for these processes.

802.1X works in conjunction with an appropriate EAP (authentication protocol) method to allow for user-based security. User-based security allows an administrator to restrict access to a WLAN and its resources by creating users in a centralized database or accessing a typical X.500 compliant server with an existing user database. Anyone trying to join the network will be required to authenticate as one of the users by supplying a valid username and password or other valid credentials. After successful authentication, the user will be able to gain access to resources to which they have been assigned appropriate permissions based on their role or organizational group.

Wireless devices that use 802.1X technology are identified using different terminology than that used in 802.11 standards-based wireless networking. This terminology is used frequently by the wireless professional in their daily work and includes three key terms:

- **Supplicant** - the wireless client device or the device requesting authentication
- **Authenticator** - the wireless AP, WLAN controller or the device/system providing access to the network
- **Authentication server** - the device or system providing the actual authentication, commonly a RADIUS server

We will later explore how the supplicant and authentication server need to be configured with matching EAP types for 802.1X to function. We'll also review several EAP types, to include strengths and weaknesses of each.

When implementing security for any sized business, it is essential that the network resources themselves are adequately secured as they are common targets of network attacks. That is, you should not rely on just the WPA2-Personal or Enterprise authentication and encryption provided by the WLAN as your only security. A cohesive, well-documented security strategy should be implemented appropriately throughout the network. This is often called defense-in-depth or layered security. Many modern enterprise WLANs are designed to tightly integrate with other network resources as part of the bigger security paradigm with the goals of policy-driven Confidentiality, Integrity, and Availability in mind.

Public Network Security

One common wireless network implementation is the publicly available Wi-Fi hotspot. This type of network is usually available at airports, hotels, restaurants, coffee shops, retail stores, on airplanes, and at many public settings. In many cases, these wireless networks are available for free, or in exchange for targeted marketing as a value-added service to the patrons of the establishment that provides goods or services. Occasionally, the hotspot provider will charge a fee for wireless network access (and risk the wrath of customers complaining on social media).

The hotspot model typically involves users connecting their devices to an unsecured wireless AP in order to use Internet resources or even to access corporate network resources across the Internet. Usually, the hotspot framework prioritizes ease-of-use over providing wireless security, and so Confidentiality, Integrity, and Availability aren't provided as they would be for users of a private wireless network. As a result, this kind of WLAN environment can be a haven for hackers working to get a variety of data using various intrusion techniques. These techniques often play on most people's ignorance of wireless security and include direct peer-to-peer attacks or connecting to another wireless station through the AP. Although it can be a hard message to get across, it is critical that devices connecting to a public wireless hotspot use appropriate and adequate security controls to minimize or eliminate potential security threats.

Here are common best practices for devices that connect to public wireless networks:

- Use a Virtual Private Network (VPN) connection whenever possible
- Secure all login accounts with strong passwords

- Ensure firewall software is installed, enabled, properly configured and up to date

- Ensure anti-virus software is installed, enabled and up to date

- Ensure that security vulnerabilities in the device operating system are patched and all service packs are installed

- Secure any open file system shares that may be enabled

- Disable file and print sharing features if not needed or used

- Be aware of people who seem a little too close to you physically, who may be watching what you type

- Be mindful of networks that should not be seen- like your corporate WLAN when you are hundreds of miles from work.

Remote Access Security

As wireless availability and use continue to grow, so does the need for remote connectivity. Many organizations allow employees to work from remote locations such as home, satellite offices, and while traveling. In some instances, there may not be an office to go to, and all employees work remotely. When a remote user connects remotely to a company network, that network, for the most part, is now extended to the remote location. Ensuring that the external connection is secure is of utmost importance, and corporate security policy must include remote access security.

While telecommuting, it is important for the employee to follow all corporate security procedures to ensure that the remote connection does not become a conduit for bringing trouble to the corporate network. Given that the remote user is more mobile and is not in the controlled office environment, remote access security can be more of a challenge.

Proper training programs should be in place as part of the corporate security policy to address remote access and remote security solutions. One common way to ensure remote connections are secure is to use a VPN solution. VPN provides users with the capability to create secure private communications over a public network such as the Internet, even if you connect from an open Wi-Fi hotspot.

Security and the OSI Model

Security techniques work at various levels of the Open System Interconnection (OSI) model from the lowest, the Physical Layer, to the highest, the Application layer. As you likely know by now, 802.11 is primarily concerned with Layers 1 and 2. At the same time, different layers may be leveraged for specific security strategies. Let's look at some of the security concerns and solutions for the most commonly discussed layers of the OSI Model.

> The OSI Model is fundamental to wired and wireless networking and is covered in more detail in both the CWNA and CWTS Official Study Guides. If you are unfamiliar with this model, consider using those resources to learn more. Additionally, the book, The TCP/IP Guide, by Charles Kozierok or the website (bit.ly/1xk2kjG) includes an excellent overview of the OSI Model.

The Application Layer

The Application layer is considered the interface to the user and is frequently referred to as Layer 7. This is where the protocols for common applications such as email, Internet web browsers, and file transfer programs reside. Some common Application layer protocols include Hypertext Transfer Protocol (HTTP), File Transfer Protocol (FTP), Post Office Protocol (POP) and Simple Network Management Protocol (SNMP) among others. These protocols provide no significant stand-alone security and many transfer information in plain or cleartext.

Clear text transmissions create severe concerns for WLAN communications when you consider that they are sent through the air using RF. Many Application layer protocols can use the Secure Sockets Layer (SSL) protocol alongside the Application layer protocols to provide for secure network communications using Internet Protocol (IP), but whether they do or not can be hit-or-miss, depending on the way the application was written. Because so many of these applications transmit in the clear, using encryption on the WLAN is essential in modern business. With wired networking and switch ports, it is far more challenging today to promiscuously sniff data on the wired network than it is on an open WLAN. The administrator should not leave the business wireless network open for such easy attack and using WPA2 or 3 will thwart the would-be eavesdropper who is hoping to leverage weak Layer 7 protocols.

Some organizations are now using Application layer firewalls to help control the traffic at Layer 7. It's important to note that these firewalls will not typically assist with WLAN security per se. This is because they control ingress and egress on the wired network borders, but do not control what passes within the network itself. Eavesdroppers could still capture information passing within the wireless network, because of the typical range of propagation of RF. For this reason, again, WPA2 or 3 should be used to protect the WLAN.

Below the Application layer, we have the Presentation, Session, and Transport layers. Though each plays an important role in networking, they aren't tremendously relevant to our discussion of wireless security. We won't discuss them individually here.

The Network Layer

Although WLAN technology operates at the Physical and Data Link Layers, the Network Layer still plays a role with respect to wireless security because of the networking protocols that reside at this layer. Layer 3, the Network Layer, uses the ubiquitous Internet Protocol (IP), which is responsible for the addressing and routing of network-carried data. When used with the Transport Control Protocol (TCP) at Layer 4, TCP/IP allows for communication across the Internet as well as across the LAN. TCP/IP itself is not a secure protocol stack and requires additional technology to ensure secure communications. One common way to secure data at the Network Layer is through the use of a VPN. There are various VPN technologies available, and as with wireless networking, some are more secure than others. Two common examples of VPN protocols are:

- Point-to-Point Tunneling Protocol (PPTP)
- Layer 2 Tunneling Protocol (L2TP)

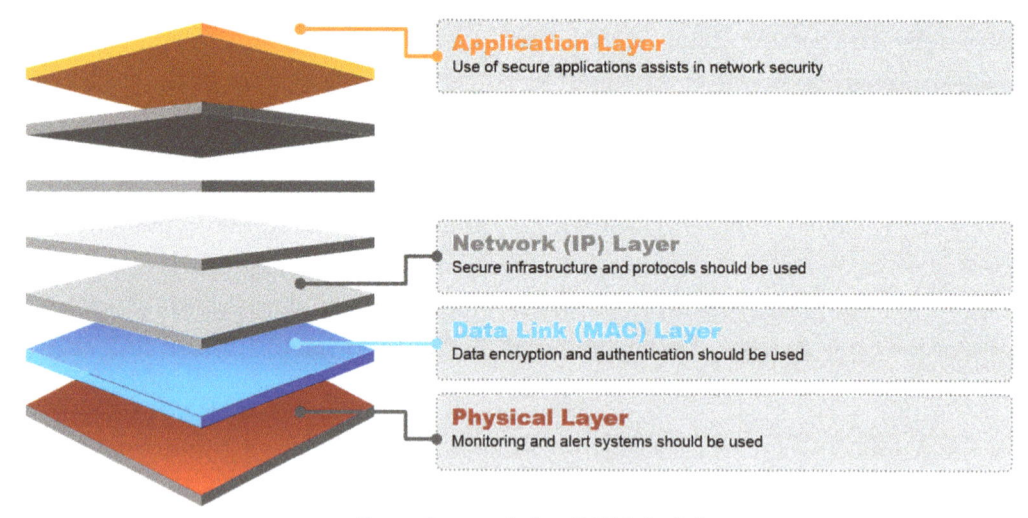

Application Layer
Use of secure applications assists in network security

Network (IP) Layer
Secure infrastructure and protocols should be used

Data Link (MAC) Layer
Data encryption and authentication should be used

Physical Layer
Monitoring and alert systems should be used

Security and the OSI Model

Somewhat analogous to how WEP and TKIP are viewed, PPTP is considered a weaker legacy VPN technology that can introduce security vulnerabilities when used with wireless networking. L2TP itself provides only a tunneling mechanism and gets its security from integration with an encryption protocol. With L2TP, the most common choice of encryption is IPSec, which provides authentication and encryption for each IP packet in the data stream.

The Data Link Layer

Layer 2 has a particular significance in the context of WLAN communications. The Data Link Layer is actually made up of two sub-layers- Media Access Control (MAC) and Logical Link Control (LLC). We'll focus our discussion here on the MAC sub-layer but know that the LLC also exists (and is pivotal to the integration between 802.11 wireless and 802.3 Ethernet). The MAC sub-layer is where bit-level communication is accomplished through MAC addressing and framing. The MAC sub-layer adds the MAC header and will allow for various WLAN security mechanisms. The header information cannot be encrypted regardless of what 802.11 methodology is employed, which is a very important point of reference. For secure wireless communications, encryption must occur within the data payload of the frames that traverse the air. Layer 2 security types include those mentioned earlier in this chapter- WEP, TKIP/RC4, CCMP/AES, GCMP-256 in WPA3 and 802.1X/EAP, though some of these have been deprecated. Using legacy or unsecured Layer 2 security mechanisms has been the cause of many real-world WLAN security-related issues.

> 802.11w-2009 (now part of 802.11-2016) introduced management frame protection. However, this does not encrypt the MAC headers of frames, and it only applies to specific management frames. The only frames protected are deauthentication, disassociation, and robust action frames. Data frames do not include this protection. The new WPA3 security certification requires that devices implement management frame protection

The Physical Layer

Layer 1 provides physical connections to the network, between devices, using various methods and connector types. For Ethernet, this is the realm of optical fiber, patch cables, and UTP station wiring. For the wireless side of 802.11 networks, the Physical Layer is RF. Since open-air is the operating domain used with wireless communications, WLAN Layer 1 security concerns are vastly different than those of wired networks. Potential vulnerabilities include eavesdropping on unsecured communications, as previously mentioned, and intentional RF interference (known as jamming). Jamming can be a denial of service attack unto itself, or it can be a component of more sophisticated attacks as we'll see later in the book.

In addition to the risks associated with Wi-Fi's unbounded medium, the wired network infrastructure at Layer 1 can also be a security concern for wireless networking. This includes unsecured physical layer wired ports that connect to Layer 2 switches, and that can be used to introduce rogue (not authorized) APs into the networking environment. Good practice includes securing unused switch ports through a variety of methods.

Security Analysis Basics

Before we close this introductory chapter, let's explore the notion of Security Analysis at a high level to provide perspective on scenarios and concepts that CWSPs might face in the real world. Threats include the individuals or groups who wish to attack your network and the systems they use to perform the attacks. Vulnerabilities are the points where your system is weak or might be able to be penetrated. You must consider both to implement and sustain an effective wireless solution. Keep in mind that your vulnerabilities, potential hackers, and penetration tools will all change over the life of a given WLAN. Several terms and concepts should be understood, which will make you a more effective security analyst. Much of the following applies to all networks, not just wireless. The first concept you must understand is the concept of the attack surface.

Attack Surface

The attack surface is inclusive of all areas that can potentially be attacked. Even small networks can have a sizeable attack surface, depending on network topology and services in use. For large network environments with sizable LAN, WLAN, and WAN components, it can take a fair amount of time and analysis to fully realize your attack surface. Additionally, just as threats change over time, so does your attack surface. Security is not a one-time thing on any network. Failure to realize the simple notion that your attack surface needs ongoing review can lead to catastrophic breaches.

Attack surface reduction is a security best practice and is the process of reducing the number of areas where your system can be attacked. This process is in line with the principle of security-as-a-process because attack surface reduction acknowledges that a system cannot have an attack surface reduced to zero and still provide a functional benefit to the organization. Put another way- you can never be without some degree of risk.

Attack surface reduction is about reducing the likelihood of attack by reducing the number of attack points. You can sum it up like this: *If you do not need a particular technology or capability for some beneficial business purpose, do not use it or leave it in place for others to use.* Think of it like the booth at the fair where you throw the darts at the balloons. The smaller balloons warrant a bigger prize (the gigantic Elmo for a four-year-old) because they are harder to hit due to the reduced surface area. In the same way, your network is harder to penetrate when you reduce the attack surface. So, go ahead and let the air out of your network's balloon by disabling unneeded services.

There are two general wireless device points of entry to consider when contemplating attack surface reduction: wireless entry and wired entry. Let's pause for a moment and remember that wireless networking is not without wires and that 802.11 infrastructure devices have 802.3 Ethernet interfaces along with their radio interfaces.

The wireless attack surface includes any components on the network with a radio interface. This includes all access points, wireless routers, wireless bridges, and other wireless devices. You need to ensure that proper security mechanisms are in place to help prevent unauthorized access through these RF entry points. One common attack method is the rogue (self-installed or unauthorized) AP or wireless router that lures users to a wireless cell that they should not be on. Even well-meaning users can install a rogue when they want wireless access, and it's not provided by the organization, or they want a less restrictive flavor wireless access in their work area. They may not mean to harm the network, but the typical default configurations and high output power of these devices make them an easy target for eavesdroppers (not to mention the performance impact from interference that comes with them).

To prevent wireless attacks, we rely on best practices for the administrative side of wireless security. These include user training, using strong encryption, and securing the management interfaces of all network devices. In more advanced implementations, consider the use of 802.1X and EAP for authentication and encryption on the wired ports, which generally involves a centralized authentication, authorization, and accounting server (RADIUS).

The AP's wired uplink is often overlooked when configuring wireless networks. To understand the potential impact of ignoring the wired side of wireless networking, consider that the Ethernet port can be used to access the AP for a number of nefarious purposes. If they can gain access, an attacker could modify configuration settings, harvest network information, and possibly exploit the underlying operating system on the AP for more sophisticated attacks. The bottom line is that the wired and wireless networks are not mutually exclusive in providing an easy attack surface for an attacker. As a CWSP, your scope will very much include the LAN on occasion.

Consider a typical pen-testing scenario. An individual posing as a copy repair person or tradesman slips into a conference room or side office in your business environment. This individual connects a PC to an open Ethernet port in the office. Immediately, the attacker notices the LED lights indicating that the port is most likely active. If the port is enabled, the attacker will likely get a DHCP address and in the process, learn a fair amount about the network. The gateway address, subnet mask, DHCP server address, and DNS server address will all be revealed at the command line.

The next step for this intruder might be to begin looking for devices to access on the network. After a brief scan with a scanning utility (like Nmap), the attacker detects more than 30 active devices. The attacker runs a script that tries to connect to port 80 using HTTP on all the discovered devices (this port is the one used by most web servers, including those on many access points or switches). Two of the IP addresses respond positively to the script. The attacker can reasonably assume that these might be infrastructure devices like switches, APs, printers, some sort of building control devices, or that they are actual web servers.

For our discussion, let's say that one of the two IP addresses that responded positively to the script was 192.168.0.250. The attacker opens a browser session for that address and sees a screen asking for a username and password. Noticing the *Cisco-Linksys* name in the dialog, the attacker remembers that the default login credentials for this brand of device are often no username with the password of *admin*. (Even if he didn't know the default credentials, the intruder could easily look them up online from his smartphone. Lists of network device default information are openly shared among the hacker community.) The attacker attempts a logon, has success and is soon looking at the configuration interface for a wireless router that is used in the production small business environment. From here, he can maliciously reconfigure the router or execute a number of other undesirable actions.

This simple- but significant- security attack was possible because the default administrative login for the wireless router had not been changed. If the principles of attack surface reduction had been employed, the attacker would not have been able to reach the wireless router in the first place. Attack surface reduction, applied to this scenario, demands that the Ethernet port in the spare office be disabled until it is needed. With the port disabled, the attacker could not have used the port to obtain an IP address and then reach the wireless router to reconfigure it. Beyond the open Ethernet port being problematic, the router using a default credential setting and even standard HTTP ports, are also weaknesses here. These may all sound fairly simple to combat, but it's easy to overlook critical basics that can lead to trouble. Hopefully, through the CWSP materials, you'll learn the importance of developing a "big picture" view of security.

In our scenario, we mentioned that the intruder scanned the network. It's fairly easy to do this sort of activity yourself to get a feel for what might go on during an intrusion. To perform a scan like this on a network where you are authorized to do so, start up a Kali Linux VM (as described in the appendix) and login as root with the password of toor and then perform the following procedure:

1. Navigate to Application > Kali Linux > Vulnerability Analysis > Misc Scanners > Zenmap.

2. In Zenmap (which is a GUI front end for Nmap), enter the IP subnet you wish to scan into the target field, for example, 192.168.10.1-254.

3. Under Profile, choose the scan type you desire.

4. Click Scan and wait until the results are displayed. You should see something similar to the following image.

> The different scan types can take different lengths of time to complete. Some scans take several minutes to scan just a single computer. When selecting a scan type, know that the Quick Scan Plus method is fast and returns open ports, which is a primary auditing tool for security analysts.

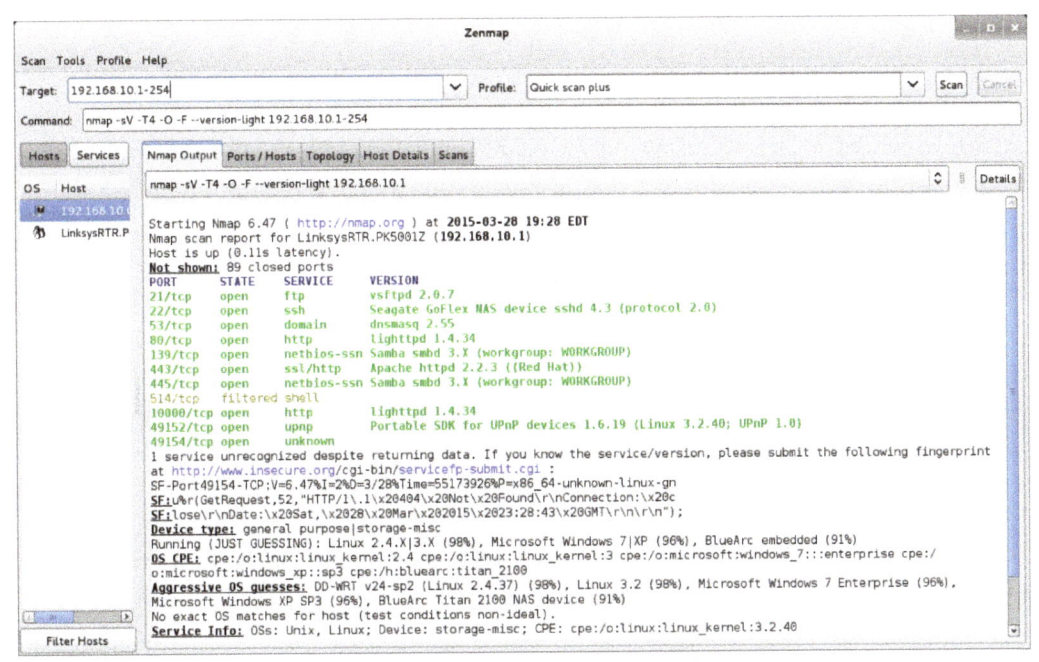

Zenmap Network Scan Results

> I am personally a big fan of Kali Linux. I use it multiple times each weak. In fact, it is the primary Linux distribution that I run as an operating system (though I still work in Windows more than anything else). In such cases, I do remove many of the hacking/cracking tools that I do not need on a regular basis, but the system is based on a stable distribution, it is updated regularly, it has excellent support for wireless adapters, and it's just a plain old great operating system configuration. If you haven't used it yet, download the VMware image from the Internet and install the free VMware Player and get started now. I'm sure you, too, will quickly come to love it. Appendix A, in the back of this book, tells you exactly how to install both VMware Player and the Kali Linux VM image. -Tom

Data Flow

Data flow analysis is the scrutiny of data as it enters, traverses, and is removed from your network. You aren't generally concerned with the departure point from the network in most wireless implementations, though that is an important concern for your overall network security strategy. Again, an environment-wide awareness will make you a better CWSP.

For wireless networking security, in particular, the focus is on the flow of data from four perspectives: the data entry point, network traversal, live storage points, and backup storage points.

The data entry point is where data starts its network journey, usually entered by a user on a laptop computer, a desktop computer, or even a web-based interface operating across the Internet. Regardless of which type of device serves as the entry point, you must focus on how that device connects to the network. If a wireless connection is in use, you must consider how to secure this data as it traverses the air.

Before we discuss the Network Traversal component of Data Flow, let's consider the types of data that are typically present on a wireless network and what we need to keep in mind about each. Remember that the level of security needed depends on the type of data in play. If you only use the wireless network for general Internet access, you might not need advanced security techniques like VPNs and 802.1X/EAP authentication. Access to secure web sites should already be encrypted with native

SSL, and insecure sites should not need encryption beyond that provided by WPA2 (more on this later).

You might categorize your data into three basic levels of data sensitivity:

- Public
- Private
- Highly private

Public data is that which anyone can see and access. You might want to limit the ability of users to modify the data, but viewing the data is not a concern (and is usually made public for mass consumption). If a client uses only public data, you do not need to be as concerned about the security of the connection. Here common-sense wireless security practices should suffice, like the use of WPA2.

Private data might include human resource information, non-sensitive trade secrets, and other data inappropriate for sharing with the public. Here you generally combine common-sense wireless security measures with non-wireless network controls to keep "inside" data restricted to access by employees only.

Highly private data is described as information that only a select few should see. This data almost always requires advanced security mechanisms such as VPNs for all wireless connections carrying the data and possibly the use of certificates, the strongest EAP methods with 802.1X, and a PKI (public key infrastructure).

How data gets protected is driven by the sensitivity and value of the data. This is a fundamental tenant of wireless security, and of IT security in general. Think of it like this: you wouldn't spend $100 to protect a common modern penny, nor would you leave a bank vault without robust, multi-level security. Likewise, you wouldn't spend thousands on security equipment and software to protect data that is worth very little or do high-value business operations over an unsecured WLAN. As a CWSP, your understanding of situational nuances will help your employer or customers arrive at the right security solution for a given set of circumstances.

We just discussed entry points for data, as well as typical data types and their implications. Now let's examine the paradigm of data moving from Point A to Point B on a network, along with associated security concerns.

Once the data leaves the point of entry (the wireless client device, in our case), it traverses the network and passes through many networked components along the way. During this data flow, transmissions can be interrupted with the potential for interception by an attacker if the traversal points are not secured properly. Part of data flow analysis is investigating these connection points as well as the medium between them for ease of access to those who might do harm.

As we talk about network traversal in the context of CWSP, the main concern is with the wireless medium. Generally, wired network paths benefit from being hidden in walls or behind (hopefully) locked doors. Consider the network represented in the following image. Assume that the user enters data in the wireless laptop client and that the data is then transferred to the database server on the network.

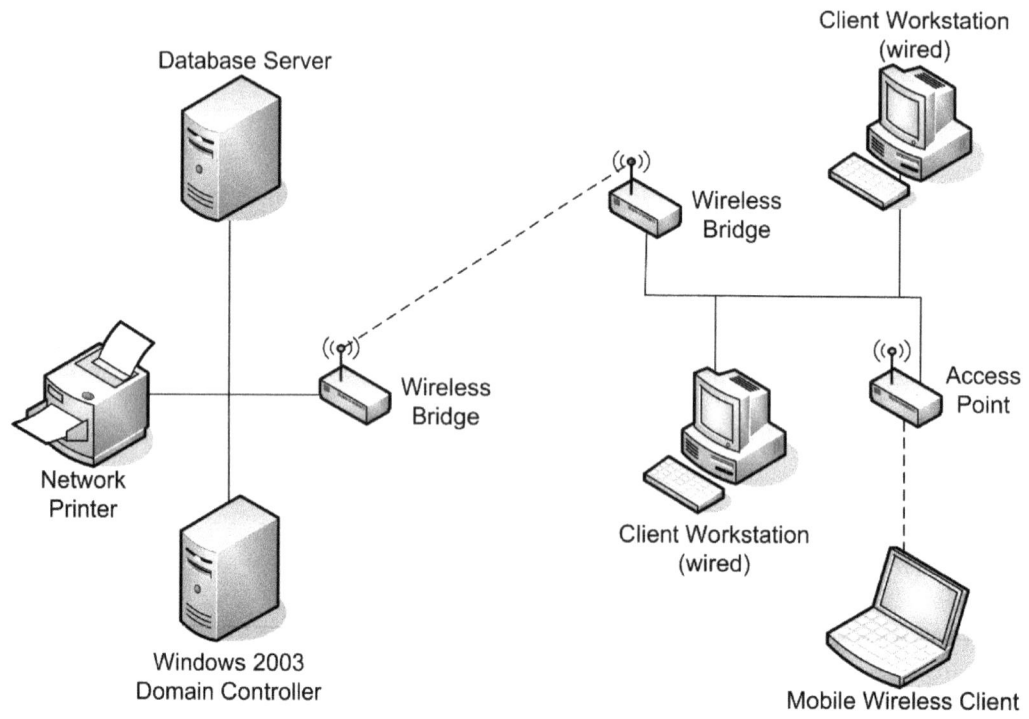

Network Traversal Diagram

Two wireless traversal points are included in this example. The first is between the wireless laptop client and the AP connected to the wired network. The second is between the two wireless bridges that link the two individual wired networks. Both of these wireless connections need to be secured to provide complete security to the data flow (assuming that the wired path is secure). If you enable WPA2-Personal on the AP but do nothing to secure the wireless bridges, you've only addressed one of the two wireless security concerns. Though bridges usually have a narrow RF propagation pattern, this characteristic is not considered a security feature as an attacker can still position his device between the two bridges and sniff the traffic from the air if it is not strongly encrypted as well.

The concept of "sniffing" the traffic means to pull the packets into your device even though they might not be intended for you. Sniffing packets is used in wired and wireless networks alike for both legitimate support and for hacking. The packets, frames, and data payloads have value to both the good guys and the bad. There are many wireless network sniffers (more formally known as network monitors, packet analyzers, or protocol analyzers) available for free. There are also powerful commercial tools available to purchase or acquire for free illegally. If the data being targeted has high enough value, the legitimate purchase of high-end wireless packet capture tools may be a small price to pay for professional or state-sponsored hackers versus the payoff of a successful breach.

The main purpose of network traversal analysis is to ensure that eavesdroppers cannot easily gain useful access to your data. Because we can't stop people that we can't see from listening to the RF, it's critical that we use encryption to protect the WLAN. Encryption renders any packets that do get captured, useless, to those who may be sniffing. This precaution helps secure data in transit, however, and does not protect against data theft during storage, which is discussed next.

After the data has traveled the network to its final destination, it is processed in some way. In many cases, the data is stored in live storage. Live storage means that data is hosted in a location that can be accessed instantly by authorized users. There are, at minimum, two points of access where this data must be secured: on the network and on the storage device itself. When you encrypt the communications between a client device and an AP, you ensure that eavesdroppers cannot view the data easily. However, if attackers discover the configuration parameters needed to associate and authenticate with the AP (maybe somebody left the WPA2 PSK written on a whiteboard easily seen through a window from the parking lot), they can access the network. This might allow them to view the data as it enters the wired network or give them the ability to access the data in live storage.

To protect against the scenario where attackers discover a method for associating and authenticating with your wireless network (in other words, they've breached the network access portion of your security), you should use secure authorization at the point of live storage on the storage device. Create users and groups, as supported by your network operating system or storage device, and then assign proper permissions to those users and groups. Because the attackers are not members of one of these groups and are not one of these individuals, they should not be able to access the data. As a CWSP, you may not have a hand in configuring storage security parameters, but you can certainly collaborate with the system administrators who do.

To truly secure against attacks on live storage, you (or those you are working with) need to understand the security mechanisms of your chosen network operating system or device in depth. You don't need to know how to do this for the CWSP exam because it is beyond the scope of Wi-Fi security, but you must learn to either do it or to work with the right staff resources for your production implementations. This understanding is critical to providing complete security to your network because skilled attackers often escalate their privileges once they gain access to the network—this means they become one of those users accessing the data "on the inside." Hopefully, you can see by now that "security" isn't accomplished in one step and that it requires a broad view of what's going on with the entire network and the host devices attached to it.

The final point of attack is the storage media you use for data backup. Many organizations use physical backup devices that are connected directly to the live storage device. Sometimes organizations transfer the data across the network to an external backup device. In these scenarios, just as when securing the wireless laptop client connections earlier, you must ensure that the traversal path is secure by securing all wireless links in the path (if there are any).

This section was not intended to serve as a complete tutorial on data sensitivity analysis. Instead, hopefully, you learned that there are many factors for a CWSP to consider depending on the construct of the network, and you should contemplate all the ways in which your data is used in order to provide proper security mechanisms. In a large organization, wireless security will be just one component of overall organizational security, and you can expect to coordinate your efforts with other staff responsible for overall organizational security.

Chapter Review

In this chapter, you learned about the foundational security concepts that are essential to understanding the remaining chapters of the book. You explored the importance of security and much of the terminology commonly used in relation to Wi-Fi security. You also reviewed the different types of networks and the security requirements they may have. Finally, you explored the relationship of the OSI Model to security.

Review Questions

1. Wireless network security is built on the foundational concept of CIA, which stands for what?
    a. Configurations, Integrity, Applications
    b. Confidentiality, Integrity, Availability
    c. Configurations, Integrations, Availability
    d. Confidentiality, Integrations, Applications

2. The original WEP encryption specification used a key of which construct?
    a. 24-bit key with 40-bit Initialization Vector
    b. 40-bit key with 24-bit Initialization Vector
    c. 64-bit key with 24-bit Initialization Vector
    d. 104-bit key with 64-bit Initialization Vector

3. The 802.11i amendment does not allow which of the following?
    a. Shared Key Authentication
    b. Open System Authentication
    c. Four-way Handshake
    d. The use of STAs for WLAN

4. The concept of RSNA came about with _____ and stands for _____.
    a. 802.1X, Reasonably Secure Network Association
    b. 802.11X, Robust Secure Network Association
    c. 802.11i, Reasonably Secure Network Association
    d. 802.11i, Robust Secure Network Association

5. Which of the following is true regarding wireless networks?
    a. Wireless networks are easily confined
    b. Wireless networks are considered to be unbounded
    c. Wireless networks are immune to eavesdropping
    d. Wireless networks can't be heard by authorized clients and intruders

6. Which of these was introduced with WPA3?
    a. OSI
    b. VPN
    c. SAE
    d. RSN

7. Why are wired networks at risk from poorly secured 802.11 networks?
    a. Users associate to both wired and wireless networks simultaneously
    b. Wired networks extend wireless networks
    c. Wireless networks extend wired networks
    d. It's very easy to eavesdrop on wired networks

8. Which organization promotes wireless networking and has been instrumental in Wi-Fi interoperability?
    a. IEEE
    b. IETF
    c. Wi-Fi Alliance
    d. Wi-Fi Institute

9. WPA2 Enterprise makes use of which non-wireless standard?
    a. 802.1X
    b. 802.3X
    c. 802.11X
    d. PPTP

10. Shared Key Authentication uses _____ management frames, while Open Authentication uses _____ management frames.
    a. 2, 4
    b. 4, 2
    c. 4, 6
    d. 2, 6

11. Which of these is a valid reason why wireless networks need robust encryption?
    a. Many application layer programs use secure protocols
    b. Undetected eavesdroppers can intercept traffic between authorized wireless stations
    c. Wireless networks are becoming very popular for server connectivity
    d. Intruders might piggyback on your wired network

12. Two examples of Virtual Private Network protocols are:
    a. WPA2, CCMP
    b. PPTP, CCMP
    c. IETF, IPSec
    d. IPSec, PPTP

13. Which security type uses GCMP-256 authenticated encryption?
    a. 802.11i
    b. WPA
    c. WPA2-Personal
    d. WPA3-Enterprise

14. A wireless network security method or practice that is no longer recommended because it is known to be weak, is said to be_____.
    a. Deprecated
    b. Retired
    c. Suspended
    d. Abandoned

15. At which OSI layer does RF in WLAN environments work?
    a. Layer 7
    b. Layer 4
    c. Layer 2
    d. Layer 1

Review Answers

1. **B.** Confidentiality, Integrity, and Availability are design tenants of wireless (and wired) network security.

2. **B.** The 40-bit key/24-bit IV combination is generally referred to as 40-bit WEP, but also sometimes as 64-bit WEP.

3. **A.** Shared Key Authentication is not allowed by 802.11i.

4. **D.** The concept of the RSNA was introduced with 802.11i.

5. **B.** Propagating in free space, wireless networks are considered unbounded.

6. **C.** SAE (Simultaneous Authentication of Equals) was introduced in WPA3 as a replacement for PSK.

7. **C.** Wireless networks extend the wired network and can provide an attack vector against the LAN.

8. **C.** The Wi-Fi Alliance has interoperability and promoting wireless use as two of its primary focuses.

9. **A.** Remember that 802.1X is not a wireless standard, but it works elegantly with secure 802.11 networks.

10. **B.** Shared key authentication has four management frames, while open authentication has two.

11. **B.** Given that there is no way to detect eavesdropping, encryption renders intercepted traffic useless to the intruder.

12. **D.** IPSec is one of the most common VPN types in service today. PPTP is valid, but its use is discouraged.

13. **D.** GCMP-256 authenticated encryption is a key component in WPA3-Enterprise.

14. **A.** The word "deprecated" is quite popular in networking when a feature or command is considered obsolete.

15. **D.** WLAN works at Layer 1 (RF) and Layer 2 (data frames).

Chapter 2: Wireless Security Challenges

Objective Covered

2.1 Identify potential vulnerabilities and threats to determine the impact on the WLAN and supporting systems and verify, mitigate, and remediate them

3.5 Secure public access and/or open networks

3.6 Implement preventative measures required for common vulnerabilities associated with wireless infrastructure devices and avoid weak security solutions

Chapter 1 hopefully provided you with a solid grasp of the objectives of the CWSP course. Now we'll start delving into specific security challenges related to 802.11 networks. This chapter will address several specific topics important to the WLAN security professional. It begins with an exploration of network discovery processes. As with Chapter 1, you may notice overlap with CWNA materials throughout this chapter, as we add depth to security-related topics from CWNA. We begin with examining the typical process that an attacker or security auditor would go through to locate WLANs.

We'll next investigate questionable recommendations that result in a false sense of security. These are called pseudo-security solutions and are as important to be aware of as legitimate security measures, as you may need to correct them in your CWSP duties. We'll consider legacy security mechanisms that should no longer be used, along with basic network attack methods.

Finally, we'll discuss a number of recommended best practices as we begin learning how to secure a given WLAN environment against the types of attacks mentioned in this chapter. There is a lot of ground to cover in Chapter 2, much of which is relevant to the daily work and worry of the typical CWSP working on wireless networks. Keep in mind that the tools and methods used by intruders are constantly being refined and improved.

Passive WLAN Discovery

802.11 wireless network discovery is foundational to using the wireless network in that you must discover it before you can associate to it. The wireless discovery process consists of passive scanning, active scanning, or both. As with many things related to 802.11 wireless, network discovery is important for its overall functionality and as a potentially exploitable juncture in WLAN operations.

Flashback to CWNA: passive discovery uses Beacon management frames, which are transmitted at regular intervals- usually every 100 "time units." Onetime unit is equal to 1,024μs (microseconds). Therefore, the average beacon interval is 100 times 1,024μs or roughly 100ms (milliseconds). Though beacon intervals are adjustable on enterprise wireless APs, the 100ms interval tends to be the de facto default value across most WLAN products and "tuning" it may do more harm than good. During passive discovery, wireless client devices simply listen for beacons. Wireless clients use beacons to identify available Wi-Fi networks and their characteristics, including the capability of each network's type of security. From a security perspective, it is important to understand what information is, and is not, broadcast in beacon management frames. The CWAP (analysis professional) course is heavy on packet and frame analysis, but we also need to talk about it here in CWSP to a certain degree.

Beacons contain a frame body which includes fixed fields and information elements. The security information elements (IE's) that appear in beacon frames for a given SSID will depend on the type of security mechanism for which the network is configured, such as TKIP/RC4 (WPA) or CCMP/AES (WPA2).

Devices that are certified for Wi-Fi Protected Access (WPA) will include a WPA information element in Beacon management frames. To client devices, this information element will identify the supported security features, including the authentication methods in use, whether passphrase or 802.1X/ EAP. For the WPA case, the information element will also specify which encryption type is in use, which would be the Temporal Key Integrity Protocol (TKIP) and the RC4 stream cipher.

Devices that are certified for Wi-Fi Protected Access 2 (WPA2) will include a Robust Secure Network (RSN) information element in beacon management frames. To client devices wishing to associate, this information element, the supported security features. These include the authentication methods, either passphrase or 802.1X/EAP, and the encryption type which is Counter Mode Cipher Block Chaining Message Authentication Code Protocol, Counter Mode CBC-MAC (CCMP) and the Advanced Encryption Standard (AES) block cipher. Keep in mind that the 802.11i amendment that defined RSN parameters also allowed TKIP/RC4 for backward compatibility in an RSN network though the industry has largely moved past TKIP. (As mentioned in Chapter 1, the notion of backward compatibility is pervasive throughout 802.11 standards.)

> **WPA should no longer be used, but you may run into it.** If a wireless device such as an AP is configured for WPA2 and WPA for backward compatibility, then both the RSN and WPA information elements will appear in Beacon management frames.

Beacons are one of the most important and prevalent frame types in Wi-Fi and can be exploited in a number of ways. The Beacon management frame, which allows for the passive scanning process by wireless clients, includes the following basic information of potential interest to attackers:

- **Capability Information** – This includes information related to operational modes, whether it is a BSS or IBSS and other capabilities.

- **SSID** – This is the name of the BSS/ESS.

- **Supported Rates** – This is the list of data rates supported, based on configuration, and is further expanded by the Extended Supported Rates field.

- **RSN-IE** – This shows the configured security capabilities of the network from a robust security perspective.

Active WLAN Discovery

Active scanning is another method of WLAN discovery and uses both a Probe Request management frame (sent by client devices) and a Probe Response management frame (sent by the AP) in the discovery process. Wireless client network adapters will scan all RF channels they support, which may include the 2.4 GHz band and the 5 GHz band (depending on model) through 802.11ax, in an effort to quickly locate WLANs that are available. The Probe Request is either aimed at a specific SSID or is very commonly used to find all SSIDs within radio range. When looking for any available network, the destination address (DA) is a broadcast address. All APs (infrastructure) or client devices (ad-hoc) on that RF channel that hear the Probe Request will answer with a Probe Response frame.

It is important to understand what information is broadcast in the Probe Request and Probe Response frames, and the differences between the two. Like Beacon management frames, the Probe Request and Response frames each contain a frame body with fixed fields and information elements. Though there are commonalities, the contents of the frame body are different for both management frames. Some of the information contained with Probe Request frames is the Service Set Identifier (SSID), supported basic data rates and extended supported rates. This frame contains limited information compared to the Beacon management frame.

The 802.11 standard requires that all devices (such as an AP in infrastructure mode or clients in an IBSS) that hear a Probe Request frame must answer with a Probe Response frame. The Probe Response frame contains much of the same information as the Beacon management frame, which identifies the specific capabilities of the service set. In addition to the SSID and supported data rates, the Probe Response frame also contains security-related information such as WPA and RSN information elements. The Probe Response frame is a directed management frame and is sent to the MAC address of the device that sent the request. There is a lot of back and forth chatter, and we are reminded that 802.11 depends on significant management traffic to work.

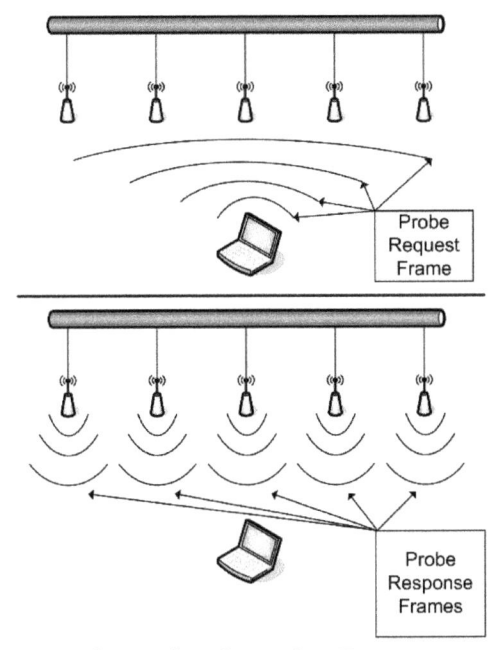

The Active Scanning Process

The Probe Request frame the client transmits may contain a specific SSID value which will identify only the networks it will associate to, or it may contain a "Broadcast SSID" as a wildcard (blank) SSID allowing the device to connect to any wireless network that responds. What determines whether a client will only talk to a specific SSID or is interested in any available SSID? The way the client device is configured makes that determination. For example, the typical off-the-shelf configuration for wireless client devices' Probe Requests is to transmit the wildcard/any SSID, but a locked-down laptop in a corporate wireless environment might only allow for a specific SSID. More on this follows later in the chapter.

Discovery Hardware

In order to connect to a wireless AP, a compatible client device with a wireless network adapter and client software is required. The types and numbers of Wi-Fi client devices on the market is constantly growing. You may find interesting mixes of device types on the WLANs that you support, and all do the discovery process. Discovery hardware can be used to legitimately detect and connect to WLANs, or it can be used to seek out unprotected wireless networks and gain uninvited access to the resources of the network that are located behind the AP. Unauthorized intruders may prefer to use lightweight, unobtrusive client platforms to perform discovery and exploitation of unprotected WLANs. Equipment such as laptops and tablet PCs make for powerful exploitation platforms, but the smaller size and convenience of handheld devices may be preferred when it comes to wireless attacks in certain scenarios.

With the proliferation of high-end smartphones and small computers like the Raspberry Pi, wireless users, IT staff, and attackers are all enjoying a range of very interesting client hardware options. Wireless support staff may use common client devices in tracking down wireless rogues, or they may use add-on hardware and/or utilities that provide Spectrum Analyzer or Packet Analysis functionality to help detect signals that shouldn't be present in their environments.

> When using a spectrum analyzer to locate rogue devices based on signal strength as an indicator of proximity, look at the *amplitude* **of the energy in the** *FFT* **view**. By contrast, when using a protocol analyzer to locate rogue devices, **use the** *signal strength value* **in frames from the target devices**. Similar information is being shown, but the spectrum analyzer works at Layer 1 while the protocol analyzer, in this case, is working at Layer 2.

At the basic level, all discovery devices require a wireless client adapter, an antenna, and discovery software. While most modern off-the-shelf client devices have built-in adapters and antennas, there's more than meets the eye in this space. There are many variations on the radio cards and antennas that are available, and custom configurations can offer an extended range and sensitivity at surprisingly low prices – an advantage to the unauthorized intruder. A high-gain antenna connected to the right wireless client adapter can significantly increase the operating range of wireless intruders to the point where they can clandestinely operate on your WLAN with pretty much no chance of being seen.

Wardriving is the term used to describe the act of performing a mass WLAN discovery activity while logging the discovered AP location information to a file for later analysis. Wardriving was all the rage when wireless networking was a newer technology and not so well understood by the masses, but still has relevance even in the current age of "Wi-Fi everywhere." The term wardriving is taken from the 1983 movie starring Matthew Broderick called WarGames, in which an automated software application was used to scan for open telephone modem connections – "war dialing." Wardriving is simply the act of performing a WLAN discovery while driving through a business park or residential area. Today's wardriving generally turns up far fewer open networks than "back in the day," but there are enough vulnerabilities with certain security types that the activity still has relevance.

Wardriving is often conducted in a secretive manner and is usually considered to be illicit activity. However, the legality of wardriving in the US is not clearly defined, and there has never been a criminal conviction for simply wardriving. Most of those who fear wardriving are under the impression that the perpetrators are in the act of accessing the wireless networks they find (piggybacking), but the nature of most wireless network scanning applications such as inSSIDer and Kismet do not allow this. These applications take control of the wireless network station adapter and do not allow them to associate with the discovered APs at the same time the discovery process is working. Therefore, wardriving can primarily be construed as a data-gathering activity. Keep in mind though that some client devices can use multiple adapters simultaneously, to both monitor and to connect to the networks being heard. And even though "wardriving" implies a vehicle is involved, the same data gathering can be done while walking, cycling, or even from a boat if RF from nearby wireless networks can be received.

The use of a Global Positioning System (GPS) device connected to a discovery PC can augment the efficacy of wardriving (it also gets used with outdoor site surveys, as a point of reference). GPS greatly increases the efficacy of location charting software by assigning a precise latitude and longitude position to each AP in the discovery listing. The GPS unit usually connects via USB and is pivotal in automating location recording when covering a large area or while operating in unfamiliar locations. While this positioning information only indicates where the GPS receiver was located when it received each AP's signal, it can still be a very useful tool for locating nearby networks and either coming back or directing others to those networks. Additional software can be used to take the raw discovery location logs and convert them into graphical map representations.

A custom-built Spectrum Analyzer that can also be used as a protocol analyzer. This example is based on a Dell Windows 8 tablet with an external USB adapter.

Wireless network scanning applications that can be used for wardriving may operate in either active mode or in RF Monitor mode. Active mode applications such as the original NetStumbler, issue probe requests to nearby listening APs using the standards-mandated broadcast (wildcard) SSID. Any APs that are not explicitly restrained from answering these probes will respond with a matching Probe Response that contains - among other critical pieces of information - the current SSID of the answering AP. Listen-only RF Monitor mode applications such as the Linux-based Kismet simply listen quietly for various types of management messages such as authentication and association exchanges, which also contain the SSIDs of the nearby networks. Active mode applications may be effective at gathering discovery information from devices that have been tailored with rudimentary security mechanisms such as SSID hiding. Where a discovered WLAN is found to be secured, the type of wireless security in use, and the skill of the intruder will determine if any SSIDs found might ultimately be exploited. Open networks are obviously at great risk.

Discovery Software

In addition to wireless cards and antennas, discovery stations require software in order to locate and connect to nearby APs. This software is typically called a client utility, or simply a client. Some computer operating systems (like Windows) include only basic wireless connectivity support from within the operating system itself, with few options beyond client credentials and basic wireless profile settings. More features tend to come with wireless adapter's manufacturer utilities that usually accompany certain new Wi-Fi adapters. The Intel ProSet utilities are a long-running example.

A growing number of Apps can also be installed on some mobile devices, providing far more features than those that are built-into the mobile operating system. One multi-OS example of these apps is inSSIDer, which is a popular Wi-Fi discovery application. It's a favorite especially among Android users in the Wi-Fi support space and brings functionality to Wi-Fi devices that the operating system itself cannot provide. Once free, inSSIDer now has a price.

Specialized client applications may be used to perform wireless network discovery, site surveys, security auditing, wireless intrusion detection and mitigation, spectrum analysis, protocol analysis, and endpoint security. Since Linux is typically the easiest to develop applications for, it is often the operating system of choice for wireless hackers. At the same time, every modern OS has some form of wireless utility that can be used for legitimate support- or nefarious purposes.

Discovery software may also fit into the category of online databases which are populated with information that's been provided by the hacker community via websites. Frequently, the information gathered during a war-drive will be published to a publicly viewable, online repository. Several such databases exist, but they have varying degrees of accuracy and may not be up to date. When performing an initial security audit in an organization that has a pre-existing WLAN installation, it may be a good practice to check the online databases to see if the organization's APs are currently listed there, and what vulnerabilities the hacker community claims are in play for those APs.

Popular Public Access Wi-Fi Databases:

- RadioCells (formerly openBmap) - https://radiocells.org/
- Skyhook - www.skyhookwireless.com/coverage-map (now requires an account)
- Wigle - www.wigle.net

Weakest Link

The overall security of any network is only as strong as the weakest link, and this certainly includes wireless networks. For example, let's say that you have 50 devices that are connected to the wireless network, and 49 devices use the strongest security available (WPA2 – CCMP/AES) while one legacy device uses WEP because it can't support better security options. The security of the entire network is assumed to be diminished to WEP because that is the lowest level of security that is in place on the network. Let's look at other potential wireless security weak links.

MAC address filtering is a popular control mechanism whose security value is frequently overestimated. Though MAC filters have their place in networking, relying on them for wireless network security is not recommended, as we'll cover later in the chapter.

Another example of weak or mistaken security is when a network's SSID is hidden to secure the network. In reality, hiding the SSID does not offer any wireless security whatsoever. Some choose to hide the SSID for various reasons, but doing so should not be considered a security strategy. Some client devices struggle when the SSID is hidden. SSID hiding is covered in more detail in the next section.

Wireless networks have challenges from both the troubleshooting and security perspectives that you may not see in a wired network infrastructure. These challenges exist because the communication medium is RF in free space, which, as we've established in Chapter 1, is an unbounded medium. Taking this into consideration is critical when it comes to understanding and implementing wireless network security.

Weak Link

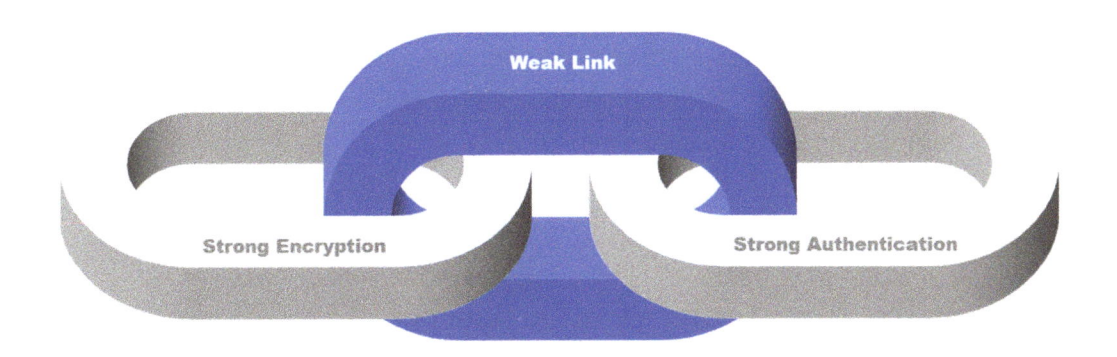

Strong Encryption — Strong Authentication

- WEP, SSID Hiding and MAC Filtering
- WPA/WPA2, 802.1X/EAP

Weak Links Break the Chain
The security of a wireless network is only as strong as the weakest link. In other words, if legacy security solutions such as Wired Equivalent Privacy (WEP) are in place you can consider that the best you will have.

At the same time, 802.11 networks are not completely different than wired networks when it comes to security concerns. It's worth mentioning again that wireless networks extend the wired network, so there are common elements in play and threat factors to consider regardless of how network clients actually connect to a given network for access. Some of the attacks that may be common with both wired and wireless networks include but are not limited to:

- Denial of Service (DoS) attacks
- Phishing attacks
- Protocol weaknesses
- Configuration error exploits

If this were an Ethernet security course, all of these would still be talked about. The specifics would be different, of course, but the over-arching concerns pervade both wired and wireless and happen anywhere that client devices and people using those devices are in play. Let's look at the wireless-specific characteristics of each of these attacks as we continue our "weak link" analysis.

Common DoS attacks can exist at both the Physical Layer and the Data Link Layer. For Wi-Fi, RF jamming (using wideband or narrowband noise) is the most obvious Layer 1 DoS. Other DoS attacks are due to exploits that have been discovered in the 802.11 protocol itself (MAC) at Layer 2. You will learn more about DoS attacks later in this chapter.

Phishing is a method used by attackers to gather valuable information. This information can be of a sensitive nature and usually includes login credentials (usernames and passwords) and other information that may provide access to financial institutions, credit card accounts, medical records, and other sensitive resources. Various methods are used in phishing attacks, including email messages, websites, telephone calls, and other electronic communications.

In the very basic sense, the 802.11 protocol was designed in a way that will allow devices to politely share the wireless medium in a one-at-a-time construct. Great care was taken when 802.11 was developed in allowing each wireless device to have an unimpeded turn, with many management and control frames required to keep it all orderly and precisely timed. Unfortunately, this methodology comes with its share of security concerns due to exploits that have been developed over the years. Some 802.11 management frames such as deauthentication and disassociation are required for basic protocol operation but can also be exploited for malicious intent. For example, the hijacking of authorized user devices and certain denial of service attacks both make use of deauthentication and disassociation frames. We will discuss how the concept of Management Frame Protection (MFP), which is optional through WPA2 but mandatory in WPA3, can somewhat help prevent against management frame attacks later.

Incorrect configuration of infrastructure devices tends to be due to human error and leads to other potential security concerns. Human beings are very frequently the weakest link in the security chain. Though misconfiguration is more common in home and small business networks because the individual installing the device may lack the knowledge required to correctly configure the devices, the stakes get much higher in business network settings. At home and in small business settings, simply following directions provided by the manufacturer of the infrastructure devices may be enough to provide a basic security posture. On corporate networks though, it's reasonable for professionals to lock network infrastructure devices down to the strongest possible configurations, and then relax settings as operationally appropriate. The configuration of infrastructure devices used with enterprise networks will be primarily driven by the corporate security policy, which helps to lessen the possibility of misconfiguration. Policy alone cannot do much good if it's not put into practice by skilled administrators who have the authority to enforce it, though.

> Creating and using a checklist is a great way to ensure that all bases are covered when it comes to securing network devices. A common checklist used by all team members will help to lessen the possibility of misconfigurations. The book, *The Checklist Manifesto – How to Get Things Right*, showed clearly that checklists work in projects of all sizes. Many high-importance tasks from military operations to flying an aircraft, to building a house, rely on checklists which can be thought of as mini project-plans. -Tom

SSID Hiding

Recall from CWNA that an SSID is more than just a network name. The Service Set Identifier (SSID) is used for wireless network identification and segmentation and allows the naming of service sets much like Microsoft Windows uses workgroups to group computers and other devices that need to function together. Other characteristics include:

- The SSID is included within several different management frames.

- Legacy security tactics may suggest hiding the SSID from intruders, although doing so provides no security.

- Current security methods adequately protect WLANs, making the hiding of SSIDs further unnecessary. SSID hiding is a technique implemented by WLAN device manufacturers that will remove the information found in the SSID information element from Beacon management frames. Depending on the implementation, SSID hiding may also remove it from Probe Response frames sent from the AP. Hiding the SSID is intended to keep casual users from noticing a wireless network but may be problematic for certain client devices. This does not offer any legitimate protection because many software utilities, and all protocol analyzers, can find the SSID in 802.11 management frames other than beacons. There simply is no "hiding"! Reducing the transparency of the SSID is the best that can be done.

WARNING: Disabling the broadcasting of the SSID within the Beacon frames is not an effective deterrent and adds no value to WLAN security.

Some organizations will hide the SSID on all WLAN profiles except for the profile that is used for guest access. This decision is not for security purposes but rather to prevent users that do not belong to the organization from attempting to connect to wireless networks in which they do not have the proper credentials. This, in turn, may help lessen unnecessary technical support calls and can thus be seen as a convenience measure (users who need to connect to hidden WLANs can be provided with the configuration profiles). Although this is not a recommended security procedure, it is a configuration option that may be used in some installations for management purposes.

Consider the flawed explanation provided by one SOHO AP manufacturer on the topic:

"It is possible to make your wireless network nearly invisible. By turning off the broadcast of the SSID, your network will not appear in a site survey. Site Survey is a feature of many wireless network adapters on the market today. It will scan the "air" for any available network and allow the computer to select the network from the site survey. Turning off the broadcast of the SSID will help increase security."

This type of messaging, though well-meaning, can mislead inexperienced users to enable SSID hiding as a standalone security mechanism. Thankfully, most manufacturers have advanced past this sort of poor advice in their product literature.

SSID Field in Other Frames

In addition to Beacon management frames, the SSID information is also included in several other 802.11 common management frames which are:

- Probe Request
- Probe Response
- Association Request
- Reassociation Request

For protocol functional purposes, the Association Request and Reassociation Request frames will always contain the SSID. The SSID cannot be removed from these frames, or the 802.11 protocol would cease to function.

Hiding the SSID will only remove it from the Beacon management frame. The information element in this frame is still intact; however, the SSID value is removed from the frame itself.

Authentication Request, Authentication Response, and Association Response frames do not contain the SSID Information Element.

Many current discovery software applications and most protocol analyzers will be able to identify the SSID regardless of whether it is hidden in Beacon management frames. Once a user has associated with an AP, a discovery utility can gather the SSID from other management frames. An intruder could simply wait for a new association to occur or actively force users to deauthenticate and then quietly learn the SSID when the devices reassociate to an AP. With that said, we can see that even if the SSID is not broadcast in the Beacon frame, most enterprise-quality packet analyzer tools (and some of the free ones) have the capability to learn what the SSID is from the other management frames and will display the SSID value.

Broadcast SSID in Probes

The 802.11-2016 standard defines a Broadcast SSID as a wildcard (blank) SSID. 802.11-2016 requires APs to respond to all Probe Requests that contain either a matching SSID or a blank SSID. Many manufacturers provide the configuration option to prevent APs from responding to Probe Request frames, even though 802.11-2016 still requires it. Curiously, the Wi-Fi Alliance does not deny certification to manufacturers who disable Broadcast SSID responses.

From IEEE 802.11:

STAs, subject to criteria below, receiving Probe Request frames **shall respond with a probe response** *only if*

> a) The Address 1 field in the probe request is the broadcast address or the specific MAC address of the STA, and either item b) or item c) below.
>
> b) The STA is a mesh STA and the Mesh ID in the probe request is the wildcard Mesh ID or the specific Mesh ID of the STA.

    c) *The STA is not a mesh STA and*

        1) *The SSID in the probe request is the wildcard SSID, the SSID in the probe request is the specific SSID of the STA, or the specific SSID of the STA is included in the SSID List element, and*

        2) *The Address 3 field in the probe request is the wildcard BSSID or the BSSID of the STA.*

Probe Response frames shall be sent as directed frames to the address of the STA that generated the probe request. The SSID List element shall not be included in a Probe Request frame in an IBSS.

Hopefully, you now understand that SSID hiding should not be considered a security solution by any measure. In review- for some scenarios, the hiding of one or more SSIDs may play the role of assisting with management or usability of a given WLAN environment, but it's never considered a security solution, as the SSID is still present in a number of management frames. Free utilities combined with low skill will show hidden SSIDs. Bottom line: no security is gained by hiding the SSID, period.

MAC Address Filtering

We touched on this at the beginning of the chapter, now let's see what is problematic with MAC Address filtering. Media Access Control (MAC) filtering is considered by some to be an effective deterrent to prevent casual or unintentional system access to a wireless network. Before we go any further, understand that CWSP-quality security philosophy dictates that MAC address filtering should not be used as a stand-alone security solution in anything beyond a home network. Even then, there is no real reason to use MAC filtering with the availability of WPA2-Personal in all consumer equipment today.

Since 802.11 WLAN device technology operates at the Physical and Data Link Layers of the OSI model, the MAC address is a big part of the wireless networking process. The MAC address, which is defined at the MAC sublayer of the Data Link Layer (Layer 2), identifies the network interface by the use of a manufacturer-assigned unique physical address or potentially a software-assigned address. MAC addresses on wireless devices are sometimes (incorrectly) called Ethernet addresses.

The purpose of MAC address filters is to allow or disallow access to the wireless network by restricting which MAC addresses can authenticate and associate to the network using 802.11 technology. Procedurally, MAC addresses are manually entered into the wireless AP or are populated by instructing the AP to add all currently connected MAC addresses, which will identify the specific devices that will be allowed or denied access to the wireless network. This is called a white list.

> In some vendors' 802.11 mesh topologies and Network Access Control (NAC) systems, there may be dynamic MAC filtering happening as part of the overall workflows that control the nuances of those systems. This is different than using MAC filtering as a security solution.

From the perspective of manageability, MAC address filters can be reasonably enabled for small numbers of client devices but can be tedious and prone to entry mistakes when the list gets large. If MAC address filters are the only deterrent, intruders can easily discover the MAC addresses that are permitted by eavesdropping and then readdress their own station adapters with an allowed MAC address to defeat the filter and gain access. This process is called MAC spoofing and is the primary reason why MAC filtering should not be used as a WLAN security solution. Next, we will discuss how easy it is to find MAC addresses and to carry out MAC spoofing.

Finding Valid MAC Addresses

MAC address-based access control lists (ACLs) can certainly provide a degree of admission control for client devices, but the process is subject to extremely simple exploitation. Discovering which MAC addresses are authorized on a system is simple when using a wireless scanner or protocol analyzer software program. Any device that is successfully passing data traffic to an AP on the wireless network is considered an authorized device, allowed through the MAC filter, if in use. An allowed MAC address can be copied and used for connectivity to the wireless network. MAC addresses can be easily spoofed using operating system techniques or third-party freeware utilities.

Keep in mind that the MAC addresses cannot be encrypted, as specified in the 802.11 standard. This physical identifier (MAC address) is broadcast in plain clear text and is shown in protocol analyzer frame decodes. Therefore, it is very easy for an intruder or anyone with a limited amount of technical knowledge and the proper software tools to identify authorized wireless networking devices from a simple scan of the unbounded RF medium used by the WLAN. Seem too easy? As mentioned in Chapter 1, the authors of the original 802.11 standard just couldn't foresee how big of a target Wi-Fi networks would become to those with nefarious intent.

One analogy to consider when contemplating the MAC address construct is the physical address of a home or building on a street. Each building is marked with a unique physical address to provide an identity for the building, for example, 123 Main Street. The street name is comparable to the SSID of the wireless network since all connected devices share the same SSID. So, 123 would be comparable to the MAC address of a connected device which is the unique identifier.

Anyone who wanted to visit this building could easily identify it from the marking of the numbers 123. If these identifying addresses from all of the buildings on the street were missing, encrypted, or scrambled in any way, there would be no way for a visitor to find the correct building. But since they are not, the buildings can be found easily. On WLANs, the same is true. The addresses are not encrypted and can be identified easily.

374	Ethernet Broadcast	00:0F:B5:6E:32:4A	00:0F:B5:6E:32:4A	802.11 Beacon	11	1.0
375	Ethernet Broadcast	Intel:4F:CE:66	Ethernet Broadcast	802.11 Probe Req	11	1.0
376	Ethernet Broadcast	Intel:4F:CE:66	Ethernet Broadcast	802.11 Probe Req	11	1.0
377	Ethernet Broadcast	00:0F:B5:6E:32:4A	00:0F:B5:6E:32:4A	802.11 Beacon	11	1.0
378	Ethernet Broadcast	00:0F:B5:6E:32:4A	00:0F:B5:6E:32:4A	802.11 Beacon	11	1.0
379	00:0F:B5:6E:32:4A	00:11:F5:1B:E4:79	00:0F:B5:6E:32:4A	802.11 Auth	11	1.0
380	00:11:F5:1B:E4:79			802.11 Ack	11	1.0
381	00:11:F5:1B:E4:79	00:0F:B5:6E:32:4A	00:0F:B5:6E:32:4A	802.11 Auth	11	1.0
382	00:0F:B5:6E:32:4A			802.11 Ack	11	1.0
383	00:0F:B5:6E:32:4A	00:11:F5:1B:E4:79	00:0F:B5:6E:32:4A	802.11 Auth	11	1.0
384	00:11:F5:1B:E4:79			802.11 Ack	11	1.0
385	00:11:F5:1B:E4:79	00:0F:B5:6E:32:4A	00:0F:B5:6E:32:4A	802.11 Auth	11	1.0
386	00:0F:B5:6E:32:4A			802.11 Ack	11	1.0
387	00:0F:B5:6E:32:4A	00:11:F5:1B:E4:79	00:0F:B5:6E:32:4A	802.11 Assoc Req	11	1.0
388	00:11:F5:1B:E4:79			802.11 Ack	11	1.0
389	00:11:F5:1B:E4:79	00:0F:B5:6E:32:4A	00:0F:B5:6E:32:4A	802.11 Assoc Rsp	11	1.0
390	00:0F:B5:6E:32:4A			802.11 Ack	11	1.0
391	Ethernet Broadcast	Intel:4F:CE:66	Ethernet Broadcast	802.11 Probe Req	11	1.0
392	Ethernet Broadcast	00:0F:B5:6E:32:4A	00:0F:B5:6E:32:4A	802.11 Beacon	11	1.0
393	Ethernet Broadcast	Intel:4F:CE:66	Ethernet Broadcast	802.11 Probe Req	11	1.0
394	Ethernet Broadcast	00:0F:B5:6E:32:4A	00:0F:B5:6E:32:4A	802.11 Beacon	11	1.0
395	FF:6E:FF:7F:B7:FF	20:87:23:C7:5B:B7	9B:CB:F6:FF:FB:D9	802.11 Probe Req	11	1.0
396	Ethernet Broadcast	00:0F:B5:6E:32:4A	00:0F:B5:6E:32:4A	802.11 Beacon	11	1.0
397	Ethernet Broadcast	Intel:4F:CE:66	Ethernet Broadcast	802.11 Probe Req	11	1.0
398	Ethernet Broadcast	00:0F:B5:6E:32:4A	00:0F:B5:6E:32:4A	802.11 Beacon	11	1.0
399	Ethernet Broadcast	00:0F:B5:6E:32:4A	00:0F:B5:6E:32:4A	802.11 Beacon	11	1.0
400	Ethernet Broadcast	Intel:4F:CE:66	Ethernet Broadcast	802.11 Probe Req	11	1.0
401	Ethernet Broadcast	00:0F:B5:6E:32:4A	00:0F:B5:6E:32:4A	802.11 Beacon	11	1.0
402	Ethernet Broadcast	00:0F:B5:6E:32:4A	00:0F:B5:6E:32:4A	802.11 Beacon	11	1.0
403	Ethernet Broadcast	Intel:4F:CE:66	Ethernet Broadcast	802.11 Probe Req	11	1.0
404	FF:FF:FF:7F:B7:FF	Ethernet Broadcast	Intel:4F:CE:66	802.11 Management	11	1.0
405	Ethernet Broadcast	00:0F:B5:6E:32:4A	00:0F:B5:6E:32:4A	802.11 Beacon	11	1.0
406	Ethernet Broadcast	00:0F:B5:6E:32:4A	00:0F:B5:6E:32:4A	802.11 Beacon	11	1.0

Protocol Capture Showing MAC Addresses

MAC Address Spoofing

MAC address spoofing is simply the process of altering a MAC address in a computer, so it matches a valid MAC address on the network. Each client device is given its MAC address at the time of manufacture, and you may sometimes hear of a MAC address referred to as burned-in address (BIA) or physical address. This terminology might imply that MAC addresses are permanently written in some unalterable chip within the network adapter, but this is not true. The reality is that MAC address values can be rewritten in software, so the network sees a different value than what the factory assigned. Several MAC Spoofing utilities have been freely available through the history of 802.11, including:

- SMAC
- MAC Makeup
- A-MAC Address
- Nmap ("Network Mapper")
- Systems Lizard

MAC addresses may also be reset with simple tools that are available by default on most computer OSs.

Linux: ifconfig eth0 hw ether 03:a0:04:d3:00:11

FreeBSD: ifconfig bge0 link 03:a0:04:d3:00:11

MS Windows: On Microsoft Windows systems, the MAC address is stored in a registry key. The location of that key varies from one MS Windows version to the next, but simple Internet searches will help you find this value so you can edit it yourself. There are, of course, numerous free utilities you can download to make this change. Additionally, for some NICs, you can modify the MAC address in the Device Manager using the following procedures:

1) Open the Device Manager.
2) Press Windows Key + S.

3) Type Device Manager and press Enter to select it in the results list.

4) Find the WLAN NIC in the Network Adapters node and double-click on it to open the Properties dialog.

5) On the Advanced tab look for a property called Network Address, MAC Address, Locally Administered Address or something similar.

6) Change it there.

> The MAC spoofing process does not always work on Windows systems, even when the feature is apparently available. The wireless drivers for some adapters just don't accommodate the capability.

MAC Address spoofing is also an effective way to bypass other MAC-based security or control mechanisms, such as paid access to hotspots. Some service providers log allowed devices by MAC address. Those wishing to bypass this type of filter on a paid network can simply spoof their MAC address with an authorized device's MAC address. This type of activity is likely illegal along the lines of "theft of services" and, as with other common hacker capabilities you may learn, is not recommended. Hotspot providers should be aware of the limitations of their system. At the same time, there is great value in learning how to spoof a MAC, so you have a sense of how the bad guys do it.

Open System Authentication

802.11 Open System authentication must be performed every time a device connects to a wireless network, or anytime it transitions from one AP to another- even on secure networks. This process is a fundamental step in the basic operation of 802.11 wireless connectivity. Without performing this task, a wireless device would not be able to associate to the AP.

From IEEE 802.11:

Open System authentication is a null authentication algorithm. Any STA requesting Open System authentication may be authenticated if dot11AuthenticationAlgorithm at the recipient STA is set to Open System authentication. A STA may decline to authenticate with another requesting STA. Open System authentication is the default authentication algorithm for pre-RSNA equipment.

Open System authentication utilizes a two-message authentication transaction sequence. The first message asserts identity and requests authentication. The second message returns the authentication result. If the result is 'successful,' the STAs shall be declared mutually authenticated.

Based on this description, it should be obvious that Open System authentication alone provides no security whatsoever. Despite including the word authentication, there really isn't any in this case, as any and every device (STA) is welcome to communicate with the AP. Don't let the terminology trip you up.

Though "a STA may decline to authenticate with another requesting STA," it's pretty much a given that devices of like technical capabilities will authenticate with each other. Despite the lack of security, it is important to know that Open System authentication is used as part of the workflow by all strong enterprise security solutions today. For example, when using WPA2-Enterprise, Open System authentication is performed first, and then the EAP authentication occurs followed by the 4-way handshake, which is used to generate the unicast encryption keys used between a single STA and the AP. You will learn about these concepts in greater detail later in this book, but it is important, for now, to understand the foundational importance of Open System Authentication.

Wired Equivalent Privacy (WEP)

By now you hopefully know that WEP should never be used today. It really is that simple. If you desire confidentiality and integrity for your data, you will not use WEP. All hardware released for the past several years will support at least WPA-Personal, which is an order of magnitude stronger in security than WEP. Better still, WPA2-Personal is now pervasive and is the minimum that you should aim for. The weaknesses of WEP will be briefly reviewed in this section in a historical context. To better understand why currently accepted WPA variants are so much stronger, it's worth digging into the flaws of WEP just a bit more. In other words, we study WEP as CWSPs to learn from the past.

From IEEE 802.11:

WEP-40 was defined as a means of protecting (using a 40-bit key) the confidentiality of data exchanged among authorized users of a WLAN from casual eavesdropping. Implementation of WEP is optional. The same algorithms have been widely used with a 104-bit key instead of a 40-bit key in fielded implementations; this is called WEP-104. The WEP cryptographic encapsulation and decapsulation mechanics are the same whether a 40-bit or a 104-bit key is used. Therefore, subsequently, WEP can refer to either WEP-40 or WEP-104.

The characteristics of WEP include:

- RC4 Stream Cipher
- Static Preshared Keys
- Manual Key Management
- Weak Implementation

WEP is unsafe for use under any circumstances or at either key size (mandatory 40-bit or optional 104-bit) because it suffers from multiple weaknesses. Ensuring that no wireless networks in your care ever use WEP should be a top priority.

WEP Weaknesses

WEP required the use of static keys. The selected key would have to be manually entered on all devices that were part of the same service set. In most cases, once the key was determined and entered on all the devices, it was never changed; even as devices left service, or, were lost or stolen along with the configured keys. In theory, changing the key periodically, or at a specific regular interval, would help to provide a more secure network. The important words here are "in theory."

You learned earlier that the 802.11 standard defined a 40-bit WEP key. In addition, a 104-bit key could be used, but not all devices supported that option. 40-bit and 104-bit are the actual key lengths. In addition to the key, WEP also used a 24-bit initialization vector (IV) as part of the encryption and decryption process. Therefore, with the addition of the IV, the key length would be 64-bit or 128-bit. The key can be made up of either hexadecimal or ASCII characters. The length for each is shown in the following table:

Key Length	# of Hexadecimal Characters	# of ASCII Characters
64-bit	13	5
128-bit	26	10

Table 2.1: WEP Key Lengths

The 24-bit IV transmitted across the wireless medium in cleartext makes the WEP key vulnerable to exploitation. This reality was a primary flaw in the WEP implementation. RC4 is not an inherently bad encryption algorithm, but the keys must be implemented in a way that avoids reuse in any reasonable window of time. Unfortunately, WEP was not implemented in this way, leaving WEP vulnerable to key reuse attacks. It's somewhat heady stuff, but further explanation follows.

Two primary problems exist with how the WEP IV mechanism was implemented. For one, the 24-bit IV was transmitted across the air in clear or plain text. Secondly, it was used as an encryption seed in conjunction with the WEP key and the RC4 stream cipher to create a key stream and finally, the encrypted ciphertext message. This was accomplished using an exclusive OR process with the Integrity Check Value (ICV), providing an encrypted frame body for the wireless data frame. Though this process resulted in basic encryption of data, these two items created an ultimately bad combination. If someone was to capture enough of the encrypted frames, the WEP key could be found using any number of key cracking programs.

In addition to the weak IV scheme, the WEP process also suffered from weak integrity protection or ICV. The WEP ICV was computed using the CRC-32 and calculated over the plaintext MAC Protocol Data Unit (MPDU) field. This made the ICV vulnerable to what is known as a "bit-flip attack" which gave someone the capability to capture frames and flip bits in the data payload of the frame. Then the ICV would be modified, and the frame would be retransmitted with the modified data payload. Unknown to the receiver, the data was modified in transit, therefore creating an additional vulnerability and losing integrity – the very purpose of the ICV. It may sound complicated, but these attacks are all too easily accomplished with just a bit of hacking skills, and WEP was not in service long before it was compromised.

In the early days of wireless, cracking WEP became a sport for the hacker community as their obsession with Wi-Fi increased with the popularity of wireless networking. Wardriving and WEP cracking also happened to complement each other for the early ambitious hackers. Since the advent of the 802.11n standard, no APs going forward should even allow WEP as an option. However, on poorly implemented hardware, you may still occasionally see WEP as a choice. You don't have to memorize the explicit details covered here in the examination of WEP's weaknesses, but it's good to know the background on why more robust encryption options were ultimately needed and to appreciate the evolution of encryption methods in wireless security. WEP is not a primary testing topic on the CWSP exam.

Shared Key Authentication

We've already established that 802.11 Shared Key authentication is a deprecated authentication mechanism, but let's look a bit deeper at its flaws. Unlike Open System authentication, which used WEP only for data encryption, Shared Key authentication required the use of WEP for both 802.11 authentication and for data encryption. While it may seem that adding an authentication exchange would enhance a network's security, Shared Key authentication may accelerate the exposure of a static WEP key. Shared Key Authentication was another one of those examples of the naivety of the early standards writers when it came to security.

Notice that in the image, Shared Key authentication uses four authentication management frames that are exchanged between two stations, usually a client station and an AP. Recall that Open System authentication only uses two authentication management frames. For Shared Key authentication to function, the same WEP key must also be installed on all stations that are part of the wireless service set, so you can see that it wasn't particularly scalable.

In this example, the first frame was sent from the client station to the wireless AP, which initiated the Shared Key authentication process. The AP responded to the requesting client station with a clear/plain text challenge message. This challenge text was seen by anyone monitoring the wireless medium with eavesdropping software (protocol analyzers and dedicated cracking tools such as those in Kali Linux). The third frame was sent back to the AP from the client station, with a message that was encrypted using the WEP key assigned to the client station. Keep in mind that this was the same key installed on all devices that were part of the same service set, including the AP. The AP validated the encrypted message and responded to the client device with the fourth frame showing a failed or successful authentication. Once this process successfully completed, the 802.11 association process ensued, and data communications then commenced.

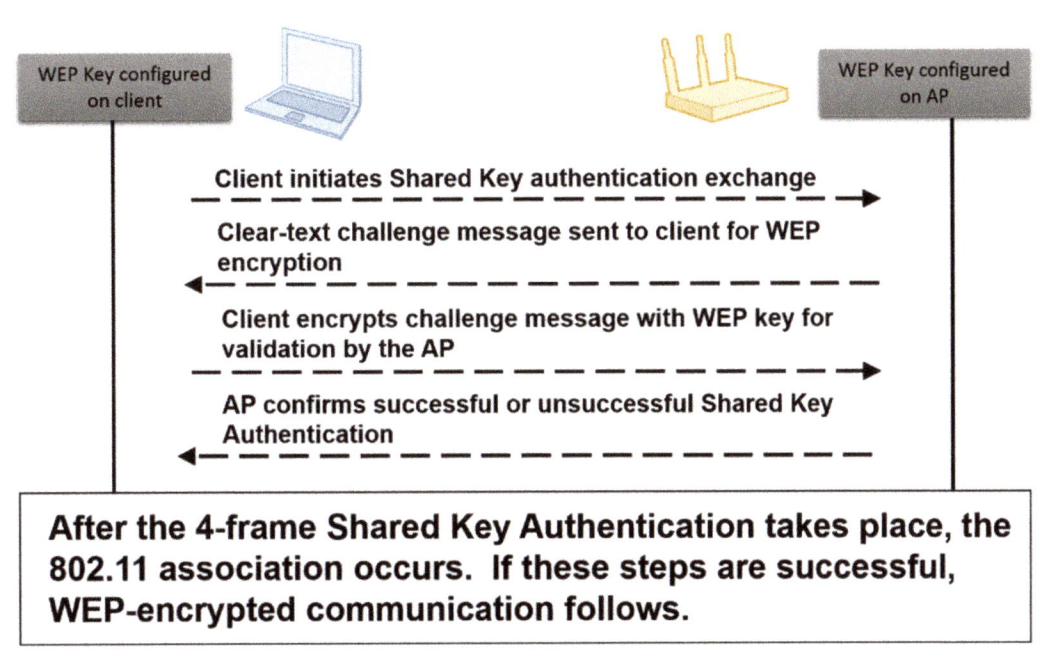

Shared Key Authentication Process

We don't need to rehash all that's wrong with WEP in the context of long-since deprecated Shared Key Authentication. As with WEP, Shared Key Authentication has academic value as we contemplate the bigger WLAN security body of knowledge needed to be successful at CWSP and in our jobs. But, also like WEP, the CWSP exam requires no more than a working familiarity with Shared Key Authentication.

Extensible Authentication Protocol (EAP)

You learned in the first chapter that 802.1X, which defines port-based access control, helps to provide a secure, scalable, and manageable security solution for enterprise wireless networks. It's also important to note that 802.1X is a framework that works in conjunction with an appropriate EAP type to allow for user-based security. There are many EAP variants that can be used to secure WLAN communications, and each is a trade-off between security (and complexity) and ease-of-implementation. Let's explore some of the EAP types that are vulnerable to intrusion and those that should not be used to secure a wireless network. If used, any of these would certainly qualify as "weak links in the security chain."

EAP-MD5

EAP-MD5 is perhaps the most glaring example of a weak EAP type. It was developed for use on the wired network to test basic connectivity between EAP participants. It does not provide dynamic encryption key management, mutual authentication (client trusts the authentication server, and the server trusts the individual clients), or any operational characteristic that would provide security for a wireless network. Because it creates numerous vulnerabilities, EAP-MD5 should never be used to secure an 802.11 network. Unfortunately, it shows up as an option in many configurations and the savvy wireless professional blows right past it.

Proprietary LEAP

Earlier in this chapter, you learned about 802.11 Open System authentication, Shared Key authentication, and WEP. You saw that all these methods are inadequate for providing trustable, secure wireless communications on an 802.11 network. At one point on the 802.11 timeline, the realization that early security methods were not sufficient (by a long shot), hit the wireless industry as the popularity of Wi-Fi exploded and a number of well-publicized LAN breaches occurred. There was an urgent recognizable need for creative methods that would provide stronger wireless security, or Wi-Fi would be stunted in where and how it could be used. Something had to be done!

With its strong wireless security mechanisms, including CCMP/AES, the answer to the security dilemma would eventually be addressed in the 802.11i amendment to the standard. However, in the early 2000 timeframe, the ratification of the 802.11i amendment was still some time away (it would not be ratified until 2004). In the interim, attempts were made to provide alternative wireless security options, like LEAP.

Cisco Systems developed its own EAP type known as Lightweight Extensible Authentication Protocol (LEAP). This proprietary EAP method was very popular because it provided more secure wireless communications and was widely deployed with Cisco networks- a large part of the overall enterprise WLAN market. Keep in mind that LEAP required the exclusive use of a Cisco infrastructure, which included Cisco client devices and wireless APs. One exception to this was via the use of Cisco Compatible Extensions (CCX) technology, which allowed for non-Cisco manufacturers to develop code that allowed their devices to use LEAP on the client device side. There have been several versions of CCX through the years, but non-Cisco adoption has been far from universal. Though LEAP was a decent attempt at bettering the wireless security situation, it too would be found to have shortcomings.

ASLEAP Capturing LEAP Information

LEAP included a vulnerability in which the username of the person attempting to authenticate was passed in clear text across the wireless medium and did not use any tunneling mechanisms to secure the communications. Theoretically, since it used a variant of the MSCHAPv2 hash for the exchange of client credentials, this behavior made captured authentication traffic susceptible to offline dictionary attacks on weak passwords. Joshua Wright is a long-time WLAN security expert who created a software program (named ASLEAP) which improved LEAP's flaws.

> In addition to developing wireless security tools, Joshua Wright has published a wealth of articles on Wi-Fi security and has taught classes on the topic. He's very active in the wireless community and is one of those people that CWSPs would do well to follow as a resource for current WLAN security trends and concerns.

After LEAP's vulnerabilities were discovered and published, Cisco Systems then introduced EAP-FAST, which has since been replaced in many deployments by newer non-Cisco-specific EAP types, such as PEAP, EAP-TLS, and EAP-TTLS.

Eavesdropping

By now, you likely realize that unencrypted wireless traffic is easily intercepted by anyone with a protocol analyzer in reception range. Any client device capable of receiving the WLAN traffic will be able to collect information that traverses the wireless medium. Modern protocol analyzers make it easy to harvest and inspect unencrypted traffic. These wireless protocol analyzer applications use a special network device driver that will allow the wireless adapter on the PC that they run on, to operate in *promiscuous mode*, which in turn makes the analyzer a passive device. You can think of it as being in "listen-only" mode. The monitoring analyzer will then be unnoticed by any intrusion prevention methods because it transmits nothing. Not all wireless adapters can be placed into monitor or promiscuous mode, but there are USB variants that can be purchased specifically for the task of wireless eavesdropping in monitor mode. Protocol analyzer software vendors recommend and sell specific adapters for this purpose. This software and hardware are marketed for analysis of WLANs aimed at performance and functional improvements, such as those covered in the CWAP® certification, but they are also often used by hackers.

802.11-based encryption obscures Layer 3-7 data from protocol analysis and is the basic deterrent to eavesdropping. Using adequate encryption of WLAN traffic is imperative to ensure security and privacy. Unauthorized protocol analysis is the most common form of eavesdropping, but its effectiveness varies with the security of the WLAN being monitored.

Because of the passive methods used by wireless protocol analyzers, there is no way to really detect or prevent this type of eavesdropping, short of physically catching one in the act. The amount of information that can be gathered by eavesdropping on a WLAN with weak security is amazing. What can be learned by passive listening is fairly shocking the first time you see network traffic exposed in this manner. Even with encrypted data payloads, you can still learn significant information, such as:

- Supported data rates

- Allowed MAC addresses
- Security cipher suites and encryption algorithms used
- PHYs supported in the BSS
- MAC and PHY features supported
- The amount of data traversing the network

One popular and easy eavesdropping method is to use the Kali Linux distribution and a compatible USB adapter. Kali can run in a virtual machine (VM) or off of a live disc so that it need not even be your native OS. It comes with all the required tools for wireless eavesdropping preinstalled and is updated often.

Getting Started with Kali for Eavesdropping

Because the best learning comes from hands-on experience, the appendix to this text provides instructions for installing a Kali Linux VM using VirtualBox. It can also be used in VMware Player – a free virtualization environment that can only run a single VM at a time. VirtualBox can run multiple VMs and, for this reason, it is the tool referenced in the appendix. However, here at CWNP, we have found that running Kali Linux in VMware Player typically results in better performance on the same machine.

Assuming you have a compatible USB adapter connected to your computer and passed through to VMware player or VirtualBox running Kali Linux, use the following instructions to perform a capture on any channel you desire:

1. In the Kali Linux VM, logon as a user you have created or as root (the default password is toor).
2. On the desktop, click the Terminal icon to load a terminal (console) session.
3. In the console, execute iwconfig to determine the WLAN adapter name, for example, wlan0 or wlan1 is common.

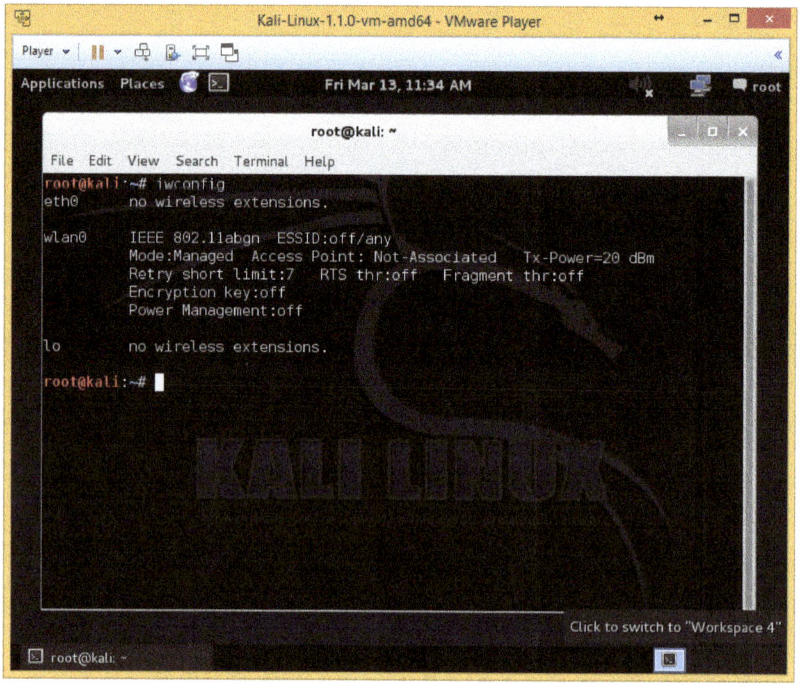
Kali-Linux-1.1.0-vm-amd64 - VMware Player
Player
Applications Places
Fri Mar 13, 11:34 AM
root
root@kali: ~
File Edit View Search Terminal Help
root@kali:~# iwconfig
eth0 no wireless extensions.
wlan0 IEEE 802.11abgn ESSID:off/any
Mode:Managed Access Point: Not-Associated Tx-Power=20 dBm
Retry short limit:7 RTS thr:off Fragment thr:off
Encryption key:off
Power Management:off
lo no wireless extensions.
root@kali:~#
KALI LINUX
Click to switch to "Workspace 4"
root@kali: ~

4. Turn on monitor mode by running airmon-ng start wlan#, while replacing # with the identifier of your adapter.

5. Next execute airodump-ng mon# --band g to look for SSIDs on 2.4 GHz or airodump-ng mon0# --band a to look for SSIDs on 5 GHz, while replacing # with the appropriate identifier for your monitor interface created by airmon-ng, for example mon0.

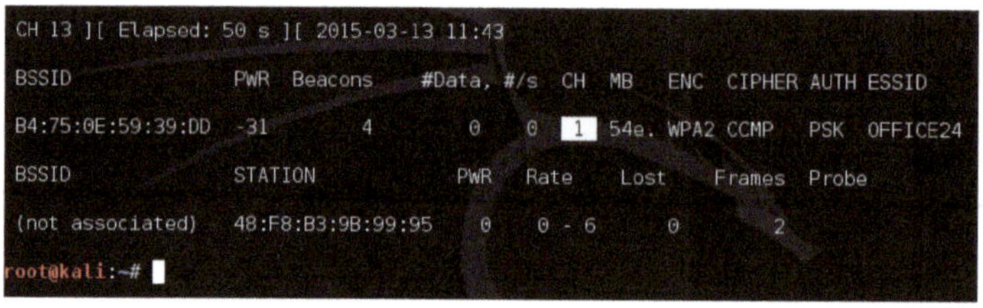
CH 13][Elapsed: 50 s][2015-03-13 11:43
BSSID PWR Beacons #Data, #/s CH MB ENC CIPHER AUTH ESSID
B4:75:0E:59:39:DD -31 4 0 0 1 54e. WPA2 CCMP PSK OFFICE24
BSSID STATION PWR Rate Lost Frames Probe
(not associated) 48:F8:B3:9B:99:95 0 0 - 6 0 2
root@kali:~#

6. Now execute airmon-ng stop mon#, while replacing # with the appropriate identifier(for example airmon-ng stop mon0). You will need to create a new monitor interface on the channel you want to scan.

7. Now execute airmon-ng start wlan# 1 to instantiate a monitor interface on channel 1 (simply change 1 to any other channel you desire, including 5 GHz channels), while changing the # as needed.

8. Run Wireshark by simply executing wireshark from the console.

9. Ignore any root errors; after all, you are performing wireless discovery.

10. After you are in Wireshark, click the mon# interface and then click Start to begin capturing, while replacing # with the identifier of your monitor interface created with airmon-ng.

11. Depending on your adapter and its supported drivers, you will get varying results.

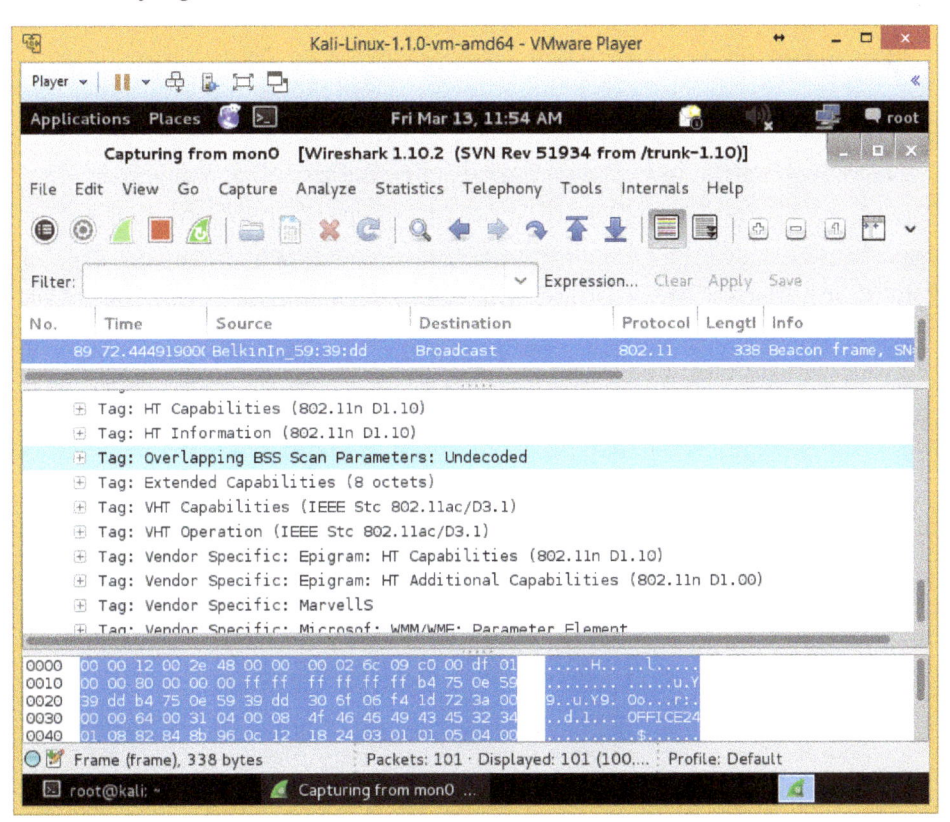

Capturing 802.11 Beacon Frames on Kali Linux

If you are unfamiliar with Wireshark, this is a great time to get your feet wet with it. It is an absolute favorite of wireless and Ethernet network engineers and support staff. It's a powerful multi-OS tool for showing packet and frame-based activity that frequently leads to solving network problems (or exonerating the network when applications are poorly written). It's also a favorite for hackers. "The packets never lie" is a popular saying among those skilled in protocol analysis and as you get used to what Wireshark can tell you, you'll understand why that expression is true. For more detailed information on protocol analysis (as well as spectrum analysis), see the CWAP Official Study Guide published by CertiTrek Publishing.

Social Engineering

Social engineering is a collection of methods used by intruders to gather information which may, in turn, facilitate the ability to circumvent an installed wireless security solution. Social engineering is perhaps one of the easiest ways for someone to bypass even the best security solutions as it takes advantage of the human component in network environments. The people who use and support a network are often quite vulnerable to exploitation by virtue of basic human nature. Most individuals are trusting to a certain degree, so network users can be easily deceived by practiced intruders. As with eavesdropping, there are also tools available to assist with social engineering.

> Without question, social engineering is the easiest method an attacker can use to gain access to, well, anything. It may not be the coolest method for young hackers and crackers, but it is the easiest, as long as you can accomplish one thing: learn to be persuasive. Of course, in this context, we are really talking about being manipulative, but manipulation and persuasion use the same tools – it's the motive that's different. Historically, the strongest of security solutions have been circumvented using social engineering. I see no reason for this to change in the future. Of course, if artificial intelligence (AI) evolves to become intelligent but NEVER emotional, social engineering may not work against AI gatekeepers. But until that becomes true, it will continue to be the most powerful tool in the hacking world. By the way, I just gave you a strong hint, emotional appeals are among the most powerful social engineering tactics. -Tom

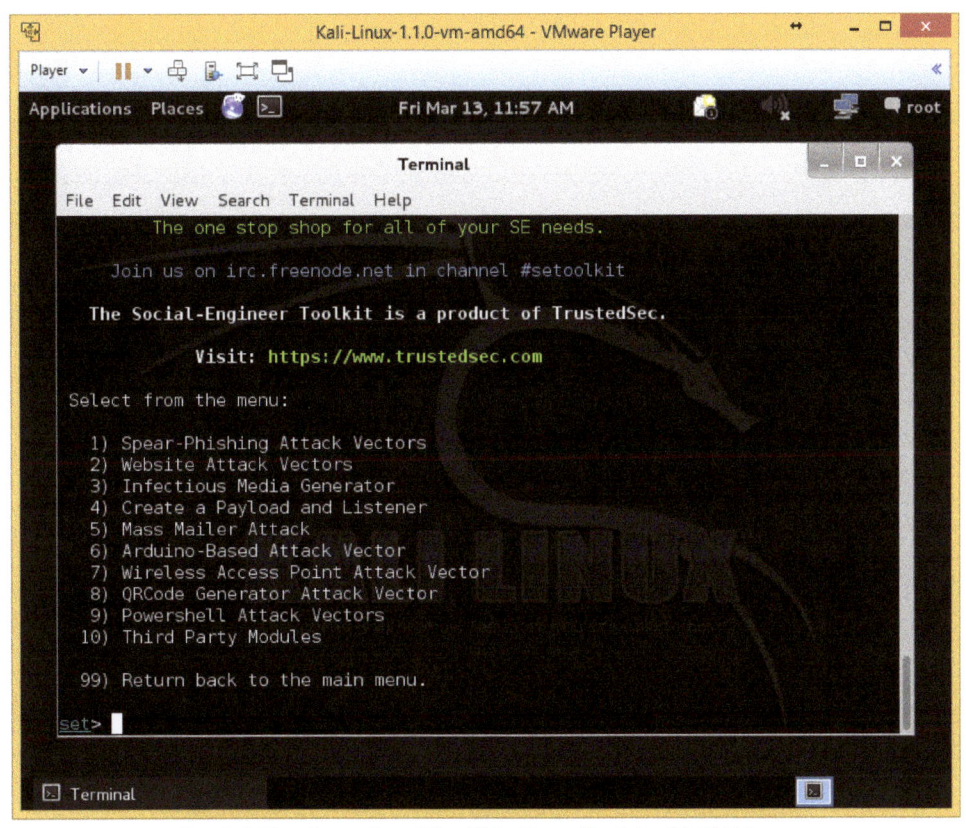

Social Engineering Toolkit Bundled in Kali Linux

Let's consider a simple social engineering example focused on the company help desk. The purpose of the company help desk is to assist users with technical problems. In many computer network installations, the help desk is commonly the first place a user will turn when they are experiencing wireless network problems and are seeking assistance. Help desk staff are in the business of answering questions and helping people as quickly as they can. But if not properly trained and aware of social engineering practices, the help desk personnel can be targets for potential intruders. Some social engineering tactics include calling the help desk and befriending the person that is assisting them through charm to get information such as WLAN passphrases. Another method used is when the intruder places a call into the help desk and requests a password reset for an authorized user account so the intruder can access that account. Social engineering variations are many and include various phishing methods, talking-the-talk with the right people, dumpster diving and other low-tech trickery.

A well-known hacker named Kevin Mitnick often addresses the vulnerability of social engineering. Many of his greatest network attacks occurred by exploiting this weak link in the security chain:

"My message today is primarily the same... I usually go around speaking on the threat of the human element, particularly on social engineering."

– Kevin Mitnick

The Social Engineering Toolkit in Kali, shown in the image on the preceding page, includes several attack vectors:

- Spear-Phishing – send emails with attached file payloads.

- Website Attacks – utilize multiple web-based attacks to compromise site visitors.

- Infection Media Generator – create USB, CD, and DVD autorun modules with a Metasploit payload.

- Creating a Payload and Listener – setup a payload to provide re-entry to compromised systems.

- Mass Mail – send emails to massive numbers through a private mail server or a junk Gmail account.

- Arduino Attacks – used to program Arduino hardware for attack purposes.

- Wireless AP – used to set up fake APs and captive portals to capture user information or infect user machines.

- QRCode Generator – simply generates QRCodes that can redirect people to attack sites.

- Powershell – take advantage of Powershell's power in modern Windows systems to perform attacks.

- Third-Party – several add-ons to extend the features of the toolkit.

This toolkit makes social engineering attacks easy, but most attacks involve direct human interaction. For this, no software can give an attacker the ability to do what some gifted and charismatic people are able to do. Whether you claim to be the CTO (or his assistant) when you demand a password over the phone or pose as a copier repair person to plant a rogue AP, it's the human interactions associated with social engineering that can do the most damage.

Social engineering should be addressed in the corporate security policy for any type of computer network with specifics aimed at the wireless side of the network. Training of all company personnel in awareness of social engineering should be explicitly defined and be made mandatory. Training will help to explain and identify the techniques and methods used in social engineering attacks and to help provide company-wide awareness as the primary countermeasure to social engineering attacks.

RF DoS

Just as the name implies, a Denial of Service (DoS) attack prevents access to a service. With regards to wireless networking, one such attack is an RF DoS. This occurs when the radio frequency necessary for wireless network communications is negatively impacted by external RF sources. This type of attack impacts Layer 1 (the Physical Layer) and will generally fall under one of two different categories, intentional or unintentional.

With standards-based wireless networking, the 802.11 PHYs specify raw RF Energy Detect (ED) thresholds, which will cause the STA to defer transmissions (they will wait to send any data) on a given RF channel. If alternative channels on other APs are not available because the jamming spreads across multiple channels, or because there aren't other APs, a complete network outage may occur as a result of excess RF noise. This is known as a PHY DoS. It's extremely difficult to defend against a PHY DoS in the wireless space.

The first category of this kind of attack is the *unintentional* RF DoS. This denial of service is usually caused by devices that are operating in the same radio frequency space as a given wireless network. The RF could be modulated or unmodulated radio frequency information, which means it may or may not understand or implement 802.11 wireless network communications. An unintentional RF DoS attack could be caused by various devices that use radio frequency, including:

- Microwave ovens
- Cordless telephones
- Baby monitors
- Wireless cameras
- Other 802.11 wireless networks

The second is the *intentional* RF DoS. This type of denial of service attack, which is typically classified as an RF jamming attack, is used to interrupt valid, active RF communications with malicious intent. Intentional jamming can have serious implications on a wireless network, as all RF communications in the range of the jamming device can be stopped. The attack could be used by an intruder to force an authorized wireless network device to reauthenticate and roam to a rogue AP, or to shut down an RF channel or channels – effectively shutting down a wireless network. Such an attack can be performed by devices like:

- RF Jammer, narrowband or wideband
- RF signal generator
- 802.11 wireless adapters using specialized software programs

> Intentional jamming is typically performed on 2.4 GHz spectrum as jammers are easily acquired (though absolutely illegal). If an intentional jamming attack were to occur on a modern WLAN that uses dual-band APs, the chances are that only one band or the other would be impacted as any jamming tool likely to be used wouldn't impact both 2.4 GHz and 5 GHz. A jamming attack can be extremely disruptive, but the range of jamming tools is limited. The same jamming that could lay waste to an SMB Wi-Fi environment would likely only impact a part of a larger WLAN, as even jamming signals are subject to free space path loss.

The best way to protect against an intentional RF DoS is to realize that it's happening and then to employ proper physical security techniques. The best tool to identify this type of attack is an RF spectrum analyzer that covers the correct frequency spectrum used, or a wireless intrusion prevention system (WIPS). While such solutions can detect the attack, they cannot stop it, if it is wideband. To stop the jamming attack, someone will need to find the source and shut it down. A narrowband attack, focused on just a channel or two, may be mitigated through solutions like Radio Resource Management (RRM), which will move an AP to an alternate channel.

Layer 2 (MAC) DoS

Let's remind ourselves that 802.11 wireless networks operate at both the Physical Layer and the MAC sublayer of the Data Link Layer in the OSI model. In addition to the Layer 1 PHY DoS attacks mentioned above, 802.11 wireless networks are also vulnerable to MAC sublayer attacks. This is a result of the manner in which the 802.11 protocol functions, with built-in vulnerabilities as we'll discuss next.

Several different documented MAC sublayer DoS attacks can be used for wireless network exploitation. Due to the half-duplex, shared-media nature of Wi-Fi, 802.11 protocols specify behaviors that require Wi-Fi devices to play nicely together. These same protocols, specified for the operational good of the network, may also be used to carry out DOS attacks.

Common Layer 2 wireless MAC sublayer DoS attacks include using Deauthentication and Disassociation management frames to disrupt normal client connectivity.

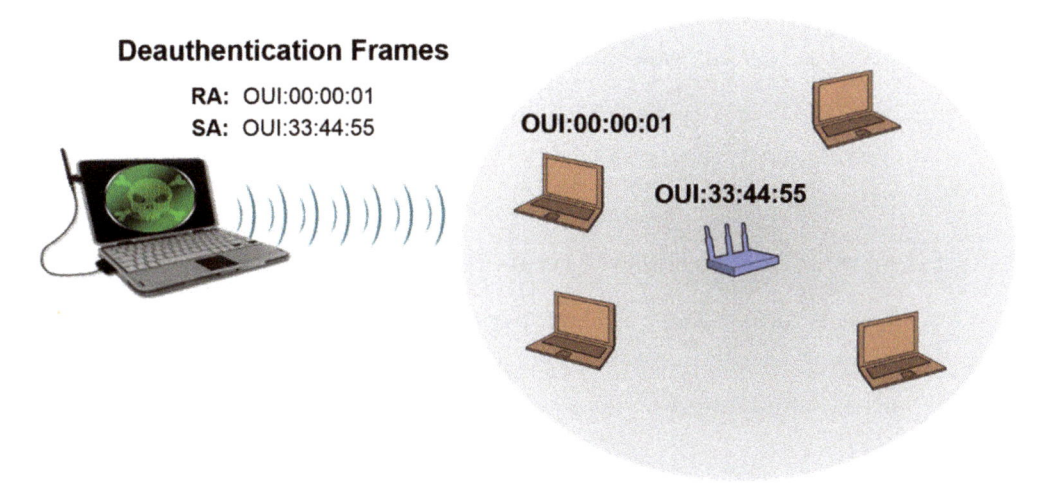

A Deauth DoS Attack

With common tools like protocol analyzers and a centrally managed wireless intrusion prevention system (WIPS), Layer 2 DoS attacks can be identified, and in many cases mitigated. Unlike the PHY DoS attacks, configuration changes can be made to circumvent such attacks more easily.

Deauth DoS with Kali

This is a great exercise in seeing how the bad guys do it, but please limit your experimentation to lab environments under your own control. A Deauth DoS can be performed utilizing Kali Linux with a compatible adapter using the following procedure:

1. Launch the Kali Linux VM (the appendix provides instructions for creating the VM).

2. Logon as root with a password of toor.

3. Launch a Terminal.

4. Determine the proper WLAN adapter identifier based on the instructions on pages 44-46.

5. Execute airmon-ng start wlan0, replacing wlan0 with the proper identifier for your WLAN adapter.

6. Execute airodump-ng mon0 --band a, replacing mon0 with the proper identifier of your monitor interface and --band a with --band g if you prefer to look for attack targets in 2.4 GHz instead of 5 GHz.

7. Note the BSSID of the AP you wish to target. This is the MAC address of the target AP. You can highlight it and then right-click and select COPY to make things easier.

8. Press CTRL+C to end the airodump-ng scan.

9. Execute aireplay-ng -0 0 –a B4:75:0E:59:39:DE –c mon0, replacing B4:75:0E:59:39:DE with your target MAC address and mon0 with the identifier of your monitor mode interface.

Peer-to-Peer

We learned in CWNA that Peer-to-Peer network communications are when one wireless client device connects to another wireless client device (as opposed to talking to a server). These communications in the 802.11 space use the Independent Basic

Service Set (IBSS) (also known as an ad-hoc WLAN) network in which client devices will connect directly to each other, or in infrastructure mode where wireless client devices will connect to each other through an AP. It is important to understand that ad-hoc networks are typically against most corporate security policies; however, they may be allowed in some cases. If they are used, proper security precautions must be taken. Even if they are not "allowed," it is important to remember that, if they are not restricted on the organization's devices, they can still be created by the users. They can also be created on user-owned devices present in the corporate WLAN footprint. Therefore, monitoring for them with WIPS solutions is important if the organizational policy forbids them.

If infrastructure mode peer-to-peer connections are not required, then peer-to-peer blocking should be enabled on the WLAN infrastructure (at the APs or controllers). WLAN equipment manufacturers use different methods to perform this task, but it's usually a simple configuration step. For networks that require this type of communication, like wireless voice handsets or Apple Facetime, peer-to-peer communications will need to be enabled. Many corporate WLAN admins have found that blocking infrastructure peer-to-peer mode just isn't practical given the way many contemporary apps work.

Multiple types of peer-to-peer attacks exist, and they are most common in open public access networks (wireless hotspots) where unsuspecting users unknowingly leave themselves vulnerable to attackers. If the establishment that is hosting the open wireless network did not implement the proper security measures such as peer-to-peer blocking, many experienced attackers would be able to leverage this type of wireless network for a variety of attacks. These attacks include data theft and accessing the client device directly because of weak security on the client system.

Man in the Middle (MITM)

A wireless man in the middle (MITM) attack is the result of an intruder placing an unauthorized wireless device between a legitimate wireless AP and its client devices. The intruder gains the capability to capture and exploit all information that is passed between the authorized wireless client devices and the wireless AP as that information traverses the MITM device. The possibilities of how to leverage the ill-gotten data are endless once a MITM attack has been successful. MITM attacks are not trivial to accomplish but are certainly within the capabilities of experienced hackers. Several steps must be taken to perform a MITM attack, and slight variations exist to the basic construct. We'll look at a common scenario here.

With minimal equipment and easily acquired programs, a MITM attack can be fairly straightforward to those who understand the basic pieces required. It is important to understand what technology is used in this type of attack in order to be able to protect your network from it.

One common method is to use a client device (usually a laptop) with two wireless adapters. One adapter will be used as a "soft AP" (or software AP) for illicit client connectivity, and the other will connect to the authorized AP. The intruder will force the unsuspecting authorized user to connect to the software AP, typically with a deauthentication attack, and will then retransmit to the authorized AP using the second wireless network adapter. The success of this process assumes the attack is performed on an open wireless network (public hotspot) or that the intruder has gained the proper credentials to connect to a secured network. Acquiring these credentials may have been the result of another attack, such as social engineering or shoulder surfing (looking over the shoulder of a target to view the username and password entered).

To perform a successful MITM attack, the intruder must hijack the authorized wireless client device after getting his MITM device set up properly. Hijacking is performed by forcing the authorized wireless client device to connect to the intruder's unauthorized wireless device. Given that a client will usually connect to the AP with the strongest signal strength from the beacon frame (or probe response frame), a deauth attack will typically cause the client to connect to the attacker's software AP (assuming the software AP radio is closer to the target and provides a stronger signal than the authentic AP). Once an authorized client device is hijacked, several other attacks can be conducted.

MITM attacks may lead to very serious security issues, or they may simply be used for eavesdropping purposes in order to gather specific information about the network and the connected wireless devices for planning future exploits. As an example, the attacker may be able to harvest website credentials used by the user, including credentials required to authenticate to a captive portal or other login destination. With adequate protection and proper security measures in place, man in the middle attacks can be prevented. Using 802.1X/EAP authentication makes it very challenging (but not impossible) for an attacker to launch such an attack. Using memes to indicate to the user that they are at a legitimate captive portal login page (such as those used by banking websites) can help, but these depend on the user actually observing that the proper meme (graphic selected by the user at account creation) is in view and not logging in when it is not.

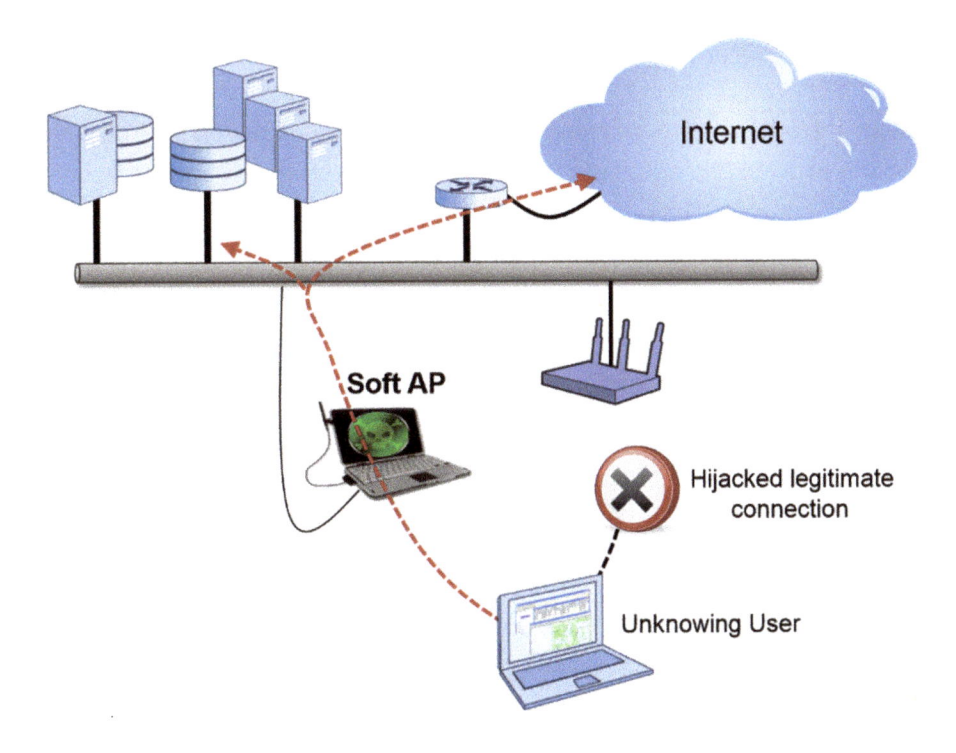

Hijacking Attack

You can effectively perform a MITM attack using Kali Linux on an Open System authentication-only network with three tools: arpspoof, driftnet, and urlsnarf. All three tools are included in the Kalie Linux distribution. Instructions can be found in many online forums, and there are even videos online that will walk you through the process. It's worth mentioning that specialized devices like the various Pineapple models from Hak5 allow for unattended attacks- including MITM. A battery-powered small hacking platform with multiple wireless adapters can be dropped somewhere, and either pre-configured for an attack or reached remotely by the intruder. They are fascinating devices with a lot of capabilities and are a class of hardware you should become familiar with as a wireless security professional.

Management Interface Exploits

Many home and small business network users do not realize the dangers of default configurations as security is frequently not their first concern when setting up new hardware. Wireless equipment manufacturers publish their default configurations and login credentials to ease the setup process, but when these parameters are not changed, they are easy to exploit. One of the first steps in staging an AP (or a managed switch or router) prior to placing it into service is to change any and all default configurations. These changes include the login credentials (username and password), remote access configurations, securing all required access control protocols, and disabling all protocols that are not needed. Changing default configuration parameters should be performed with home, small business, and enterprise installations. While changing the default values, it's also good practice to check for firmware updates that might be available as new firmware may address security flaws.

When managing network devices, weak protocols like HTTP and Telnet send session authentication traffic as clear text, and an eavesdropped session could allow access to an intruder. When management interfaces are accessible to intruders, complete DoS attacks—or worse—are very easy to perform.

Enterprise network deployments will typically specify configuration parameters as part of the corporate security policy. The policy should document all required steps and help to ensure that everything related to staging and management of devices is covered, as the main defense against management interface exploits. Physical access to infrastructure devices like wireless access points is an important topic that must be considered. Unauthorized physical access to these devices can introduce many security issues including theft, device replacement, resetting to factory defaults, access to the console port that is used for configuration, and other concerns. Many solutions are available to help control physical access to APs, including special enclosures and device locks. As with configuration parameters, physical access restrictions should be identified and documented in the corporate security policy.

Some best practices recommend managing wireless infrastructure devices from a wired network connection only, and never from a wireless connection. In the event this is not possible, proper security must be used to ensure eavesdropping will provide no security credentials or parameters that may pose a security risk to the network infrastructure.

> Management interface exploits can be prevented by implementing proper staging and management procedures. Staging includes all tasks during the initial setup of the equipment. Management procedures should be performed over secure channels. Always use encrypted protocols like HTTPS, sFTP, and SSH.

Authentication Cracking

We've covered that some authentication protocols that are (and once were) commonly used with 802.11 wireless networks may be weak and vulnerable to authentication cracking. As a reminder, these include but are not limited to:

- 802.11 Shared Key authentication
- WPA & WPA2 personal mode
- Lightweight Extensible Authentication Protocol (LEAP)
- EAP-MD5
- Point-to-point tunneling protocol (PPTP)
- PAP, CHAP, MSCHAP, MSCHAPv2

In most cases, if these authentication methods are cracked, the process used to encrypt the data will also be cracked. The result is the exposure of network and user data to intruders. It should also be noted that some of these authentication methods can be implemented in a way that makes it difficult to crack them today, particularly WPA2-Personal with a very strong PSK, and MSCHAPv2 when properly implemented in a secure tunnel that is itself established through a mutual authentication process. Now a bit more on each of these authentication protocols, through the lens of cracking:

802.11 Shared Key authentication – the shared key challenge hash is easily recovered and even accelerates recovery of the WEP key, allowing an intruder to authenticate to the wireless network and have access to all encrypted data. Cracking is easy- and thankfully this one is obsolete.

WPA-Personal and WPA2-Personal – though both methods can be reasonably secured with sufficiently strong passphrases (used to create the pre-shared key or PSK), weak passphrases may jeopardize the network security. Weak passphrases can be discovered by capturing the 4-way handshake and the use of dictionary attack software. With the intruder knowing the passphrase, they have the ability to connect to the AP and potentially see user data. Even more devastating than weak passphrases is the publication of KRACK in 2017. This attack took advantage of a weakness in the 4-way handshake process used in WPA/WPA2 and forced pretty much all AP and client device vendors to issue patches. KRACK hits both authentication and encryption. We'll cover KRACK in more detail elsewhere in the text.

LEAP – Recall that ASLEAP is a well-known software utility that can be used to recover LEAP authentication credentials. An intruder can capture the frames used for authentication that traverse the wireless medium in order to recover the user password. The ASLEAP software, in conjunction with a dictionary file, will be able to recover weak passwords.

EAP-MD5 – Simply by capturing a few frames, EAP-MD5 authentication can easily be cracked. This is because it does not provide dynamic encryption key management, mutual authentication, or any characteristic that would provide security for a wireless network. This is one of those authentication methods that should go away as an option but just hasn't yet. Avoid it.

Point-to-Point Tunneling Protocol (PPTP) and other legacy protocols like PAP, CHAP, MSCHAP, and MSCHAPv2 are also vulnerable to authentication cracking. Although PPTP is used with VPN solutions, the authentication process can be cracked and can introduce security vulnerabilities if used with wireless networking. Like LEAP, PPTP allows an intruder to capture the frames used for authentication that traverse the wireless medium in order to recover the user password. L2TP/IPSec has replaced PPTP in many installations where robust VPN security is required.

Encryption Cracking

WEP is the best-known weak encryption scheme used with standards-based wireless networking. It has been around the longest and is essentially the first model of how

NOT to do wireless security. Earlier, you learned about the specific vulnerabilities of WEP and why it should be avoided. Several encryption cracking tools are available for its exploitation. Cracking WEP is a straightforward process and can be achieved with software programs designed for that specific purpose and minimal required effort. Modern software tools even allow for WEP to be cracked without capturing any data frames, while early WEP cracking methods required an intruder to capture potentially large amounts of data traffic (usually 300 or more MB).

TKIP also has well-known encryption weaknesses. In late 2009, it was widely publicized that "new" TKIP weaknesses were discovered and much was made about these vulnerabilities. However, the weaknesses with the TKIP MIC (Message Integrity Check) function were actually known from its inception and were allowed to exist because TKIP was never intended as a long-term solution. Rather, it was introduced as a stop-gap solution for WEP/RC4 while 802.11i was being ratified, and eventually implemented with CCMP/AES. The published TKIP weaknesses may allow an attacker to inject traffic to probe for wired side vulnerabilities and possibly conduct DoS attacks. As TKIP/RC4 has also been deprecated, it should have been largely removed from networks by now. It was a transitional solution, and CCMP/AES is where production networks of any size should currently be.

Although there are strong encryption methods used with 802.11 wireless networks, encrypted data still has the potential to be viewed by intruders. This weakness is based on how the technology operates. One example is with both WPA-Personal and WPA2-Personal modes. Although WPA2 can use strong encryption methods such as CCMP/AES, which is considered uncrackable as an encryption method, the way WPA2 personal mode is designed may allow an intruder to view encrypted user data. This is because the passphrase used with WPA/WPA2 to secure the service set also is used as a seed to ultimately create a unique session encryption key known as the pairwise transient key (PTK). The PTK is basically used to encrypt unicast traffic that traverses the wireless medium. The PTK is created during the 4-way handshake after 802.11 authentication and association and if the passphrase is known, capturing the 4-way handshake will yield enough information to view encrypted data assuming the correct software tools are used. As mentioned above, the KRACK attack has also caused great trouble for even WPA2 networks, and partially the reason why WPA3 will not use the famous 4-way handshake.

> When using WPA- and WPA2-Personal, you must use a strong and long passphrase, preferably one that is 20 characters and not comprised of words. Use a mixture of uppercase letters, lowercase letters, and digits. If you add in special characters, it will just be that much stronger. WPA3 has enhancements that help greatly, even when weak passphrases are used.

Other Common Concerns

Other attacks may take advantage of basic weaknesses in a computer network environment. We've already touched the dangers of physically accessible access points and open Ethernet ports as an ingress point for a number of malicious activities, but let's expand on the physical security notion a bit.

The need for an overall physical security plan cannot be overstated. Theft of WLAN infrastructure is one concern, for sure. But, tampering with installed devices can lead to other security issues. If an intruder were to have physical access to an installed wireless AP, there is the possibility (depending on the manufacturer) that the administrator or logon credentials could be reset to the default configuration or reconfigured with a hacker's credentials, but still, maintain the functional configuration of the device. This would allow the intruder free reign within the actual devices that he has access to and yield the ability to read configuration information and the possibility of creating unauthorized access to the wireless network for later exploitation.

Physical access to unsecured Ethernet ports provides the opportunity for placement of Rogue APs (or nasty platforms like the Pineapple). Keep in mind that a rogue AP is one that is not organizationally authorized but has still been placed into service. Rogues can be installed for either innocent (but usually misguided) or malicious purposes. Rogue APs placed on a network infrastructure with malicious intent may provide an opening to the network for more serious attacks. Rogues are classified as *intended* – installed by intruders – or *unintended* – placed on the wire by employees who didn't know better. Make sure security policies that you are involved with address both types of rogue APs.

Public Access Networks

Public access wireless networks, also commonly called wireless hotspots, have their share of vulnerabilities as we discussed in Chapter 1. This type of network can be a big draw for intruders. Many times, a business like a restaurant or coffee shop will install a wireless AP for the convenience of its customers, and to draw traffic to the business. The infrastructure devices that are installed for public hotspots are often residential or small office/home office (SOHO) grade devices and so lack enterprise security features. Just as bad, those installing these devices also tend to lack the skills to tell the difference between suitable business hardware and those best left at home. Given their limited feature sets, low-end wireless devices may not have the capability to provide additional security features such as peer-to-peer blocking- where it is most needed! Other public access networks may use enterprise-grade equipment and be managed by the corporate IT group or outsourced to a provider with the capability to professionally manage the devices.

Most public access wireless networks will not have any wireless security authentication and/or encryption features enabled because ease of use generally wins out over security for hotspots. They are configured to support standards-based 802.11 Open System authentication only because they want users to connect with the least amount of problems and without the need for support. The CIA philosophy talked about in Chapter 1 is usually set aside in the public wireless model. Security is typically left up to the end-user when it comes to hotspots, which means it's every man or woman for themselves. In many cases, the end user's lack of security knowledge pretty much guarantees problems can happen on public Wi-Fi. Many users have no idea that their network communications traversing the wireless medium can be intercepted and viewed by intruders. End-user education is one defense, but that can be extremely difficult to accomplish.

There is hope coming for security on public wireless networks in the form of Opportunistic Wireless Encryption (OWE), which is part of the Wi-Fi Alliance's Wi-Fi CERTIFIED Enhanced Open™ program. We'll talk more about OWE elsewhere, but it's essentially encryption that happens with no user configuration whatsoever.

The current risks of using unprotected public wireless networks include:

- Spam transmission

- Malware injection
- Information theft
- Peer-to-peer attacks
- Various Internet attacks

If the operator of the public access network knows how to select and configure the APs correctly, it will help to lessen these potential threats to the network and the clients who use it. If they do not know how to perform these tasks or cannot outsource to professional wireless help, then these threats will likely become security incidents. Some of the configurations that will help to lessen these types of threats include:

- Firewall configuration
- Port blocking
- Protocol blocking
- Peer-to-peer blocking

Arguably, the onus for wireless security should not be singularly put on the establishment that is hosting the public access network. Users also have the responsibility of securing the devices they use on the network, though most can be utterly clueless. User responsibilities include using up-to-date anti-virus software, proper firewall configurations, strong passwords, securing any file shares, and using VPN technology.

Hotspot 2.0 and Passpoint may eventually provide solutions to the public hotspot dilemma. They offer the ability for hotspot providers to grant Internet access, but do it using backend authentication systems to which the users are subscribed. Additionally, Opportunistic Wireless Encryption (OWE), which is making its way into most major vendor equipment, can provide a solution as it allows for encryption to be implemented without any kind of really user identity authentication. At the time of this writing in 2019, one can only say that time will tell.

General Recommended Practices

Legacy security is wishful thinking based on a simpler time when the WLAN industry as a whole, was far more naïve about the threats to wireless networks. In the legacy security mindset, an administrator's only hope was to try to hide from intruders or throw a few obstacles in the way of intrusion. This philosophy says that if you put enough inconveniences in the intruder's way, he will find an easier target. CWSPs know not to bet their livelihood on this sort of strategy - especially since modern 802.11 security mechanisms if implemented correctly, will provide strong protection, deterrence, and threat mitigation. Conveniently, modern security steps are actually easier to implement than things like MAC filtering and other weak solutions.

Recommended best practices include the following:

- Use the latest security suite
- Replace legacy solutions and clients
- Upgrade firmware
- Implement Lifecycle Management procedures for WLAN security

All modern Wi-Fi certified equipment should now support WPA2. However, some WLAN implementations may have older specialty devices that are not WPA2-capable and only support WPA. Ideally, these devices should be upgraded, but this may not always be an option due to budgetary constraints, proprietary software implementations, specific use cases, or a variety of other situations.

Replacing legacy WEP security solutions is long past the point where it is a requirement. This is backed up corporate security policies, business models, and regulatory compliance with Health Insurance Portability and Accountability Act (HIPAA), and guideline compliance with Payment Card Industry Data Security Standard (PCI-DSS) and others.

Regularly upgrading the firmware in all devices (this goes for all networks, not just wireless) will provide many benefits. These include adding new features, resolving issues that may have been discovered after the older firmware was released, and opening up enhanced security features such as WPA2. In some cases, firmware updates may not be available because of embedded operating system form factors, end-of-life equipment or other reasons. In situations such as this, it is highly recommended to work toward an appropriate replacement path even if it's inconvenient or expensive to do so. As a CWSP, it's up to you to help bring about needed changes in these situations.

As part of lifecycle management of the WLAN and the devices that use the wireless network, it's important that new devices not just "show up." Ideally, you would identify new wireless devices before they go online and have the opportunity to assess their security requirements and capabilities. There should be no rude surprises at audit time. Lifecycle management helps with that by properly introducing new devices while aging out those no longer needed and adjusting security functionality appropriately.

Chapter Summary

In this chapter, we explored the basics of WLAN security concerns. This began with a review of the 802.11 passive and active discovery procedures, discovery hardware and software, ill-advised concepts like SSID hiding and MAC filtering, weak authentication systems, and basic attack methods.

Review Questions

1. In one WLAN discovery scenario, Probe Requests and Probe Responses are exchanged between the wireless client device and access points. Which type of discovery is this?
    a. Synchronized discovery
    b. Active discovery
    c. Passive discovery
    d. RSN discovery

2. Which of the following SSID descriptions in a Probe Request will elicit a Probe Response from all APs within range?
    a. Blank SSID
    b. Wildcard SSID
    c. Broadcast SSID
    d. All of these, as they mean the same thing

3. An intruder wants to listen to your WLAN and stay undetected as she eavesdrops. What operational mode is she likely to employ on her Wi-Fi adapter?
    a. Listen mode
    b. Stealth mode
    c. Monitor mode
    d. Silent mode

4. Wireless DOS attacks can occur at which layers of the OSI model?
    a. Layer 1, Layer 2
    b. Layer 1, Layer 7
    c. Layer 2, Layer 7
    d. Layer 2, Layer 3

5. Besides being present in the Beacon management frame, which one of the following frames also contains the SSID Information Element?
    a. Association Request
    b. Association Response
    c. Authentication Request
    d. Authentication Response

6. Which of the following regarding 802.11 wireless MAC addresses is true?
    a. They are associated with the Physical Layer
    b. They cannot be encrypted
    c. They cannot be spoofed
    d. They are assigned locally by DHCP

7. Open System Authentication has been part of the 802.11 standard from the very beginning. On modern WLAN, how is Open System Authentication acceptably used?
    a. As part of all strong encryption methods
    b. As a stand-alone security method
    c. As part of the 4-way handshake
    d. There is no acceptable use for Open System Authentication

8. WEP has a number of characteristics which make it weak for wireless network security. Which one of the following is not a characteristic of WEP?
    a. RC4 stream cipher
    b. Manual key management
    c. 40-bit key
    d. Dynamic pre-shared keys

9. When using WPA2 Enterprise as your wireless network security solution, which of the following properly describes the order of functional events?
    a. 4-way handshake, EAP authentication, Open System Authentication
    b. 4-way handshake, Shared Key Authentication, EAP authentication
    c. Shared Key Authentication, EAP Authentication, 4-way handshake
    d. Open System Authentication, EAP Authentication, 4-way handshake

10. When using 802.1X, which of the following EAP types work at Layer 2?
    a. All EAP types work at Layer 2
    b. EAP-FAST
    c. LEAP
    d. PEAP

11. Opportunistic Wireless Encryption (OWE) will provide improved security in which scenario?
    a. Public access wireless networks
    b. VPN
    c. WEP
    d. Shared Key Networks

12. LEAP is a _____ EAP type, crackable by the _____ utility.
    a. Proprietary, Netstumbler
    b. Private, LEAPCRACK
    c. Public, ASLEAP
    d. Proprietary, ASLEAP

13. Encryption methods specified by 802.11 standards obscure data from which layers of the OSI model?
    a. 1-3
    b. 2, 3
    c. 1-7
    d. 3-7

14. Which two EAP types were developed by Cisco for use with 802.1X-based secure wireless networks?
    a. PEAP, LEAP
    b. LEAP, EAP-FAST
    c. PEAP, EAP-FAST
    d. TLS, LEAP

15. An intruder has a number of cracks he plans on executing against a local business WLAN. His first step is to conduct eavesdropping from a nearby parking lot using a high-gain antenna and a Wi-Fi adapter in monitor mode. What can the business do to protect itself from this kind of network discovery?
    a. Use a WIPS overlay
    b. Use integrated WIPS
    c. Secure their switch ports
    d. There is no real defense

Review Answers

1. **B.** Active discovery uses probe requests from the wireless client and probe responses from the access point

2. **D.** All of these choices are synonymous and will result in a probe response from all APs within range

3. **C.** Monitor mode is a "listen-only" capability

4. **A.** Wireless DOS attacks can happen against the RF at Layer 1, or at Layer 2 against the 802.11 frames

5. **A.** The SSID IE is also present in the Association Request, the Probe Request, the Probe Response, and the Reassociation Request

6. **C.** MAC addresses are not encrypted by 802.11 standards-based security methods

7. **A.** Open System Authentication is largely an operational formality but is present in all strong encryption methods

8. **D.** Basic WEP did not make use of dynamic pre-shared keys

9. **D.** Remember that Open System Authentication comes first

10. **A.** 802.11 wireless works at Layer 2, therefore EAP works at Layer 2

11. **A.** OWE is part of the Wi-Fi Alliance's Enhanced Open™ program and provides encryption with no configuration required by the user

12. **D.** Created by Cisco, LEAP can be cracked by the ASLEAP utility

13. **D.** 802.11-based security encrypts from Layer 3 up

14. **B.** LEAP and EAP-Fast are both Cisco-authored EAP types

15. **D.** When it comes to passive eavesdropping, of the options given, there is no defense (though Faraday cages and RF paint/wallpaper technically can help defend against eavesdropping, these measures are rarely practical to implement)

Chapter 3: Security Policy

Objectives Covered

1.1   Define WLAN security requirements

1.2   Develop WLAN security policies

1.3   Ensure proper training is administered for all stakeholders related to security policies and ongoing security awareness

2.2   Describe and perform risk analysis and risk mitigation procedures

Security policies are extremely important in networked environments of any size. Policy drives the solutions we implement, provides a roadmap for dealing with security incidents, and provides the framework for conduct in which the users of the network operate. Documented policies define minimum requirements for security in an environment based on the operational needs of that environment. Without policies, even the best IT teams are "winging it" for systems configurations and daily operations. By contrast, with sound policies, you can implement the security you truly need for your organization and provide a documented reference that keeps the entire organization grounded in a common security mindset. The foundational importance of developing and maintaining an effective, well-informed security policy cannot be overstated.

In this chapter, we will explore the purpose and goals of various granular security policies including those governing passwords, acceptable use of the network and attached resources, WLAN access, personal devices, and device management for both infrastructure devices and client devices. We will also address the development of policies and related concepts like security awareness training for users and administrators.

Security Policy Defined

A policy is a documented set of operational parameters in the form of a management-endorsed agreement. The policy includes rules, regulations, and courses of action in response to different situations based on organizational needs and requirements. An organization will have various policies based on their explicit business model and internal use cases. From an IT perspective, a corporate security policy is an especially important written document which contains detailed information about protecting the integrity of the entire scope of corporate computer networking operations.

The content of a corporate security policy will vary based on the type of organization or vertical market for which it is written. For example, educational institutions, financial firms, government, healthcare facilities, retail establishments, transportation companies, and other organizations will have specific policies based on their individual business models. Although many organizations have operational areas that are common amongst them, different types of organizations require some sections of their policies to be tailored to their specific business needs. As an illustration, a healthcare organization and a company that deals in transportation both require a network password policy, but the healthcare organization must have a policy for specific industry regulatory compliance such as Health Insurance Portability and Accountability Act (HIPAA). Given that HIPAA generally does not impact transportation firms, the password policies of each company likely won't be identical.

Every organization should have an IT security policy in place with a specific focus on networking. If the policy is found to be missing or outdated, establishing or revising policy documents should be considered high priority, followed by immediate implementation. This is very much an area in which CWSPs often participate in, playing the role of policy consultant, implementer, advisor, or even author.

Some organizations may already have an existing security policy; however, it may not specifically address wireless networking. This is especially true with organizations that may be new to the WLAN technology arena. Perhaps an environment that has long had Wi-Fi is just going down the wireless-connected IoT path- creating the need for updated policy. By now, you realize that WLAN concerns are unique enough to warrant their own operational guidance, and refreshed security policy is occasionally part of that.

One of the hardest parts of creating a security policy is figuring out where to begin. If a good overall network security policy exists, it may be easy to simply add a section for wireless access. On the other hand, you may find that wireless is integrated deep enough into the network that other parts of the overall policy need to change as a result.

If a security policy does not yet exist, to get started, it is natural to look for existing policies from other organizations to use as templates. There are also organizations that offer templates that could help to streamline the creation process. One such organization is the SANS Institute, at www.sans.org.

To view and modify a sample wireless security policy, follow this procedure:

1. In your Internet browser, navigate to www.sans.org.
2. Select Resources > Security Policy Project from the menu.
3. In the Find the Policy Template You Need section, choose Network Security, and then Wireless Communications Policy.

4. Download the DOC version so you can open it in Microsoft Word.

5. After downloading, open the document.

6. In the document search for and replace <Company Name> with your company name.

7. Note that this policy depends on the Wireless Communications Standard document. This is quite common in a security policy document set. It allows for less redundancy in information delivery.

8. When you are finished viewing the policy, close the Word document.

> Without question, the most important factor in security policy development and implementation is gaining management support and buy-in. If the management (executives) do not support the enforcement of policies, users will not abide by them.

Regulations

The operation of a network in unlicensed or licensed frequencies will always be subject to the governing authorities in the network's regulatory domain. It is important to ensure that your network policy acknowledges and complies with the appropriate regulations of the governing authorities. These include agencies like the FCC in the United States, the ISC in Canada, and equivalent agencies around the world. You will not be required to memorize all such organizations for the CWSP exam, but you should understand (and recall from CWNA) that they dictate various parameters related to network usage. If you do wireless work in other countries, you will need to be aware of specific regulatory constraints and which agency enforces them.

Industry-specific compliance regulations and guidelines are also becoming increasingly more important and demanding. Publicized data breaches have led compliance groups like the Payment Card Industry (PCI) to tighten their restrictive belts and have opened the eyes of many a security manager to the gravity of network-related data loss. PCI has developed the PCI Data Security Standard (PCI-DSS), which is becoming a mainstream part of retail implementations and an integral part of a growing number of security frameworks. Other compliance requirements, like FIPS and additional government-related requirements are already very strict and need to be integrated into both organizational policy and daily operations. Most companies realize the importance of security and compliance when it comes to the network, but budget constraints often keep administrators from achieving their intended security goals. Organizations with a global presence also have the challenge of ensuring their policies are compliant with the regulations of the countries in which they operate.

Industry and other regulations and guidelines to be aware of include:

- PCI DSS (Credit card processing)
- HIPAA (Health care)
- Sarbanes-Oxley (Publicly traded corporations)
- FIPS 140-2 (Government network security/encryption)
- DoD Directive 8100.2 (Use of wireless in DoD spaces)
- GDPR (General Data Protection Regulation- EU law on data protection and privacy)

Legal Considerations

Several legal considerations related to WLAN security policy should be considered. Of course, regulatory compliance requirements are of interest to a company's legal or Risk Management departments. IT management might leverage the legal department to help make a case for expenditure on staff and system resources to ensure that regulatory compliance is met.

Any effective security policy must have executive support behind it. If employees are to be reprimanded or terminated for a breach of security policy, it is important to have legal counsel's guidance for these decisions, and for these disciplinary actions to be coordinated with senior management. Similarly, when a breach is detected, and forensics are being gathered and analyzed as evidence, it is important that the legal team be involved. These steps, among others (like active rogue mitigation), may require input from a legal representative to make sure that policy does not become a liability instead of the asset that it should be.

Policy Importance

As wireless professionals, it can be tempting to wave away the larger security policy discussion as someone else's problem. Please remember that CWSP is a professional-level certification and that you may well find yourself as both the administrative and technical point of contact for security policy at some point in your career. It's imperative that you develop an appreciation for the importance of policy and the impact it has on wireless solutions.

Let's examine several important reasons to create a well-defined security policy. First, a policy is required if network security is to be maintained with consistency. Second, the IT staff requires the documented authority in order to enforce the policies they have defined and for which executive buy-in has been achieved. Additionally, with large IT groups, it is important to have a single, central source of documentation that defines practices and procedures for everyone to follow to prevent a patchwork of departmental policies that do not align.

> In complex network environments, wireless policies may need to cover a range of situations. Consider the case where PCI activities are present, an open guest network is offered, telecommuters are allowed to connect from open networks while traveling, and general company business is conducted over the secure wireless network. Top it all off with a new generation of wireless lighting controls in the environment that fall under the heading of IoT. A written policy will need to uniquely address each of these.

The following reasons for creating a security policy are important:

- Maintain the desired level of network security
- Uphold compliance
- Legal protection
- Asset documentation
- Procedural continuity

- Authority

Traditionally, documented policies have been overlooked or neglected because their importance is not understood (until there is an incident). Responsibility for policy creation is often confused between IT and management staff, and it's not uncommon for the two groups to not see eye-to-eye on policy specifics. Times have changed, and the sheer importance of the network to most organizations has given network security new importance. As mobility and pervasive wireless networking take deeper root in both the business and personal spaces, the stakes have never been higher for security.

When deciding which 802.11 security mechanisms to use, it is important to consider the circumstances, requirements, and various use cases of the organization implementing the network. It is essential that the organization's document is planned for a secure wireless environment within a wireless network security policy. Due to the speed with which changes have occurred in the wireless industry, it is desirable to create a security policy that is easily modified after periodic audits and technology evolution to take advantage of on-going security enhancements. Even where open wireless networks are used, policy needs to be developed that sets the limitations on what can be done over that open WLAN and how any security-related issues will be responded to.

The steps involved in creating a wireless security policy can be defined as follows:

1) Perform a risk assessment.

2) Define and document vulnerabilities and countermeasures.

3) Obtain support from senior management.

4) Provide communications between the departments or individuals that will be involved.

5) Provide ongoing monitoring and security auditing.

6) Plan response, forensics, enforcement, and reporting tactics in advance of a security policy breach. Identify key security team players for policy breach situations.

7) Revise and fine-tune the policy as needed.

8) Publish all changes to the security policy and provide an educational forum to keep users apprised of current status.

> Both IT and organizational management must support the security policy development process, and they must also support enforcement of the policies for the policies to have a significant impact and to result in a more secure environment. Without critical executive buy-in, a policy can be worthless and unenforceable.

A WLAN addendum or special wireless section should be added to the general corporate security policy. The author or contributor to the policy needs to be identified, and senior management needs to be aware that policy is being developed or revised. You may be the primary writer or contributor to a wireless security policy as a CWSP. Minimally, you will be an implementer of the policy.

Continuous monitoring and periodic security auditing are crucial in determining security policy adherence. All companies should perform continuous monitoring - especially those that have a "no WLAN" policy. It's just not enough to hope or assume that intruders will leave your network alone. Continuous monitoring often has the added benefit of identifying conditions that may impact network performance if the underlying cause isn't a security concern.

As mentioned before, as an important underpinning of CWSP processes, the level of WLAN defensive countermeasures enacted should be proportional to the determinations of your risk assessment. Stated plainly, you should not implement more security than is required and you should not implement less, assuming that security solutions equate to costs. If complete and total security were free (or even possible), we would all gladly implement it. But even open-source solutions are not truly free, as there is a time cost factor associated with the implementation of any technology.

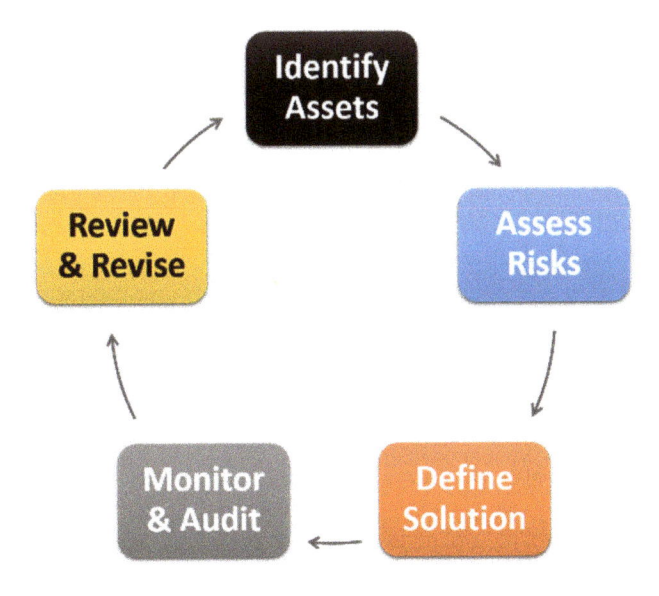

Risk Assessment

An important first step is to perform a risk assessment of the organizational assets that the policy will cover. This assessment is used to determine the level of vulnerability that exists, and to attempt to quantify the consequences that could result from an intrusion into the LAN and connected hosts from the wireless segment.

Risk and Impact

Risk assessment and impact analysis are somewhat related concepts. Together they comprise the evaluation of assets and vulnerabilities, and they help formulate the level of necessary security. This is an early step in the policy creation process.

Risk Assessment Process

The risk assessment is basically an audit of the security in place, and it will help to determine if the controls in use are adequate to provide the necessary security required by the specific organization. This can be performed in-house (typically by an unrelated department), or it can be outsourced to a different organization who specializes in assessing network security. Standard penetration testing procedures will help provide the information needed to complete a full risk assessment and impact analysis. All issues, concerns, and lack of compliance (whether incidental or intentional) should be clearly defined and documented as part of the risk assessment process. Understanding the balance between potential security solution costs and how the organization will benefit from specific security controls should be evaluated. Also, the type of business or organization will usually determine whether or not they are a significant target for intrusion and the likelihood of an attack. For example, a financial institution or a government entity is expected to have a higher risk of attack over a company who manufacturers general-purpose widgets that have no valuable intellectual property to target. At the same time, the widget maker also needs policy even if the chance of them being targeted is smaller.

Understanding the impact of successful intrusions is also part of the risk assessment process. A few years back, the wireless network typically functioned as an extension to the wired network and allowed access to only a limited amount of company resources. The potential to gain access to specific types of information or to the network infrastructure was minimal, and so the typical attack surface was usually much smaller. Today the wireless network is the primary access method, or is "the network" for many organizations, allowing access to all available resources and infrastructures. Therefore, the potential impact of a successful intrusion has grown tremendously in recent years.

The risk and impact analysis processes and procedures are not a onetime endeavor, as shown in the Risk Assessment Process graphic. As technology evolves, so do the security risks and how they will affect an organization. Think of this from the client device security perspective. In this example, you can purchase and install computer anti-virus software and run a full system scan. Any potential current threats will be discovered and mitigated. But new and more sophisticated computer viruses are created every day. If the anti-virus software is not updated on a regular basis and scans are not performed periodically, the potential of a virus infecting the computer is greatly increased. The same premise holds true for the risk and impact analysis procedures. New threats are introduced constantly, so ensuring an organization is adequately protected is an ongoing process. A security policy should define the frequency of conducting risk analysis and what the subsequent analysis processes will entail.

Assets are the resources you desired to protect. A complete asset inventory is required to do proper risk analysis. If you don't know what your assets are, you don't know their vulnerabilities or value.

> Loss expectancy is a standardized calculation used in risk management. Annualized Loss Expectancy (ALE) is the result of the Single Loss Expectancy (SLE), which is the loss incurred from a single incident event, multiplied by the Annual Rate of Occurrence (ARO), which is the number of times an incident event is expected to occur. Though this formula may sound odd in a wireless security book, it helps illustrate how non-IT people measure loss from network insecurities.

Document and Define

After risk assessment, it's time to document and define the discoveries and deficiencies. Once the policy document is created, based on the risk assessment, ensure the following:

- It is accessible to all relevant parties via a public file share or on one another's computer.
- It is marketed/promoted/distributed within the company.
- It is kept up to date.
- Its importance is defined and communicated to new and current employees alike.

Buy-In and Training

We've mentioned that for the security policy to have authority, senior management buy-in is necessary. For effectiveness, training must also be performed, and awareness must be promoted. If a strong and authoritative security policy is written, but employees are not aware of it, how can you expect them to follow it? Similarly, if you draft a policy, but management does not buy into it, what good is it? Security is as much of organizational culture (and individual state of mind) as it is a technical and procedural framework.

Management buy-in offers the following benefits:

- Provides authority, as the policy is endorsed by top management
- Allows for enforcement of technical policy requirements
- Enables commitment of resources (people and money)
- Commitment to disciplinary behavior when violations occur

Training of end-users and administrators should include:

- Security awareness training should be provided to end-users and administrators
- New hires need immediate training; all staff should get periodic training
- How to identify and report social engineering
- The expectation that all will abide by the password policy
- Prevention of rogue APs and clients
- Understanding repercussions to policy violations
- Acceptable Use and Abuse policies covering organizational and BYOD devices
- Remote networking procedures
- Creation of overall security awareness

When management buy-in and training are implemented with these considerations, your policies will result in a more secure environment.

Response

The wireless security policy discussion is incomplete without addressing incident response. Incident response is the process or action taken when a security incident occurs. This response should be documented in an incident response plan that also benefits from management buy-in and training for those staff members that would be expected to respond. Such a plan may be as simple as a one-page Visio flowchart or as complex as a multipage bound document. Either way, you should have a plan.

The response plan should address items like the following:

- Forensic data analysis
- How to respond to rogue APs
- Analyzing system logs, and agreement on how long log data will be retained
- Accounting services
- What immediate reaction is taken with a compromised network infrastructure?
- What authorities are notified and involved
- To whom do end-users and admins report security violations
- Who is responsible for owning the response process and seeing it through to conclusion?

Enforcement

The security policy is of little value if it cannot be practiced and enforced because it is too obtuse or impractical, or if no one has been empowered to enforce the policy. As such, it is necessary to formulate a workable and functional set of rules that can be realistically administered. This set of rules determines how the wireless network will be used. Following are criteria that should be included within the security policy functions:

- Use of passwords
- Amount and frequency of training focused on the use of the chosen security model and awareness of social engineering attacks
- The methods to be used in order to provide awareness of security risks and vulnerabilities of WLAN implementation
- Definition of acceptable and unacceptable uses of the WLAN
- Employees should be made aware that any or all of their WLAN traffic may be captured, filtered, and analyzed

- The procedures used to implement and enforce the security policy must be consistent
- The creation and maintenance of a WLAN security checklist and a change management program
- Endpoint security, personal firewalls, and virus checking software may be mandated by the security policy for employee devices when:
  - Used on the corporate campus
  - Traveling and remotely connecting to the corporate network
  - Corporate information is contained in the employee's mobile computing devices.
- Management of WLAN devices, including security applications installation, maintenance, and support, may be required by the IT support department.

The ongoing enforcement of functional policy is crucial. This can be a time-consuming task, and in some cases depending on the size of the organization, a full-time job. One tool that can help with policy enforcement is a Wireless Intrusion Prevention System (WIPS). Properly implemented, configured, and tested, using a WIPS can be very beneficial to an organization. Though it can add significant cost to the total cost of ownership (TCO) of a WLAN environment, the use of WIPS can be a force-multiplier in ongoing support and will typically cost less than the cost suffered as a result of a data breach. WIPS solutions tend to be more likely used in certain verticals, like the finance sector.

Advancements in technology such as multifunction mobile client devices bring many new security concerns to the forefront. In addition to a WIPS solution, a mobile device management (MDM) solution should be considered to aid in security policy enforcement for mobile device technology, if it is allowed and used on the corporate network. MDM solutions help to administer and control how mobile devices such as smartphones, tablets, laptop computers, and even desktop computers can be used on a network and how they access company resources. The MDM market has become mature and impressive in the many options available.

> Definitions of enforcement are the statements within policies that define or explain prescribed disciplinary actions for violations of the security policy. Such definitions guide management in the proper manner of dealing with a security incident.

Monitor and Audit

The initial security audit provides a baseline of all active wireless devices and is used to classify those devices as to their role. To ensure that the security audit baseline remains current, it is necessary to provide on-going monitoring. This can be done manually in smaller environments or through the use of automated sensing systems such as a WIPS.

Several wireless security manufacturers offer WIPS solutions that perform automated, around-the-clock monitoring, alarm notification, and reporting without administrator intervention. Many of these systems are equipped with the ability to isolate and nullify the actions of threatening wireless devices. This activity is referred to as "threat mitigation."

A WIPS solution will use distributed sensors which are either separate, stand-alone infrastructure devices (referred to as an overlay), or those that are integrated into wireless APs (and likely requiring additional licensing fees to use), strategically placed around a facility, campus, or another wireless service area. The sensors are passive monitoring devices which monitor the RF environment within the system's coverage area and gather a wealth of information used to report both performance metrics and security policy violations to a central analysis engine, or to a WIPS server.

Suggestions for WLAN monitoring include:

- 24x7x365 monitoring
- Implement WIPS
- Periodic and automatic report generation
- Enable appropriate alarms and notifications

- Keep the WIPS sensors updated to match WLAN technology used by the WLAN

- Making sure the WIPS is properly "tuned" for a given environment

- Keeping staff motivated and aware of the importance of the monitoring process. It's easy to suffer "dashboard overload" in complex network environments, but when it comes to security, it's critical to not become numb to the information presented.

Audits provide spot-checks on organizational compliance with security policies and can be internal or external. An internal audit is one performed by the staff of the organization. An external audit is performed by an outside agency. Depending on your industry or the situation, an external audit may be required.

During the auditing process:

- Test for known weaknesses
    - Authentication cracking
    - Social engineering
    - Sloppy password management
    - Rogue devices
- Generate detailed audit reports
- Ensure compliance with industry regulations and guidelines

The good news is that many modern and properly tuned WIPS solutions will actually provide much, if not all, of this data to you without requiring manual actions. WIPS are further addressed elsewhere in the text.

Review and Revise

As WLAN technologies and the company's IT and business needs change, it is important to review and revise the security policy to keep it up-to-date and relevant. Having an outdated policy is often as bad as having no policy. An outdated policy can give a false sense of security when action is actually needed and create confusion for network users.

A revision and review process should be implemented to keep the policy up to date. This may be an annual process, or it may be performed with each new system upgrade. The revision and review process may be triggered by discovered vulnerabilities, but the trigger or schedule should be defined. The process usually involves, at least, the following:

1) Perform a policy review

2) Perform an internal/external audit

3) Modify the policy based on results

4) Set a time frame for policy updates to be complete

5) Communicate the policy changes to various stakeholders

Additionally, while it is important to review and revise your policies, it is also important to stay informed of regulations and guidelines that may change for your industry. For example, if you work in healthcare, you want to know if anything changes in relation to HIPAA regulations or any other healthcare standard that could drive changes to your policy and operational framework.

Password Policy

Password policies define the required parameters for user, system, and/or device passwords. When creating a password policy, consider the following elements:

- Password length (depends on the use case)
- Mixed alphanumeric with lowercase, uppercase and special characters
- Password change policies
- Password sharing policies
- Password access policies (who has access to passwords)
- Password storage policies
- Application or hardware password requirements from vendor

The following are example statements that can address the preceding list (Note that these statements will vary greatly depending on the needs of the specific organization):

- Passwords used by individuals for authentication purposes must be 10 characters or longer.

- Passwords should include mixed characters and be case sensitive.

- Passwords used by individuals for authentication purposes must be changed every 60 days. Passwords used by systems or devices must be changed every 6 months.

- Passwords used by individuals may not be shared with anyone. Passwords used by systems or devices may only be shared with documented and authorized individuals.

- No individual will have access to the plain text of any other individual's password. Only individuals with documented authorization should have access to system or device passwords.

- Passwords must be stored with sufficient encryption to avoid any currently known attack method that could extract the passwords within the password change requirement window.

Again, these are just examples of password-related policy statements and not fixed guidance. In addition to password policies, it is important that authentication requirements be addressed in the policy. Authentication requirements include the use of client credentials, sharing, and storage of passwords/secrets, the use of multi-factor authentication, and many other considerations.

> Frustratingly, you may find that password policies conflict with wireless client device capabilities, on occasion. Some devices only support passwords of limited length, and occasionally, the use of special characters can be problematic for certain authentication databases. It is what it is.

Additional Policies

The following additional policies should also be addressed. While not covered in the previous examples of policy statements, these are equally important:

- Acceptable use – defines the intended use cases for the provided system.
- WLAN access policy – defines who can access the WLAN, and how they can access it.
- Personal device or BYOD policy – defines the allowed uses of personal devices and may include requirements such as onboarding and mobile device management.
- Physical security policies – the policies related to the protection of devices and the environment and may include requirements such as security gates/doors, locks, enclosures, and cameras.

Security Baselines

Security baselines provide a starting point to work from when striving for wireless network security. They give the minimum configurations that must be deployed when staging a new device. Security policies may define baseline security parameters, such as:

- SSID naming
- Authentication mechanisms
- Supported encryption types
- Device types used
- Device labeling
- Rogue AP and client policy
- Endpoint security requirements
- Default configurations
- Remote networking
- Management protocols
- Monitoring
- Security layering

- Segmentation
- Role-Based Access Control

Device Management

Different business types will require varying device management policies. A SOHO and some SMBs often only have one wireless AP, or possibly a few. In this case, the APs will often be autonomous models requiring independent configuration for each AP. Proper staging is required to ensure that there are no security holes allowing an intruder to gain access to the wireless network. This includes items such as:

- Changing the appropriate device default configuration values
- Securing the device login credentials
- Disabling remote access capabilities unless needed. If they are needed, using adequate security methods must be addressed
- Enabling firewall settings to meet the highest security requirements
- Disabling all protocols that are not used or needed
- Placing devices where they are not easily manipulated

This is not a complete list, but it includes some of the common items which need to be considered.

Enterprise wireless network device management policy is more involved due to the size and complexity of the infrastructure topologies in play. In addition to the items just mentioned for smaller networks, infrastructure devices used in enterprise networks will have the following management considerations:

- Wireless LAN security profiles
- Management protocols and software such as SNMP and third-party solutions
- Logging requirements
- Change management procedures
- Device labeling requirements

One important best practice for infrastructure device management recommends, managing wireless infrastructure devices from trusted space on the wired network only. If this is not possible, it is important to ensure adequate wireless security mechanisms are in place for the wireless devices that are accessing the management interfaces of the infrastructure devices. If an intruder was to get the administrator credentials, they would gain the keys to the network kingdom, which can have catastrophic results.

BYOD Policy

Bring Your Own Device (BYOD) policies either forbid employee-owned devices from being used at work or more typically allow employee-owned devices to not only be brought into the workplace environment but also to access the corporate wireless network. It may sound like a simple paradigm, but BYOD adds significant complexity to network management. In addition to the potential impact on network capacity, network security is a big concern when personally owned devices are part of the network. If employees are not allowed to use their own devices, then writing a BYOD policy will be fairly straight forward. But if a given environment is among the growing number of organizations that allow BYOD on the network, then this part of the security policy can get very complex. If BYOD is not allowed, audit processes will need to ensure that users are not violating the "No BYOD" policy.

One major benefit to BYOD is the fact that a given company, school or hospital - to name a few examples - can leverage this technology for organizational business by allowing employees to use their own devices for operational processes under controlled circumstances. This could provide a large endpoint cost savings for the organization, as long as the paradigm is managed in a way that doesn't leave the door open to costly security risks. Items that should be addressed in a BYOD security policy include, but are not limited to:

- Allowed and supported device types, i.e., Android, Apple iOS, etc.
- Supported mobile operating systems and versions
- Device provisioning and enrollment methods
- Containerization to separate corporate and personal data
- Permitted apps, distribution methods, and app stores used

- Remote device management
- Location services capabilities
- Data encryption methods
- Remote access security, virtual private network (VPN) use, and public access network restrictions
- Firmware, operating systems updates and software patches or hotfixes

MDM Solutions

The growth of wireless networking has spawned a number of complementary industries, and one example is mobile device management (MDM). With the large number of mobile smart devices that exist between corporate assets and BYOD, management of these devices is vital to ensure corporate security policy is maintained and enforced. WLAN vendors might either provide built-in MDM capabilities or partner with 3rd party MDM solutions as a competitive differentiator. MDM solutions provide a way to control and administer client devices that not only include portable devices like smartphones and tablets but also laptops and even the possibility of desktop computers. MDM can provide configuration management capabilities for both corporate-owned devices and those belonging to employees.

Selecting the best MDM solution for specific networking requirements involves some careful consideration. As with any technology, the correct MDM solution must meet the needs of the organization and will require planning to validate it does so. MDM solutions are typically available in two forms: in-house (on-premise) solutions or cloud-based alternatives, which are provided as Software as a Service (SaaS) technology. This is another example where you, as a CWSP, may be involved in recommendations or implementation.

MDM solutions typically share the following common feature sets:

- Multiplatform management and support
- Application distribution
- Device registration
- Remote lock and wipe

- Password control
- Feature lockdown
- Secure communications
- Policy enforcement

Notice several items on the above list are directly related to functional security and corporate security policy. Much of what you learned about earlier in this chapter also apply here, including enforcement, monitor and audit, password policy, acceptable use, physical security, and device management. Mobile device management provides a way to ensure that, as wireless technology continues to evolve, it does so in a way that meets the security policy requirements of the organization.

Balancing mobile device security and usability is also something to consider. If a policy is too tight, it may limit productivity or interfere with a user's ability to utilize their devices correctly. If a policy is too loose, it could potentially compromise the security and integrity of the corporate network. Therefore, how to provide an acceptable balance between the two must be carefully considered. A poorly balanced mobile device security policy is directly proportional to the potential for policy failure. Consider the example of passcode or password policy. Most industry best practices agree that the longer and more complex a passcode/password is, the more secure it will be. However, if it is too long or complex, a user may have difficulty remembering it or depending on the device, it may be a challenge to type it into the device's interface. In cases like this, the policy did not meet the necessary goal. Thankfully, WPA3's password-related enhancements will largely negate the risk that comes with weak passwords.

As BYOD has become more accepted in corporate culture for businesses of all sizes, the small form factor of mobile devices has presented a new challenge. Device loss or theft is of particular concern because the sheer number of small devices containing sensitive data is skyrocketing. Consider the CIO who leaves her smartphone in a taxi in a busy city or the network administrator whose tablet gets stolen from his car in the parking lot of the local mall. These scenarios are commonplace and could result in massive data breaches when you consider how much business is done over mobile devices. MDM solutions help to make sure that highly mobile devices have a number of defenses configured in case of loss or theft. The ability to remotely lock or wipe a device through the MDM framework is the ultimate defense when one goes missing. Even if other MDM features are not employed because of administrative overhead, it's common (and wise) to put into place mechanisms that wipe a device if a high number of bad passwords are entered.

Social Networking Policy

Social networking services such as LinkedIn, Twitter, and Facebook continue to grow in popularity and have proven to have business applicability that ranges far beyond personal use. Many organizations now use social networking extensively as part of their business practices. In some cases, this is a significant part of how many large businesses provide communications channels to their customer base and to the general public.

Organizational Risks
- Brand reputation
- Privacy
- Copyright issues
- Trademark infringement

Data Risks
- Intellectual property leaks
- Unauthorized data sharing

Social Media Risks

Technology
- Viruses and worms
- Spyware
- Denial of Service (DoS)
- Misc. vulnerabilities

Personnel
- Productivity loss
- Social engineering
- Human Resource information leakage

Despite potential benefits, social networking can also introduce many potential security concerns for organizations that choose to use it as part of their business continuum. An appropriate social networking policy is important to help lessen the possibility of corporate security breaches which could come about through social media channels. A corporate security policy that addresses social media platforms is easy for companies to overlook. It is important that social media be recognized for the threat it poses and made part of security policies. Why does this matter in the context of CWSP? WLAN does not operate in a vacuum, and social media and Wi-Fi go hand-in-hand. Here are some of the items that make social networking vulnerable to potential threats within an organization:

- Authentication / login management
- Phishing attacks
- Malware threats
- Corporate intellectual property (IP) integrity
- Reputation management

At a minimum, corporate security policy must consider how to deal with these threats. Unfortunately, the chances are high that many organizations are lacking when it comes to addressing and maintaining a policy with respect to social networking.

Some WLAN manufacturers provide "Social Wi-Fi" technology (either integrated or third-party solutions) that will allow a user to login to a guest network using their social media credentials. Behind the scenes, the use of "Social Wi-Fi" trades WLAN access for a varying amount of probing of your social media accounts for data that can be monetized. This type of captive web portal has fallen in and out of industry favor over the last few years but is still prevalent enough to be considered relevant. "Social Wi-Fi" should be considered within the corporate security policy.

Chapter Summary

In this chapter, we reviewed the important factors related to security policies. First, policies were defined, and then the details of various policies were explored. We examined example policies and learned about keeping policies updated after periodic review. Additionally, the basics of the risk assessment process were addressed. Finally, modern issues such as BYOD and MDM were acknowledged with consideration of how policies should address them.

Review Questions

1. You are advising your management team on the benefits of having a process in place to deal with security incidents after they happen. Which of these are you describing?
    a. Disaster Prep Plan
    b. Incident Response Plan
    c. Security Investigation Plan
    d. Call-Down List
2. The capability of performing remote device locking and wiping comes with implementation of which system?
    a. WIPS
    b. Back Up
    c. BYOD
    d. MDM
3. In your role as Network Security manager, you are reviewing the company's security policy and realize that it is quite dated. You find that there is no policy covering a fairly recent technology that could lead to copyright infringement, theft of IP, and productivity loss. What technology are you concerned with?
    a. Social Media
    b. MDM
    c. BYOD
    d. Mobility
4. Risk assessment is _____ in relation to security policy?
    a. Easy
    b. A one-time process
    c. Optional
    d. Critical

5. You are hired to do a security audit of the Wellington Company's wireless network. You find that the executive team at Wellington wants little to do with IT operations and doesn't seem to know where to locate the wireless security policy or what's in it. What important component of wireless security policy is missing in this situation?
    a. Audit process
    b. Management buy-in
    c. Incident response plan
    d. Social Media policy
6. When it comes to wireless network security policy, which of these applies to all organizations?
    a. Policy is optional
    b. All policies will be the same
    c. Policies will differ by company types
    d. Some companies won't need risk analysis
7. For hospitals and medical facilities using wireless networking, which of these regulations will help shape security policy?
    a. HIPAA
    b. FERPA
    c. FIPS
    d. PCI-DSS
8. A given company has not updated its wireless security policy in the last ten years. What important aspects are likely missing?
    a. MDM
    b. BYOD
    c. Social Media
    d. All of these

9. What is the best approach for developing and communicating a wireless security policy?
    a. Each department should create their own
    b. There should be a single company-wide policy
    c. To save money, employees should be trusted to follow their own common sense
    d. Senior management should not be part of the policy framework as they tend to not understand IT

10. Security policy covering end devices would not include which of these?
    a. Anti-virus software
    b. Personal firewalls
    c. Strong passwords
    d. Administrative interfaces

11. What do we call the ability of a WIPS system to identify and nullify detected wireless threats?
    a. Threat mitigation
    b. Threat attribution
    c. Threat classification
    d. Threat isolation

12. Which of these is likely to not be part of a password security policy?
    a. Password length
    b. Password storage
    c. Passwords at home
    d. Password sharing

13. As you construct a security policy document, you are about to consult with your manager on the topic of Wireless Access Policy. What will this specifically cover?
    a. Infrastructure administration
    b. Members of Incident Response teams
    c. How to communicate a wireless policy to users
    d. Who can access the WLAN, and for what reason

14. Which of these are not common among Device Management policies likely to be used in small business and enterprise wireless settings?
    a. Use of SNMP
    b. Securing management interfaces
    c. Disabling unnecessary protocols
    d. Changing device defaults
15. Which of the following can be used for automated auditing of wireless security policies?
    a. MDM system
    b. BYOD sensors
    c. WIPS system
    d. RADIUS server

Review Answers

1. **B.** An incident response plan will help bring order to the chaos that can ensue after a security incident occurs.

2. **D.** A Mobile Device Management system provides the ability to remotely act on client devices.

3. **A.** Social media use brings a range of concerns that need to be addressed through policy.

4. **D.** Risk assessment is critical to understanding what you are protecting.

5. **B.** Without management's buy-in, a security policy is doomed to failure.

6. **C.** The final policy will vary by company, but the process that leads to the policy being created is generally the same for all companies.

7. **A.** The HIPAA regulations that govern hospital patient information policy will have a great impact on wireless security policy.

8. **D.** All of these have come about in the last several years.

9. **B.** A single, central policy is easier to write and keep updated and has a better chance for successful implementation.

10. **D.** Administrative interfaces are associated with network devices.

11. **A.** Threat mitigation is a unique capability associated with WIPS.

12. **C.** Though employees would do well to use strong personal passwords, the enterprise security policy is not likely to cover the use of passwords.

13. **D.** Wireless network access policy deals specifically with those using the WLAN, and for what purposes.

14. **A.** The use of SNMP is generally beyond the scope of small network environments.

15. **C.** A properly installed WIPS system will augment an organization's ability to monitor and enforce a security policy.

Chapter 4: Authentication

Objectives Covered

3.1   Select the appropriate security solution for a given implementation and ensure it is installed and configured according to policy requirements

3.2   Implement or recommend appropriate wired security configurations to support the WLAN

Authentication is the foundation of 802.11 WLAN security. To quote wireless expert Tom Carpenter, "If your authentication is weak, little else matters in your security." The AAA security model (authentication, authorization, and accounting) is built on this principle. If your authentication is weak, authorization is irrelevant from the perspective of strong security because an attacker could simply log onto the network, posing as an individual with access to the resource. The same holds true for accounting- if authentication is weak, the attacker can log onto the network with ill-gotten credentials, and the accounting records will attribute any malicious activity to the authorized user whose identity is spoofed. Strong authentication is truly the foundation of overall network security.

> It feels a bit odd, as the General Editor of this book, to have someone quote me in the content. However, the quote is an essential one. You can implement WIPS solutions, advanced authorization, fancy SIEM systems, and more, but if your authentication is weak, the rest just doesn't matter. After more than twenty-five years of implementing, testing, researching and learning in the area of security, I hold more firmly to this quote than ever. -Tom

In this chapter, we explore available 802.11 authentication methods in-depth, excluding those that are part of WPA3 (covered in Chapter 10). These methods range from simple passphrase authentication to the powerful Extensible Authentication Protocol (EAP).

But first, authentication itself must be introduced in more detail.

Authentication

Without a clear understanding of authentication and identity management, you will have difficulty with much of the CWSP material, and, also, with configuring a secure wireless network to meet the requirements of different security policies.

Both basic and advanced authentication methods exist, and many wireless systems include the ability to support both. For example, Windows Server systems allow for advanced authentication mechanisms through the Network Policy Server (NPS, Microsoft's RADIUS implementation) for environments needing strong security. The same Windows Server can also accommodate basic authentication using simple passwords against the Active Directory database. Both methods serve valid purposes and are best for different scenarios. Determining which method is right for meeting policy requirements in your scenario is the first step in implementing secure authentication.

Beyond selecting either advanced or basic authentication methods (or both), you must determine whom or what to authenticate. Do you need to authenticate the clients only from the authentication server side? Or, is authentication of the authentication server by the wireless clients required as well? When both the client and the authentication server are authenticated, it is called *mutual authentication*. Mutual authentication helps prevent the introduction of rogue authentication devices into your network. Client-only authentication allows the network to have protection from unauthorized clients and is generally easier to implement, whereas mutual authentication allows confidence and trust in both the client and the network, but with more administrative work (and possibly cost) required to accomplish. Which one do you use? Again, it depends on the security policy for that WLAN.

When considering authentication, it is wise to understand where it fits in the paradigm of AAA (called "triple-A"). Authentication should not be confused with authorization. Authentication is defined as *proving a person or object is who or what they claim to be*. Authorization is defined as *controlling or granting access to a resource by a person or object*. Authorization assumes the identity has been authenticated. If authentication is spoofed or impersonated, then authorization schemes fail. From this, you can see why authentication is such an integral and important part of network and information security. In contrast to authentication, accounting processes log who or what performed a given action. It assumes the authentication of the acting identity was performed accurately and securely and logs authentication failures. Again, authentication is the foundation of strong security and figures prominently in the duties of many CWSPs.

Advanced authentication systems generally utilize stronger user credentials and better protection of those credentials than is the case with basic authentication systems. The strength and protection of the user credential are determined by the effort it takes to exploit it. Password-protected credentials are usually considered weak when compared with biometric protected credentials, for example. This could be a misconception, however, because strength of authentication really depends on how the authentication information (the credential and proof of ownership) is sent across the network. If you were to implement a biometric system, such as a thumb scanner, and the client sent the credentials and proof of ownership (a unique number built from the identity points on the user's thumb) to the server in clear text, it would be no more secure than a standard password-based system. Thankfully, no known biometric authentication systems work in this way. The point is, seemingly secure methods can be defeated by lack of careful implementation.

The primary methodology of securing the authentication path is encryption; that is the *hashing* of the user credentials or at least the proof of identity information. This protection can be accomplished with VPN technology or with well-designed authentication systems. One example of such an authentication system is 802.1X with a strong EAP type. 802.1X and EAP types are discussed in their respective sections in this chapter, and in later chapters as well.

Before we dig deeper, consider that you use authentication every day in normal life. For example, when you are at a seminar, and the speaker says he is a subject matter expert, you use authentication mechanisms to verify this information. In other words, you listen to the information he delivers and use it to determine if he is truly an expert. In addition, suppose someone walks up to you and says, "Hi, my name is Bill, and I am tall." You would look at him and compare his height with what you consider to be tall, and authenticate in your mind whether he is truly tall or not. If he is not tall, by your standards, he will lose credibility with you.

Remember the word *credentials?* Consider additional important "cred" words: credit, credibility, and credentials. Do you see how they are all related? They all have to do with having proof of something. When you have good credit, you have proof of your trustworthiness to pay debts. When you have credibility, you have proof that you are genuine, persuasive, and dynamic. When you have credentials, you have a title, certification, or the experience that proves your skill or identity. All of these words originate in the Latin word *credere*, which meant "to believe or trust." Authentication is the process used to verify credentials and establish trust.

Advanced authentication is more secure than basic authentication because advanced mechanisms are used to protect the credentials. This usually means protecting a username and password pair, but it can also include protecting a user/certificate combination, a user/machine combination, or any other user/object combination used to identify a specific user. In addition to the extra protection offered by advanced authentication systems, when 802.1X-based systems are used, you have the benefit of widely supported standards-based security technology. Hardware and software from many different vendors are likely to support the authentication process when that process is built on standards. Most wireless clients used in enterprise WLAN will support both simple and advanced authentication. Unfortunately, most consumer-grade wireless clients, most printers, and IoT devices only support simple authentication, and many are finding their way into enterprise settings. This makes for some interesting challenges for the CWSP.

Importance of Authentication in Wireless Networks

Authentication is important in any network, but it is particularly important in wireless networks for several reasons. Consider first, the situation where an intruder or attacker may be in the WLAN environment. Many wireless networks generally do not provide for simple location identification. For example, with a wired network, a smart switch can report to you the exact Ethernet port being used by a particular client. From there, a cabling record can tell you exactly where the premise wiring run, (connected to the switch port) goes, so physically locating the connected client is possible. In a wireless network, the best that most systems can report is that a specific wireless client is connected to a certain AP. The client may be located anywhere within the three-dimensional transmit range of the AP, extending to multiple floors and several thousand square feet. For this reason, authenticating users becomes very important in wireless networks.

Triangulation capabilities are available on some wireless systems allowing a smaller window to the wireless client's locale but will be of varying accuracy depending on the quality of approach when the WLAN was implemented. Due to the mobility of 802.11-based connections, you may determine the location of the client only to arrive at that location and discover the attacker has physically moved, yet is still connected. A smart attacker is not going to stay connected from the same location for very long unless he knows there is no triangulation system in play. Further, a truly wily intruder might employ a partial Faraday cage, which could be placed around a portion of the antenna, or a highly directional antenna, in such a way that wireless sensors can only see it from one direction. This would effectively disable or cripple many triangulation systems – they would no longer properly calculate signals related to the antenna that they are analyzing because they often assume the use of omnidirectional antennas. Such a scenario would allow the attacker to communicate in the direction of the AP but not communicate—or be seen—from any other direction for proper triangulation.

Another issue making a case for trustable authentication is that the default Open System authentication mechanism of wireless networks results in anyone being able to connect. Network managers are, therefore, unable to track and control who is accessing their networks. Now contrast Open System to more robust methods. Through combinations of computer certificates and user credentials, an advanced authentication system can provide both the identity of the user and the exact machine he or she is using to connect to the wireless network. Effective authentication systems help enforce network policies, such as who can gain access and what network resources they can use. These authentication systems also make it very difficult for an attacker to gain access to your network in the first place. The attacker would have to steal both a computer certificate and user credentials to fully gain access. Although some networks do not desire or require the strongest authentication systems possible, the good news is that they are available for wireless networks when required. There is also middle ground between no authentication and the strength of 802.1X's most robust EAP types.

Choosing the Right Credentials

Many different credential solutions are available for securing your wireless networks. It's important to select the right system for your needs, as defined by the security policy. During the selection process, you will consider the primary features of a credential solution, and whether or not you need a multifactor authentication framework. As a CWSP, you should be aware of the various credential types available to you.

A credential solution should provide a means of user or client device identification that is proportional in strength to the security needs. You do not want a credential solution that places unnecessary burdens on your wireless network users or that results in greater costs (of both time and money) than the value of the information assets you are protecting. You will need to determine whether the selected authentication solution provides for redundancy and integration with other systems such as Active Directory or LDAP servers as you evaluate how it would fit into the larger network picture for a given environment. The system should obviously also support the needed credential types, such as smart cards and/or biometrics.

The following are also important when selecting a credential solution:

- The method used to protect the credentials
- The storage location of the credentials
- The access method of the credential store

The method used to protect the credentials: If an authentication system sends the credentials as clear text, there is no protection of the credentials. Advanced authentication systems will protect the user credentials by encrypting them or avoiding the transmission of the actual credentials in the first place. Instead of transmitting the actual credentials, many systems use a hashing process to encode at least the password. Hashing the passwords means that the password is passed through a one-way algorithm resulting in a fixed-length number. The resulting number is known as the hash of the password, or as the *message digest*. The hash is stored in the authentication database and can be used as an encryption key for challenge text in a challenge/response authentication system.

The storage location of the credentials: The credentials, both username, and password (or hash) or certificates, must be stored in a secure and responsive location, accessible by other resources on the network which needs to interact with it. It must be secure to protect against brute-force attacks, and it must be responsive to service authentication requests with minimal delay. Certificates are usually stored in a centralized certificate store (known as a certificate server or certificate authority/CA) as well as on the client, which uses the certificate for authentication. Both locations must be secure, or the benefit of using certificates is diminished. In addition to the standard certificate store, users may choose to back up their certificates to disk. Backups are usually password-protected, but brute-force attacks against the media store may reveal the certificate, given enough time. For this reason, network administrators should be well-educated in this area, and understand the vulnerability presented by the existence of such backups.

The access method for the credential store: Access methods vary by the authentication system and storage paradigm, but standards exist that define credential access methods. One widely used example is LDAP (Lightweight Directory Access Protocol). LDAP is a standard method for accessing directory service information. LDAP information can include many objects but is usually inclusive of authentication credentials. LDAP is or can be, used by Lotus Notes, Open LDAP, Slapd, and

Microsoft's Active Directory, among others. LDAP is one of the most important parts of many enterprise security environments.

Credential Categories

In those cases where a single credential doesn't provide robust enough security, multifactor authentication has advantages. Multifactor authentication is a form of authentication that uses more than one set of credentials. The generic credentials available are:

- Something you know (information: passwords, PINs, etc.)
- Something you have (physical objects: smart cards, keys, etc.)
- Something you are (biometric: thumb scanners, retina scanners, etc.)

An example of a multifactor authentication process would be the use of both passwords and thumb scanners. Typically, the user would place her thumb on the thumb scanner and then be prompted for a password or PIN (personal identification number) code. The password may be used for network authentication, or it may only be used for localized authentication before the thumb data is used for network authentication. However, in most cases, the password and thumb data are used to authenticate to the local machine and then the network or just to the network alone. A common example of multifactor authentication would be an ATM card. The card is used, and a PIN (something you have and something you know) must be entered. You'll also see the term "two-factor" commonly used when applicable.

Credential Types

Many credential types exist. They include:

- Username and password
- Certificates
- PACs (Privilege Attribute Certificates)
- Biometrics
- Tokens

Username and password pairs are currently the most popular type of credential. They are used by most network operating systems, including Mac OS X, Linux, UNIX, and

Windows. Due to the human factor involved in the selection of the password, they often introduce a false sense of security. Because the chosen password is often too weak to withstand dictionary attacks and brute-force attacks, many users proceed without realizing that their credentials could be compromised with common attacks. Additionally, passwords are often written down or stored in plain text files on a network-accessible device and then changed infrequently, resulting in a longer attack opportunity window.

An alternative to username and password pairs is the digital certificate. In order to use certificates throughout an organization, a certificate authority (CA) must exist. The CA can be operated by the organization, or by an independent third party. A CA is basically a server or service that issues and manages the lifecycle of a certificate. In either case, the costs are often prohibitive for widespread use due to the need for an extra server or even a hierarchy of servers. Small and medium-sized organizations usually opt for server-only certificates, or no certificates at all, because of the cost of implementation. A full PKI (Public Key Infrastructure) would usually consist of more than one CA. Each CA would be a single server or cluster of servers. They would be implemented in a hierarchy with one CA, called the root CA, at the top. The PKI is the mechanism used for generation, renewal, distribution, verification, and destruction of user and machine certificates.

The Privilege Attribute Certificate (PAC) is used by the Kerberos authentication protocol in Windows 2000 through Windows 10 and all Windows Server versions in that same interval. The PAC contains authorization data for the user and includes group memberships and user rights. This feature means that the user's group and rights assignments are transferred as a portion of the ticket-granting ticket (TGT)—a feature of the Kerberos authentication protocol.

Yet another authentication credential is the individual, him or herself. Biometrics-based authentication leverages the uniqueness of every human being for authentication purposes. For example, a thumbprint can be used as a unique identifier, as can your retina via a scanning device. The balancing of cost and security is important with biometric credentials. While hair analysis could potentially be used to authenticate a user, the cost and time involved are still too high for practical use (remember the fundamental security axiom, that overall cost of security should not exceed the value of data being protected). Both thumb scanners and retina scanners have become quite popular.

There are two common types of authentication *tokens*: software-based and hardware-based. Software-based authentication token systems often run on tablets or cell phones. This allows users to launch the application, on their mobile device, and retrieve an authentication code. This code is commonly used in conjunction with a password or PIN, essentially creating a two-factor authentication system. Google Authenticator is an excellent, free example of a software token that you can try out if you have yet to see this type of authentication. Hardware-based token systems usually provide a keychain-sized device that, with the press of a button, will show the current time-limited authentication code. This code, again, is used with a password or PIN. Most token systems work off of a time-synchronized (or some other synchronization point) algorithm that generates proper codes in the software system or hardware device. This functionality means that a valid code today will not work tomorrow and provides greater security.

Hardware-Based Token Example

Passphrase-based Security

The preferred cipher suite from the current 802.11-2016 standard is CCMP (Counter Mode Cipher Block Chaining Message Authentication Code Protocol, Counter Mode CBC-MAC Protocol – yes, the first C in CCMP stands for nine words). CCMP is based on the advanced encryption standard (AES) encryption algorithm and is available through any 802.11 equipment that supports the WPA2 interoperability certification. The personal version of WPA2 allows the use of a static passphrase to be entered by the administrator in lieu of using an EAP key generation and management technique. The use of a passphrase as the master session key (MSK) can be regarded as very strong, but only as long as the passphrase is selected using an unpredictable or unlikely to be repeated methodology. As mentioned elsewhere, WPA3 has different operational aspects and is covered in Chapter 10.

Many manufacturers allow for the entry of passphrases as either ASCII text or hexadecimal characters. An important point to remember is that ASCII-based passphrases will be converted to a 256-bit, pre-shared key (PSK) using a conversion hash. The 802.11-2016 standard provides a passphrase-to-PSK mapping process.

Weak passphrases can be a security risk, and CWSPs need to understand why. Software is available allowing specific information captured over the unbounded wireless medium, to be challenged against a "dictionary file." The dictionary, in this case, is a file that contains common words and phrases, and these files are available online. The software used to perform this attack will compare the passphrase to every item in the dictionary, looking for a possible match. Continuous testing will allow the intruder to try millions of combinations until they successfully discover one that works. With late-generation computing power, even a modest laptop can run this sort of analysis with ease. If a match is found, the security has been compromised. For this reason, if passphrases are to be used, it is critical to select one using a maximum-entropy technique and then keep it secret. This means that long, unique passphrases that you carefully control are recommended to help provide adequate security.

Additional attacks can be performed against WPS PIN-based security in Kali Linux using the following included tools:

- airmon-ng
- airodump-ng
- wash
- reaver

Reaver, in particular, can crack WPA and WPA2 PSKs when WPS is also in use. To perform attacks with such utilities, you will need to be able to inject frames onto the wireless network. Let's take a hands-on look at how this is accomplished.

To test your adapter for this capability, perform this procedure in Kali Linux:

1. Open a Terminal window.

2. Execute the airmon-ng start wlan0 command. Change wlan0 to match your adapter identifier, if needed.

3. Execute aireplay-ng -9 mon0 command. Change mon0 to match your monitor interface identifier, if needed.

4. You will see output showing both "Injection is working!" and the percentage of directed probe requests successfully sent to various APs on the configured channel, if injection works with your adapter.

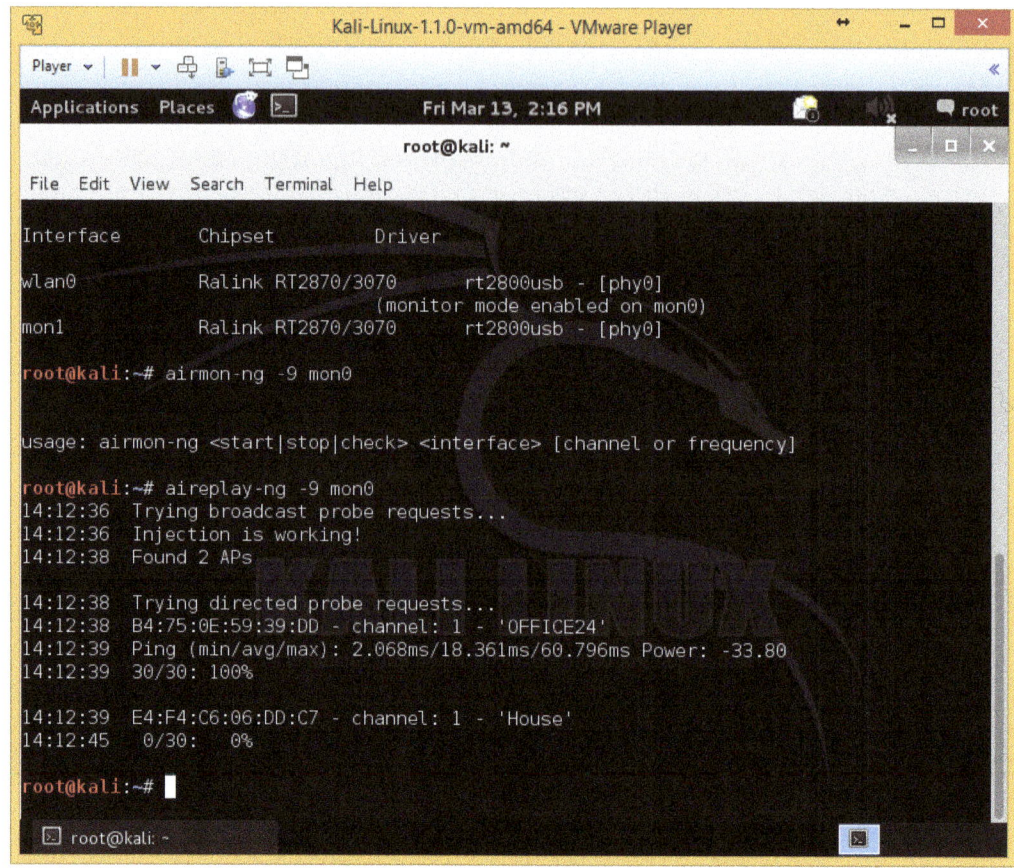

Testing Injection Ability in Kali Linux

Using default passphrases creates another issue. Most manufacturers publish the default credentials used to log into and manage their devices, on their online support pages. They also, frequently, publicly share the default passphrases that may be used to secure the wireless service set. As a reminder, security policy should address the

142

issue of using default settings to include management interface credentials and proper passphrase security.

If an AP is not configured correctly with respect to passphrase security, it will likely cause authentication issues. The same passphrase must be installed on all devices that are part of the same service set. Common errors in configuration include:

- Wrong passphrase entered
- Incorrect letter case used
- Errant spaces

It is important that all devices use the same exact passphrase and that they utilize the correct alphabetic letter case. Otherwise, the device will not be able to complete the authentication process because of a syntax mismatch between the devices. Controller-based WLAN systems are easier in this regard because you must only provision the controller, and it then provisions the APs with the passphrase. When multiple controllers are used, the passphrase syntax must match on all controllers that host APs in the same service set. Regardless of whether the APs are controller-based or autonomous, the passphrase for a given SSID must be consistently deployed across APs.

An important part of a CWSPs body of knowledge is the various keys that are used in WLAN authentication. We will present key-related generation and use details in the next several pages. This is an important part of the book to read over as many times as it takes to grasp the significant details.

WPA & WPA2 personal mode authentication, both happen by means of a shared pairwise master (PMK). Both the supplicant (client device) and authenticator (AP) are configured with the same passphrase (or HEX ASCII), which is converted into a PMK. The PMK is then used for dynamic encryption key generation during the 4-way handshake. Because the PMK is used as an input to dynamic encryption keys, if the PMK on both devices does not match, shared encryption keys will not be generated, and the 4-way handshake will fail.

When a passphrase is used, rather than directly entering a hexadecimal PSK, the passphrase is converted into a PSK by the following method:

PMK = PRF(passphrase, SSID, ssidLength, 4096, 256)

In this formula, PRF refers to a pseudo-random function that is calculated against the string comprised of the passphrase, SSID, and the SSIDlength. It is hashed 4096 times to generate a value of 256 bits, which then becomes the very important PMK. The PMK is never transmitted across the unbounded wireless medium.

The PMK is used to generate a subsequent key known as the pairwise transient key (PTK). The derivation of the PTK is done by exchanging MAC addresses and randomly generated tokens (known as *nonces*) between the supplicant and the authenticator, using the RSNA-defined 4-way handshake. The first two frames in the 4-way handshake contain all of the information required to create the PTK, except the PMK. Did you get all that? If not, read it again, as its importance cannot be overstated. (We'll also re-visit this keying process later in the book when we talk about secure roaming, so we need to grasp it.) Now let's go even deeper with the 802.11 encryption and key paradigms.

WPA or WPA2 Personal

The original 802.11 standard specified WEP for a security level which was only intended to prevent casual eavesdropping (which is about all it did). Hopefully, by now, you accept that WEP is unable to provide sufficient security on modern WLANs because it is easily compromised, as mentioned earlier in the book. Thankfully, most hardware simply no longer allows you to enable it. Let's review what we've already covered about WEP again, so we can expand on it. In the early 802.11 days, the perspective was that stronger WLAN security would be available with the eventual 802.11i amendment to the standard. Around the year 2001, however, the 802.11i amendment to the standard was still some time away. Stronger security was needed for standards-based WLAN technology to continue to evolve. Many organizations were reluctant to install 802.11 wireless networks because they lacked adequate security measures at the time. As happens today, there were early, well-published wireless breaches on business networks that highlighted the dangers of insufficient security on the WLAN. The Wi-Fi Alliance developed a fix for WEP called TKIP, which was based on the pre-ratification draft of the 802.11i amendment. This interim solution served well until stronger security mechanisms, also based on the 802.11i amendment, were finally ratified. IEEE 802.11i would ultimately provide very strong security that works well in home, small and medium-sized business and enterprise wireless networks. That is, 802.11i would provide strong security where properly implemented.

TKIP provided security enhancements that improved the way WEP operated by fixing some of the inherent security flaws native to WEP. These enhancements included a longer initialization vector (IV), stretched to 48 bits instead of the original 24. TKIP also provided an improved integrity check. In most cases, all that was required for legacy infrastructure devices to support TKIP was a firmware upgrade. For legacy client devices, driver updates and an OS patch or two brought TKIP support to capable hardware. TKIP has also become "legacy" and should not be used if client devices support AES.

Another important milestone in WLAN security was the creation of the Wi-Fi Alliance. Formed in 1999, this organization gained popularity, and provided an important service, by providing interoperability certification testing for member companies that manufactured 802.11 standards-based WLAN devices. Early device certifications were for simple communications functionality and operation. With the demand for better wireless security and the lack of a ratified amendment, the Wi-Fi alliance developed the Wi-Fi Protected Access (WPA) interoperability certification, which became available in 2003. WPA provided an interoperability certification for TKIP technology that allowed manufacturers of enterprise-grade wireless infrastructure devices to successfully build and market devices that provided a stronger wireless security solution that served well until the 802.11i amendment to the standard was approved. Once the 802.11i amendment was ratified, it would offer even stronger WLAN security by using CCMP/AES, which is certified by the Wi-Fi Alliance in WPA2.

> TKIP uses a per-MPDU TKIP sequence counter (TSC) to prevent replay attacks. A replay attack occurs when a frame is retransmitted, with or without modification.

Defining WPA and WPA2

TKIP (temporal key integrity protocol technology and the WPA certification provided a great interim wireless security solution ahead of the advent of 802.11i. However, the WLAN industry and its customers still required better wireless security before wireless could become the widely preferred access method of choice in business settings.

When the 802.11i amendment was ratified in 2004, it incorporated the use of CCMP/AES technology and introduced the Robust Security Network (RSN) concept. Although TKIP/RC4 helped fix some of the problems associated with WEP, the arrival of 802.11i and CCMP offered the best and strongest security available for creating the RSN. It may have been slow in coming, but it was worth the wait, as 802.11i is the backbone of enterprise wireless security today. Only WPA3 is stronger, and that is just getting started as per Chapter 10.

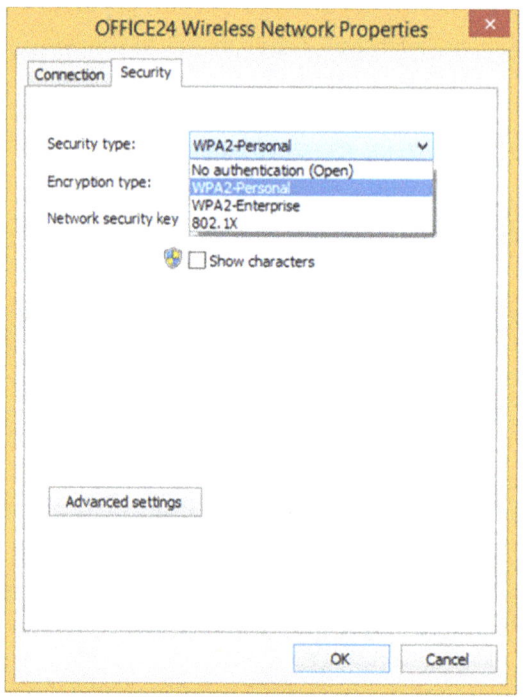

Configuring WPA2 in Windows

Given that certifying pre-802.11i WPA security was such a success, the Wi-Fi Alliance decided to also certify equipment based on the 802.11i ratified amendment. This new certification was named WPA2 and was made available in late 2004. Since then, manufacturers of enterprise-grade wireless equipment have been able to design, build, market, and sell 802.11 WLAN devices that supported the stronger CCMP/AES security mechanisms.

Enterprise equipment for the last decade has supported CCMP/AES and should be used over TKIP/AES as a rule. In some cases, organizations with older specialty client devices, may not be able to upgrade their devices for whatever reason, so a mixed environment of TKIP (WPA) and CCMP (WPA2) may still be required in corner cases.

Authentication of WPA-Personal and WPA2-Personal is basically the same, leaving the encryption method as the only differentiation.

The RSN information element (IE) is also very important to the CWSP, and it's contained within certain wireless management frames. This IE defines an Authentication Key Management Suite List field, which specifies the type of authentication supported in a network. If the field is populated with "00-0F-AC:02", it indicates PSK-based authentication. We will examine other values in a just a bit.

One distinguishing characteristic of both WPA (which should be avoided) and WPA2 is the presence of what is known as the 4-way handshake. Made up of four unicast management frames, the handshake is used to derive the necessary encryption keys that will secure both unicast and broadcast/multicast wireless traffic on a per-user basis. You simply cannot progress in wireless security without understanding the basics of the 4-way handshake.

The 4-way handshake provides an exchange of specific and required information, allowing the wireless AP and the wireless client device (in either an infrastructure or personal network) to create the same encryption keys. These keys will be used to encrypt and decrypt information sent over the wireless medium, and so are critical to WLAN security.

With WPA personal and WPA2 personal modes, "authentication" occurs during the 4-way handshake process via the fact that the client STA has the pre-shared key. After deriving the pairwise transient key (PTK), the supplicant (client device) message integrity check (MIC) protects the 2nd frame of this exchange. The hash used to calculate the MIC value includes the key confirmation key (KCK), which is a part of the PTK. If the authenticator's (AP or controller) PTK and the supplicant's (client device) PTK do not match, the MIC will fail, and the authenticator will silently discard the frame, ceasing the 4-way handshake exchange. The same process happens for frames 3 & 4, so mutual possession of the PTK is confirmed. The significant point here is that the pre-shared key, when properly configured to match on the AP and the client devices, results in a successful 4-way handshake. If the key is different, the handshake fails and, therefore, authentication and key generation has failed.

As important as the 4-way handshake has been to wireless security through WPA2, the KRACK attack exposed weaknesses that would ultimately lead to the development of WPA3. But, WPA2 will be with us for quite a few years as WPA3 takes root, and so key generation, key hierarchy, and the 4-way handshake remain crucial wireless security topics.

Per-user PSK (PPSK)

Typical implementations of WPA and WPA2 personal mode passphrase security use a single common passphrase that is shared by all users for a given SSID, as per the original intent of 802.11i. Potential security issues arise with standard uniform PSK because all users of the service set will know the passphrase, or at least have the same passphrase configured. The passphrase is used to create a 256-bit pre-shared key that will restrict access to the wireless network and will be used to generate keys and secure individual user data. If the PSK is updated, then all devices in the WLAN need to be updated- which can be laborious. But there is another non-standard form of PSK that you may encounter. Manufacturer proprietary mechanisms allow for unique Per-user PSK, which limits the ability for users to gain one another's passphrases. While the security offered by such a solution may not be as robust as secure implementations of 802.1X/EAP, this option provides many advantages over traditional one-for-all passphrases which include:

- Allowing granular, user-specific control of network privileges (not provided by traditional 802.11-based shared passphrases)
- Alleviating management burden when a passphrase must be changed due to employees leaving or passwords being compromised
- Providing enhanced accounting functionality
- Enhancing security between users of the same network by preventing decryption of unicast traffic
- Does not pose a conflict with 802.11 protocol operations

It is important to understand that PPSK is implemented differently by each vendor that offers it and that it is only available from a limited number of WLAN manufacturers.

Although it can enhance the way passphrase technology is used, it should not be used as a substitute for enterprise 802.1X/EAP security solutions.

Entropy

The 802.11-2016 standard provides for the use of a 256-bit pre-shared key to be entered directly as the pairwise master key (PMK). In addition, the standard allows the use of a more user-friendly 8-63-character passphrase from which the actual PSK can be derived using a key mapping technique. The 802.11 standard says, "The RSNA PSK consists of 256 bits, or 64 octets when represented in hex." The strength of the passphrase confidentiality mechanism can be considered sufficient for any non-governmental or non-military usage as long as the administrator enters the PSK directly, using a 64-octet hexadecimal pre-shared key.

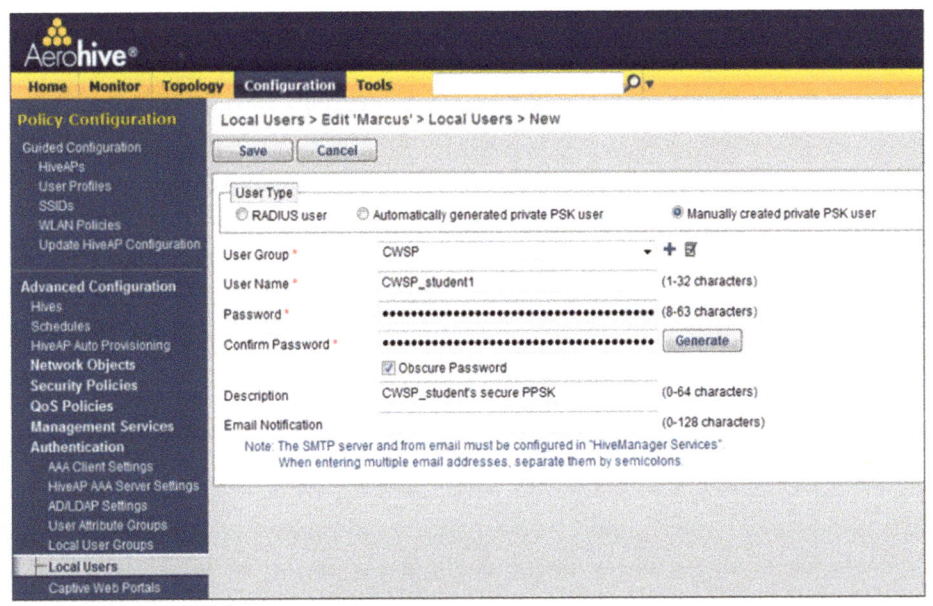

Per-User PSK Configuration in Aerohive Management System

The primary vulnerability in either the TKIP enhancement to WEP or the CCMP replacement for WEP becomes evident when a weak passphrase is used to create the 256-bit hexadecimal pre-shared key through the IEEE-recommended mapping routine. The 802.11 task group felt it was too difficult to expect users to enter long hex keys as part of their configuration duties.

From IEEE 802.11i-2004:

"This clause defines a pass-phase-to-PSK mapping that is the recommended practice for use with RSNAs. This pass-phrase mapping was introduced to encourage users unfamiliar with cryptographic concepts to enable the security features of their WLAN.

*A pass-phrase typically has about 2.5 bits of security per character, so the pass-phrase mapping converts an [n] octet password into a key with about 2.5[n] + 12 bits of security. Hence, it provides a **relatively low level of security**, with keys **generated from short passwords** subject to **dictionary attack**. Use of the key hash is recommended **only where it is impractical to make use of a stronger form of user authentication**. A key generated from a pass-phrase of less than about **20 characters** is **unlikely to deter attacks**.*

> If it was true in 2004 that the passphrase-based security in WPA and WPA2 networks was unlikely to deter attacks, if the passphrase was *less than about 20 characters*, how much truer is it today? Computers are dozens of times faster. Distributed computing, thanks to cloud services and malware, is far easier than ever. Before easily giving into the desire to "just use PSK because it's easier," remember that the recommendation is to use PSK only if stronger forms of authentication are impractical. This recommendation comes straight from the source – the 802.11 standard itself (and it's still there in 802.11-2016). -Tom

To counteract the tendency of using repetitive patterns when creating cryptographic keys, it is necessary to intentionally add an additional measure of uncertainty into the selection process. In digital communications, this uncertainty or randomness is called *entropy*, and it is measured in bits.

To visualize entropy, think about the act of flipping a coin. The coin has two possible states: heads or tails. We can be certain that the coin will come to rest in one of these two states, but we cannot say for certain whether it will end up as heads or tails. A two option scenario results in one bit of entropy. The following figure helps in understanding the randomness with which entropy is involved. Note that adding a single new slider always increases the possibilities by a factor of ten. The result is that a four-slider lock provides 100 times more possibilities than a two-slider lock – not twice as many possibilities.

$10^2 = 100$ possibilities $10^3 = 1000$ possibilities $10^4 = 10000$ possibilities

Locks Representing Possibilities and Randomness

Simply defined, entropy is lack of order or predictability – or randomness. Technically speaking, passwords or passphrases themselves do not contain entropy or rather, they contain an entropy value of zero. It is the method you use to select the passphrase that contains the entropy. So, entropy is an estimate of how difficult it would be to deduce your passphrase. The more entropy (measured in bits) contained within the method you use to create your passphrase, the more difficult it will be for someone else to deduce it.

Current information technology security best practices state that 72-bits of entropy should be safe for the foreseeable future while 128-bits is definitely safe for a very long time. Again, this assumes entropy and not some specific length alone. Longer passwords comprised from a pool of fewer random options are not as strong as shorter passwords comprised from a pool of more random options. This is why, using a minimum 20-character PSK that is not a common phrase and includes complexity (upper case, lower case, digits, and special characters) results in strong WPA- and WPA2-Personal implementations. In Chapter 10, you'll read about WPA3's mechanisms for defending against weak passphrases.

Strong Passphrases

Is wireless passphrase technology (WPA/WPA2 personal mode) strong and safe enough for use in a branch or remote office? If the preceding recommendations concerning the safe selection and storage of maximum-entropy passphrases have been followed, then there is only one remaining weakness that may be of concern for corporate usage. When using a passphrase/pre-shared key, there is one common pairwise master key (PMK) which is shared among all of the wireless devices that are part of the wireless service set. A potential security issue may arise from this configuration. Anyone who has knowledge of the PSK/PMK can decrypt encrypted data between a pair of stations (e.g., a wireless mobile device and an AP) if they also capture the nonces from the 4-way handshake (sent in clear text across the unbounded wireless medium). Therefore, vulnerability to internal attacks is still possible. Use of profiles for configuration of WPA- and WPA2-Personal or direct entry by administrative staff will limit the number of users who know the PSK.

Additionally, from the information contained in the captured 4-way handshake and with the help of additional dictionary attack software, there is a possibility that weak passphrases can be discovered. These captured frames and hacked passphrases will then provide the 256-bit pre-shared key, which is in turn used to create the PMK for the service set. It is critical for users of this technology to use strong passphrases in order to protect the network. Several tools are available that will aid in creating strong, secure passphrases.

Password generation utilities and sites are available to assist in adding entropy to the generation process. Examples include:

- LogmeOnce Password Security
    - Password strength tester available as well
    - https://www.logmeonce.com/online-password-generator-to-generate-passwords
- Gibson Research Free Online Passphrase Generator
    - Ultra-high entropy, perfect for WPA/WPA2-Personal
    - https://www.grc.com/passwords.htm
- DiceWare
    - Let the dice create your maximum entropy passphrases in the form of easy to remember word groups
    - http://world.std.com/~reinhold/diceware.html

To generate a random password in Kali Linux, use the following procedure:

1. Open a Terminal window.

2. Execute the openssl rand –base64 20 command.

3. Note the password-like text generated, which is likely far more random than you would generate manually.

4. If you desire shorter or longer passwords, simply change the last value, 20, to the number you desire.

If PSK vulnerabilities are of concern to an organization, they should be addressed in the security policy as well as in their implementation. The policy should clearly state requirements for the creation of PSK passphrases and identify the individuals who may be informed of the passphrases.

When using 802.1X/EAP as the authentication and key management (AKM) technology for 802.11 wireless networks, each individual 802.11 association has a unique PMK and a subsequent set of unique temporal keys. This unique key hierarchy will not allow for any keys to be discovered by capturing the contents of the 4-way handshake. (More on this when we discuss WPA3.)

> Like the IEEE, I recommend a 20-character passphrase if you are using PSK. I know, it's frustrating and takes longer to enter and people forget it. But if you want good security with PSK, the passphrase should be 20-characters and not contain a common short sentence or even words commonly used in your vocabulary. Use unusual words, like imbroglio, coupled with numbers and special characters to add strength. Yes, you can use shorter passphrases. No, they will not be as secure. -Tom

AAA

The abbreviation AAA (as stated previously) stands for Authentication, Authorization, and Accounting. "Triple-A" plays an immensely important role in WLAN security for enterprise settings. The AAA framework/protocol can be thought of as a set of services provided within a network to securely manage and track access to network resources. The following helps to define this further:

Authentication – The authentication process refers to the verification of an entity's identity. 802.1X/EAP is a common authentication protocol used with standards-based 802.11 wireless networks and will validate that an entity is who/what it claims to be. Authentication can be accomplished in a variety of ways including username/password pair and user certificates, for example. In summary, authentication can be thought of in relation to who a network user is.

Authorization - This term refers to the allocation of network resources in accordance with the privileges of a user or group. Authentication confirms a user's identity, and then authorization provides access to network resources according to policy. Proper authorization will ensure the authenticated user has access only to the network resources and services they have been explicitly assigned. In summary, authorization can be thought of in relation to what a network user can do.

Accounting - Once resources have properly been authorized and a user has performed actions on a network, it is important to track and log those actions so that an accounting trail is made available. Accounting includes monitoring, analysis, and reporting of network events. In summary, accounting can be thought of in relation to what a network user did while connected as well as reporting on passed and failed authentications.

Remote Authentication Dial-In User Service (RADIUS) is the most common AAA protocol in use with 802.11 wireless networks, and it is used in almost every network that supports 802.1X/EAP. RADIUS server software is available for Windows, Linux, and even Mac OS/X servers, and many WLAN vendors sell their own RADIUS platforms. Often RADIUS is part of larger NAC solutions in big WLAN environments.

RADIUS is a networking service that provides centralized authentication and administration of network users. In the 1980s, RADIUS started as a way to authenticate and authorize dial-up networking users to allow access on a network, hence the name. A remote user would dial-up to a network using the public switched telephone network (PSTN) and a modem. A modem from a modem pool on the receiver side would answer the call. The user would then be prompted by a remote access server to enter a username and password in order to authenticate. Once the credentials were validated, the user would then have access to any resources for which they had permissions.

As computer networks grew in size and complexity and remote access technology improved, there was a need to optimize the process on the remote access server-side. This is where RADIUS provides a solution. RADIUS took decentralized remote access services databases and combined them into one central location, allowing for centralized user administration and centralized management. (If you can remember that far back, think of AOL's millions of customers and how difficult it might have been to manage that subscriber environment without a central database that RADIUS could query.) It eased the burden of having to manage several databases and optimized administration of remote access services. RADIUS is commonly used as an authentication server for wireless networks in 802.1X/EAP implementations.

> RADIUS servers can use return list attributes to set group membership, and this can be used to implement appropriate security profiles and even network configurations for authenticated users of the WLAN. This makes RADIUS very powerful beyond simple authentication.

Mutual Authentication

The Internet Engineering Task Force (IETF) Request for Comment (RFC) specifying EAP, along with the 802.11-2016 standard, requires support for mutual authentication in the creation of a robust security network association (RSNA). Mutual authentication confirms the identity of the EAP peers, also known as the supplicant and the authentication server (AS). Most EAP methods implemented in modern standards-based wireless networks support- but may not require-mutual authentication.

Mutual authentication is a method used for two entities to authenticate each other. In 802.11 wireless security, this typically means that s a client device and an authentication server will authenticate each other. Though many EAP types are available to the wireless security professional, not all are good choices as discussed elsewhere in the book. For example, EAP-MD5 is an EAP type that does not support mutual authentication. It was not designed for wireless networks in the first place and should never be used with wireless unless it is a tunneled authentication (and even then, MS-CHAPv2 would be a better choice). Mutual authentication should always be supported for any authentication mechanism to be considered secure. Mutual authentication is also required for dynamic encryption key generation.

Mutual authentication is essential because it prevents the easy implementation of rogue authentication servers. If the server is authenticating the client (typically thought of when authentication is considered), but the client is not authenticating the server, *evil twin* attacks and other impersonation attacks are much easier. This alone should provide sufficient motivation to implement mutual authentication on the WLAN. Remember, it is no more complicated than the simple step of implementing a strong EAP type. When you do this, you have mutual authentication.

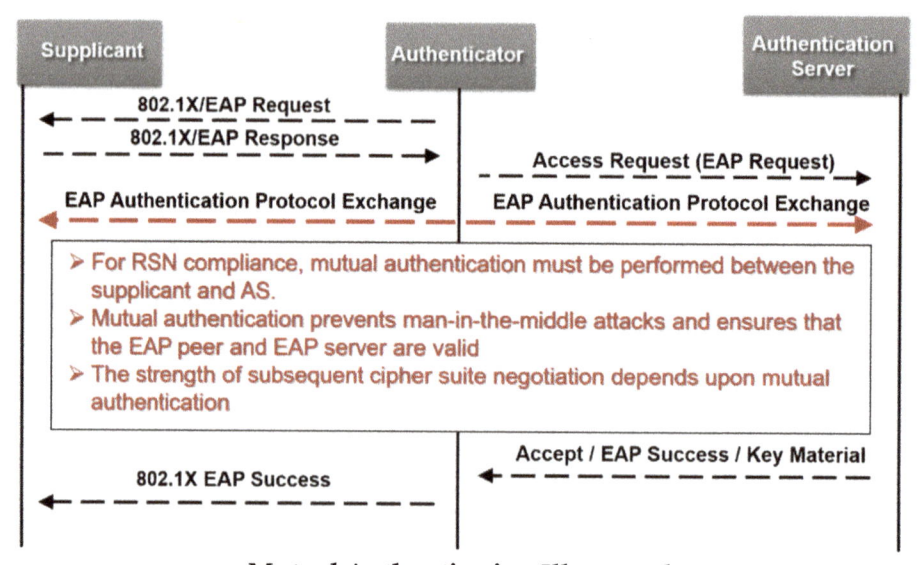

Mutual Authentication Illustrated

Some key points to consider:

- To be considered a robust, secure network, mutual authentication must be performed between the supplicant and authentication server

- Mutual authentication prevents client hijacking, which in turn will prevent man-in-the-middle attacks and ensures that the EAP peer and EAP server are both valid

- The strength of subsequent cipher suite negotiation depends upon mutual authentication

Authorization

Authorization may be performed in a few different ways, but generally speaking, policies are defined and applied to users or groups of users via a profile mapping in the user database. Authorization includes the application of policies such as ACLs, VLANs, firewall policies, bandwidth controls, location/access permissions, traffic filters, and Quality of Service (QoS) policies.

As previously mentioned, RADIUS servers may use *attributes* in a RADIUS response to designate a specific role or policy for a user/group. This allows for simplification of the distribution of authorization rules in a WLAN and amounts to a powerful way to dynamically make the network enforce the security policy.

Authorization can be allowed on a per-user or per-group basis and may include:

- **Access Control Lists (ACL)** - What can the authenticated user do on the network
- **Stateful firewalls** - Allowing or restricting network services and ports
- **Bandwidth controls** - How much data can a user transmit or receive (i.e., 5 Mbps)
- **Time controls** - What days and/or hours can the network be accessed
- **Location permissions** - What can be done based on the user login location
- **Traffic filters** - Restricting or allowing certain types of network traffic based on specified criteria
- **QoS policies** - Specifies the quality of service capabilities

> RADIUS supports ACCESS-REQUEST and ACCESS-ACCEPT packets to use for service authorization. When a RADIUS server receives an ACCESS-REQUEST packet, which includes a list of desired access rights, it must respond with an ACCESS-ACCEPT packet if all desired attributes are acceptable or an ACCESS-REJECT packet if one or more attributes are not acceptable.

It is important to understand that authorization is also known as *access control*. It may also be called Network Access Control, or NAC. You are controlling the resources to which the user has access by only authorizing access to appropriate resources.

Role-Based Access Control

Role-based access control (RBAC) refers to the general process of applying roles or groups to users. Then, filters, rules, and permissions are applied to a security policy. Finally, a security policy is mapped to a specific group or role. In the end, a user is assigned a security policy through its role. The security policy sets the rights or permissions of the user.

RBAC should be required for most enterprise wireless networks and should be specified in the security policy. RBAC requirements should include:

- Defining network access roles
- Assigning authentication parameters to each role
- Assigning authorization parameters to each role

RBAC allows for access rules based on specific roles rather than an actual user identity. This is not to say that an actual user identity does not exist, but simply that the user is assigned a role or group membership and then authorization occurs based on the role or group, rather than the user identity itself. RBAC was designed to ease the task of security administration on larger enterprise networks and shares characteristics like those of a common network administration practice. For example, creating users and grouping objects in an authentication database, which have been used in Windows and other systems for decades.

In computer network administration, to give a network user access to a network resource, such as a file share, best practices suggest creating a group object, then assigning the group permission to the resource, and finally adding the user object to the group. This method allows any user who is a member of that group, access to that specific resource. The major benefit of this model is that the administrator can easily change the group permissions and will automatically change all user permissions in the process.

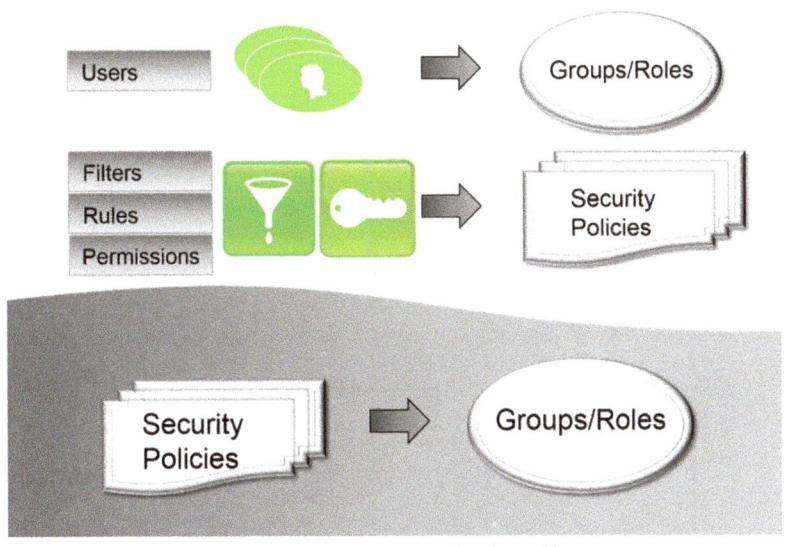

RBAC Represented Visually

RBAC, for the most part, works the same way and can be used for various activities users may perform while connected to a wireless network. Configuration options include:

- Enforcing time restrictions
- Bandwidth restrictions
- Controlling access to specific resources such as the Internet

Do some of the items on this list look familiar? Consider what was stated earlier with respect to authorization. RBAC is effectively an authorization system that is easier to manage in large-scale deployments with hundreds or thousands of users. It is not likely to be as beneficial in SOHO or small business deployments with less than 20-30 users. At the same time, cloud-managed WLAN is becoming quite popular with franchises that have small user counts, and RBAC can be rolled out to hundreds or thousands of small sites as if they were all one big enterprise via a well-designed cloud management dashboard.

RBAC also has applicability to tangential areas like Mobile Device Management (MDM), physical security systems, and even network administration and monitoring.

Accounting

The final part of the AAA process is accounting. Accounting allows a network administrator to track all or selected activity an authenticated user has performed while connected to the network. The logged information is valuable when an incident occurs as it shows who performed a given action on the network. Of course, the accuracy of this data depends on strong authentication.

RADIUS supports network accounting via default port 1813 or 1646 as specified in RFC 2866, and must be enabled on both the AAA client and the AAA server.

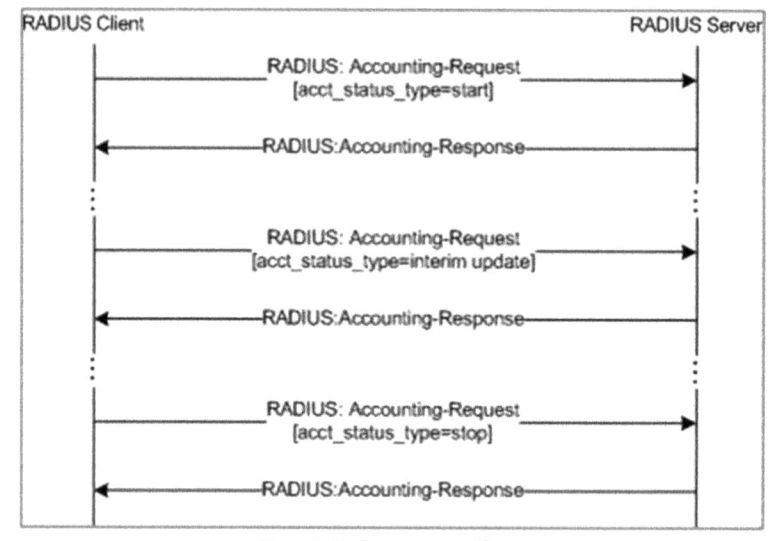

RADIUS Accounting

In addition, the wireless security professional must not lose sight of the accounting systems already available on the network. These include centralized log servers of infrastructure devices, such as Syslog servers, and operating system logs, like the Event Viewer logs in Windows environments. One useful tool is the ManageEngine EventLog Analyzer, available at:

www.manageengine.com/products/eventlog

Tools like EventLog Analyzer can be used to aggregate logs from infrastructure devices, operating systems, and more. Once aggregated, it can then generate reports against terabytes of data to show you the important information you need.

Accounting logs on the RADIUS are also extremely helpful in diagnosing why a given client failed authentication and can be invaluable in troubleshooting individual clients, the RADIUS server itself, or the integration between RADIUS and systems like Active Directory. Some RADIUS servers do accounting logging far better than others, and this should be a differentiator when you are shopping for a RADIUS solution.

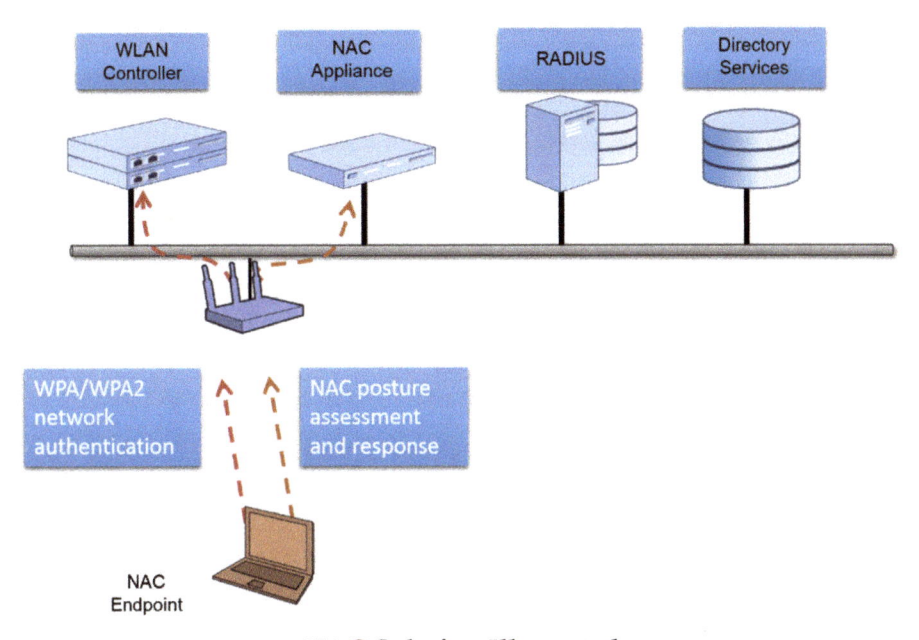

NAC Solution Illustrated

Network Access Control (NAC)

NAC is a security posture assessment tool. It may be stand-alone as appliance-based solutions, or as a set of features integrated with other key services and distributed across multiple components. Before a client joins a network, the NAC service validates that the client complies with the network policy. Actions performed by a NAC solution may include:

- Ensure all appropriate policies and security mechanisms are met by endpoints
- Ensure policies are applied to enforce security on a network

- Enforce requirements like antivirus software version and scans, OS updates, security patches, firewalls, user restrictions, etc.
- Authentication & authorization
- Posture assessment
- Quarantine
- Remediation

WPA & WPA2 Enterprise

The three primary authentication components of WPA-Enterprise and WPA2-Enterprise are:

- Remote Authentication Dial-In User Service (RADIUS)
- IEEE 802.1X - Port-based access control
- Extensible Authentication Protocol (EAP)

We have learned that RADIUS allows for centralized authentication services and acts as the authentication server (AS) in the 802.1X process. We've also established that RADIUS can be used to query a local self-contained user database, or an external user database (like Active Directory) via Lightweight Directory Access Protocol (LDAP).

IEEE 802.1X is the standard that defines *port-based access control*. It specifies the roles of the components used in the authentication process. In a wireless network, regardless of the vendor in use, these components are the "building blocks" of RADIUS:

- The supplicant - (client device to be authenticated)
- The authenticator - (the AP which provides the wireless connection, or the controller that lightweight APs are joined to)
- The authentication server - (the RADIUS server that validates user credentials)

IEEE 802.1X contains virtual controlled and uncontrolled ports to filter pre-authentication and post-authentication user traffic. It's easy to get confused at first-how do WLANs, with no physical ports- make use of port-based security? Remember that we are referencing virtual ports!

The image, WPA2-Enterprise Illustrated, shows the basic flow of client authentication in a WPA- or WPA2-Enterprise network using RADIUS, 802.1X/EAP, and an LDAP-compliant user database.

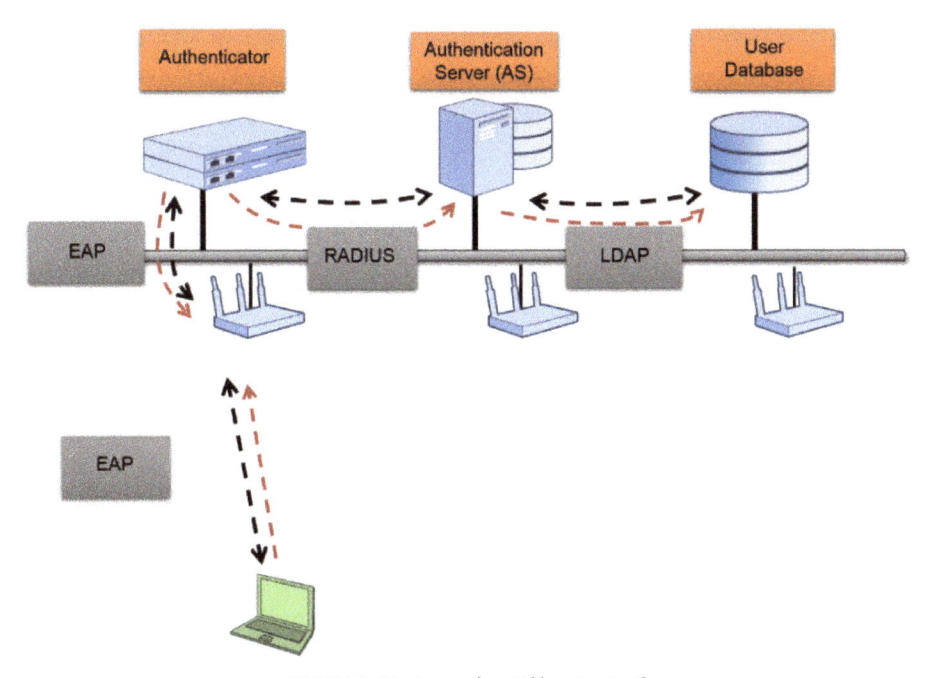

WPA2-Enterprise Illustrated

RADIUS Authentication

RADIUS services are the most often used user-based authentication services in WLAN infrastructures. RADIUS services can be implemented either natively in a WLAN controller or cloud dashboard for smaller user counts or on a stand-alone server computer (Windows, Linux, OS/X, etc.). RADIUS is based on UDP rather than TCP, though RFC 6613 supports RADIUS over TCP. When implemented on a server that runs a local firewall, it is important to open the following default ports where relevant:

- Port 1812 for authentication operations.
- Port 1813 for accounting operations.

A RADIUS server can have an integrated (native) database, or it can proxy to SQL, older Windows Domains, newer Active Directory domains, Novell eDirectory, Lightweight Directory Access Protocol (LDAP), or another RADIUS service and is supported by all WLAN infrastructure providers.

Access points and WLAN controllers point to RADIUS servers in order to authenticate users (using 802.1X/EAP AKM) and to define authorization parameters to those users.

Several steps are required on different components in order to successfully configure 802.1X/EAP. Most RADIUS servers are relatively simple to configure because the user database that provides the authentication credentials is already configured and stored externally to the RADIUS service, in most cases.

The first step in configuring a RADIUS server is licensing. The licensing paradigm will vary by vendor but is typically based on supported authenticator counts. (Where RADIUS is part of a larger NAC environment, licensing can get painfully complicated at times.) Once licensed, if the server will be using digital certificates for server-side authentication, then the certificate needs to be imported or generated. If a certificate is generated, it will need to be distributed to client devices (via email, onboarding portal, file transfer, etc.). If the certificate comes from a trusted party whose certificate is already stored on client devices, distribution of the server-side certificate is not necessary. It is important to remember that a time-limited digital certificate resides on the RADIUS server and that the certificate will need to be refreshed periodically.

> RADIUS servers have digital certificates used to authenticate the RADIUS server itself. These certificates can be self-signed or provided by a trusted third-party.

Configuring an enterprise RADIUS server geared toward standards-based WLAN security specifically is typically a straight-forward process once you've defined what users or devices are able to connect, and by using EAP types. As a reminder, solutions enforce policy, and RADIUS is no different. You need to know what you want out of the server before you can configure it!

For WPA and WPA2 Enterprise, configurations must be performed in these areas:

1. Configure the client by selecting the WLAN profile, configuring the security parameters, including EAP type, and selecting the certificate. This may be automated using an onboarding config tool.

2. Configure the WLAN Controller or AP with the IP address, correct port, and shared secret of the RADIUS server. Optionally, the configuration of RADIUS services such as accounting may also be performed.

3. Configure the RADIUS server by adding the approved APs, controllers, or network subnets where authenticator devices are located. If user accounts are to reside on the RADIUS server (native RADIUS accounts), then they should either be imported or entered manually. Specific EAP and RADIUS services must also be selected, configured, and enabled on the RADIUS server.

4. Depending upon the external user database, additional configurations may be required for database integration and connectivity.

These steps illustrate a common workflow to be performed, but there may be other steps required depending on network complexity and components involved.

RADIUS Authentication Exchange

Beyond using the RADIUS server logs, a protocol or application analyzer can be useful when trying to troubleshoot an authentication failure. Depending on when the failure occurs within the authentication exchange, an analyzer can be used to pinpoint whether the fault is due to incorrectly supplied user credentials or misconfigured user database information. In addition, an extraordinarily high number of authentication failures may indicate vulnerability probing by a hostile intruder, or a high number of misconfigured supplicants. Too many failures can cripple a RADIUS server, as the server must devote processing resources to all authentication attempts, even failures.

When using RADIUS authentication, frames are exchanged on both the wireless network connection - between the supplicant and the authenticator, and the wired side - between the authenticator and the authentication server. Therefore, appropriate tools will be required in order to gather all of the necessary information and perform adequate troubleshooting steps. A protocol analyzer must be placed between the authenticator and the RADIUS server and another between the client STA and the authenticator to capture all communications related to the RADIUS authentication process.

You can build a RADIUS server for lab purposes using the Kali Linux VM. You will need to install FreeRADIUS and configure it. To install FreeRADIUS, follow this procedure:

1. Open a Terminal window.
2. Execute the command apt-cache search freeradius to view the available FreeRADIUS packages. NOTE: You may wish to run apt-get update first to ensure you see the newest packages available to you.
3. You will see output similar to the following:

freeradius - high-performance and highly configurable RADIUS server
freeradius-common - FreeRADIUS common files
freeradius-dbg - debug symbols for the FreeRADIUS packages
freeradius-dialupadmin - set of PHP scripts for administering a FreeRADIUS server
freeradius-iodbc - iODBC module for FreeRADIUS server
freeradius-krb5 - kerberos module for FreeRADIUS server
freeradius-ldap - LDAP module for FreeRADIUS server
freeradius-mysql - MySQL module for FreeRADIUS server
freeradius-postgresql - PostgreSQL module for FreeRADIUS server
freeradius-utils - FreeRADIUS client utilities
libfreeradius-dev - FreeRADIUS shared library development files
libfreeradius2 - FreeRADIUS shared library

4. Now, run apt-get install freeradius to perform the installation.
5. When asked if you want to continue, type Y and press ENTER.
6. When the installation completes, you can add a web-based management interface by executing apt-get install freeradius-dialupadmin.

After these steps, you will need to enable Apache with PHP and then create the appropriate MySQL configuration. More detailed information can be found here:

wiki.freeradius.org/guide/Dialup-admin

LDAP Authentication

LDAPv3 (RFC 3377) databases are often used in large enterprises to hold objects such as network users. RADIUS servers may query LDAP databases to authenticate wireless users. LDAP is based on the X.500 standard, and Microsoft Active Directory and Novell eDirectory are/were LDAPv3 compliant.

802.1X/EAP Configuration

In addition to RADIUS configuration, the WLAN infrastructure must also be configured with the proper authentication server information. This configuration is typically limited to the creation of the WLAN profile which includes:

- The network name (SSID)
- 802.1X/EAP authentication
- The encryption scheme used
- Any SSID specific settings
- Configuration of a RADIUS server

RADIUS-specific parameters include:

- An IP address
- The shared secret
- Authentication and accounting ports
- Several timers, and other required and optional parameters

The shared secret is a common password shared between the authenticators (the APs or controllers) and the RADIUS server. This is one security mechanism that will help to prevent the installation of rogue APs that may be connected to the wired network infrastructure, as rogues won't be allowed to send authentications to the RADIUS server without the RADIUS shared secret. The shared secret here has nothing to do with clients directly.

802.1X/EAP Client Configuration

Each supplicant may have different 802.1X/EAP capabilities and should be checked for compatibility prior to selection. Remember that the supplicant can be native to the operating system (Windows, OS X, Linux, etc.), or in the form of the client utilities provided by the wireless Network Interface Card (NIC) vendor, or as a third-party supplicant sold separately. Also know that many devices that common sense says SHOULD support 802.1X, actually don't - like most printers and many streaming devices.

Common client configuration steps, whether done manually or with automated tools, include:

1. Configure the SSID and basic 802.1X/EAP security parameters, including encryption type
2. Verify the specific EAP type is selected and configured
3. Select the proper server and client certificates to be used, and configure some tunneled EAP parameters
4. Select advanced settings, such as user or machine authentication, as well as fast, secure roaming features if required

Certificates and Tunneled EAP

To use WPA or WPA2 enterprise, digital certificates are required with many EAP types. Following are a few examples of enterprise-grade RADIUS applications capable of generating self-signed digital certificates and acting as a trusted root certificate authority:

- FreeRADIUS (Open source)
- Microsoft – Network Policy Server NPS (Windows Server 2012 and newer)
- Cisco –Integrated Services Engine (ISE)
- Aruba (HPE)- Clearpass

Some of these applications also provide methods of distributing the server-side certificate to client stations.

Many manufacturers of enterprise WLAN equipment also provide built-in RADIUS, which is integrated into the platform or management software they provide, but again this capability is aimed at small end-user counts.

Server-side Certificates

A self-signed server certificate must be created and imported into the client's trusted root certificate store. In many deployments, self-signed certificates are installed with installation executable files or other automated processes. They may also be installed during device staging and deployment. For example, when deploying Windows systems, the certificate may be installed before the system image is created.

WPA2 (and WPA) Enterprise may provide mutual authentication of both the authentication server and the wireless client, using several forms of extensible authentication protocol (EAP). The most popular of these EAP types use a digital certificate as the authentication credential for the authentication server. In addition, some of the EAP-types also use a digital certificate as the authentication credential for the wireless client.

> The Wi-Fi Alliance no longer lists WPA as a certification program they offer for new equipment on their certification programs page; however, they are still listing it on certificates. They do not recommend using it in any environment unless an upgrade is simply impossible. We've tried to make it abundantly clear throughout the text that WPA should NOT be used, even though we discuss its theory of operation. Just because we'd all like something to go away, doesn't mean it will!

A digital certificate is a data file, exchanged between the authenticating entities. Digital certificates are created, distributed, and authenticated by trusted Certificate Authorities (CA), which, in an enterprise deployment, are part of a Public Key Infrastructure (PKI). The CA certificate should be installed in the local store of trusted roots on all client devices so certificates issued from the CA will be trusted as well.

To provide authentication, several forms of EAP rely on transaction layer security (TLS) based protocol variants. TLS is based on the secure sockets layer (SSL) protocol originally developed by Netscape. The TLS standard does not specify how security is implemented. Instead, TLS leaves the decisions on how to initiate handshaking and how to authenticate credentials such as digital certificates and secret keys, to the protocol designers. These credentials may be exchanged either during or following the TLS handshake procedure.

TLS provides the mechanism to allow the client and server to authenticate each other and to negotiate an encryption algorithm and cryptographic keys while guaranteeing privacy using asymmetric cryptography and secure message integrity. TLS negotiations are secure from eavesdropping, hijacking and man-in-the-middle intrusions, and thus have great importance in 802.11 security. It's worth pausing for a moment and letting the importance of TLS sink in.

Authentication Models

When implementing authentication for a single building or even for a single-site campus, the functional model can be quite simple. However, when implementing authentication for branch offices and remote locations, the complexity increases. In this brief section, single building/site and branch deployments will be considered.

Single Building or Campus

In a standard enterprise single-site network environment, it is most common to see one or more wireless controllers authenticating users against RADIUS, which in turn proxies to an LDAP directory service. Because all components providing authentication are contained within the local high-speed wired network, this authentication model should perform well if the network infrastructure has been designed correctly. The same authentication model used for the traditional wired networks will simply be extended to the wireless network, with the possible addition of a RADIUS server, if it is not already in place.

Branch Option #1

In a standard enterprise multi-site environment, it is common to see one or more WLAN controllers or cloud management systems authenticating users against RADIUS, which in turn proxies to an LDAP directory service. In the multi-site deployment pictured below, both the headquarters and the branch office LANs are connected using an Internet connection. This is a very common method used in many implementations. It is also important to note that each location has its own RADIUS server used for authentication and its own replica of the LDAP database containing the user-directory database.

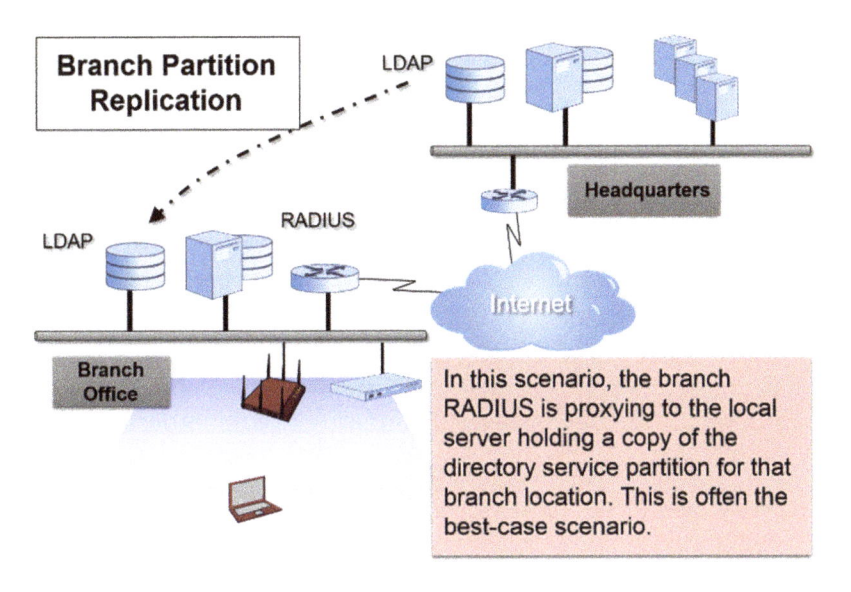

Multi-Site Directory Deployment

If there was a link failure between the sites, it would not cause authentication issues. Since they have their own RADIUS services and user database, each location is, for the most part, a stand-alone entity, of course, the branch office wouldn't be able to access any resources at headquarter locations, and updates would not be replicated until the link between sites was restored.

Branch Option #2

If the branch office does not have a local LDAP server (user directory database), RADIUS may proxy to a remote site (like the main office) to authenticate users. Doing this introduces a potentially slow and less reliable link (the Internet connection) into the equation- especially if the branch ISP connection is under-provisioned. This slow link could cause extreme latency for initial 802.1X/EAP authentications, so it's imperative that fast/secure roaming mechanisms such as OKC (discussed in chapter 8) be considered on the branch WLAN controller and wireless stations when using this authentication option.

If the branch or remote office does not have a replica or a partition of the user database, it may result in several potential problems including:

- Availability

- Latency
- Wide area network (WAN) utilization
- Fast secure roaming

This topology relies on an available connection (either leased line or the Internet) between locations. In the event of a link failure (Internet connection), the branch office RADIUS services would not be able to contact the headquarters location to authenticate users that attempt to log into the network.

The connection speed of the link between locations is also an important factor. If a slow link is in place, the authentication could suffer crippling latency since RADIUS traffic must cross the wide-area network to perform the authentication. The same holds true with a heavily utilized WAN link. This can also cause delays and long authentication times.

Depending on the fast, secure transition method used within the Wi-Fi environment at the branch location, it could cause roaming delays when a user transitions from one AP to another. If the fast roaming method used requires RADIUS server traffic to cross the WAN link, related delays could cause the connection to break, requiring reauthentication to occur.

A Deeper Look at IEEE 802.1X Port-Based Access Control

We introduced the fact that the 802.1X standard specifies port-based access control a few pages back, in the WPA2 section. Remember the building blocks of RADIUS: the supplicant (client device), the authenticator (AP or controller), and the authentication server (RADIUS server, frequently integrated to an LDAP credential store like Active Directory). Now let's go deeper.

The use of 802.1X in a wireless environment introduces several key benefits. Reasons to use EAP with 802.1X include:

- Maturity and interoperability as a solution
- User-based authentication and authorization
- Dynamic encryption key management (generation and distribution)
- Flexible authentication (many EAP types available)

The first step to creating a Robust Security Network Association (RSNA) in an infrastructure BSS is to become 802.11 authenticated and associated. As this happens, each STA receives the other's Robust Security Network (RSN) IE that describes their respective capabilities and requirements.

The second step in creating the RSNA is for the supplicant and authentication server to complete the mutual 802.1X/EAP authentication, and for the authentication server to pass the PMK to the Authenticator.

Note that an 802.1X Port consists of both a Controlled Port and an Uncontrolled Port. (Remember- this is wireless, so these are virtual ports.) This notion is very important to understand, as it plays a significant role in wireless network authentication.

- **Uncontrolled Port** – Used for authentication. The 802.1X Controlled Port is blocked from passing general data traffic between two STAs until an IEEE 802.1X authentication procedure and key management process complete successfully over the 802.1X Uncontrolled Port. This means that not even DHCP addressing for the client occurs until AFTER authentication.

- **Controlled Port** – Passes protected data. Once the Authentication & Key Management (AKM) completes successfully, data protection is enabled to prevent unauthorized access, and the 802.1X Controlled Port is unblocked to allow protected data traffic to flow for this particular STA. This is when the client gets a DHCP address and starts functioning as a network-connected device.

802.1X Supplicants and Authenticators exchange protocol information via the 802.1X Uncontrolled Port.

> When using 802.1X virtual ports, the uncontrolled port is used for authentication, and the uncontrolled port also allows data communications across the controlled port after such successful authentication.

Remember that the uncontrolled port is used for exchanging authentication frames while the controlled port allows for the communication of data frames. In an 802.11

wireless network, the controlled port is blocked during the entire EAP authentication process. Once the 4-way handshake has successfully completed, the controlled port is no longer blocked, and encrypted data will flow over the controlled port.

A Port Access Entity (PAE) operates the algorithms and protocols associated with the Port Access Control Protocol. A PAE exists for each port of a system that supports Port Access Control functionality in the supplicant role, the authenticator role, or both.

In the supplicant role, a PAE is responsible for providing information to an authenticator that will establish its credentials. A PAE that performs the supplicant role in an authentication exchange is known as a supplicant PAE.

In the authenticator role, a PAE is responsible for communication with a supplicant, and for submitting the information received from the supplicant to a suitable authentication server, in order for the credentials to be checked and for the consequent authorization state to be determined. A PAE that performs the authenticator role in an authentication exchange is known as an authenticator PAE.

Both PAE roles control the authorized/unauthorized state of the controlled port depending on the outcome of the authentication process. If a given controlled port has both authenticator PAE and supplicant PAE functionality associated with it, both PAEs must be in the authorized state in order for the controlled port to become authorized.

IEEE 802.1X/EAP Framework

Let's continue to build our knowledge of 802.1X by further examining the framework defined by this incredibly useful authentication standard.

The IETF RFC 5247, which defines EAP, does not advert to a specific implementation of EAP (other than MD5, which is useful for nothing more than basic testing of link communications and should never be used for secure authentication in any scenario). Instead, the RFC specifies a generic EAP framework that may be adapted to the specific purpose of different EAP implementations, such as EAP-PEAP (often just called PEAP or PEAPv0 for short), EAP-FAST, or any one of several others.

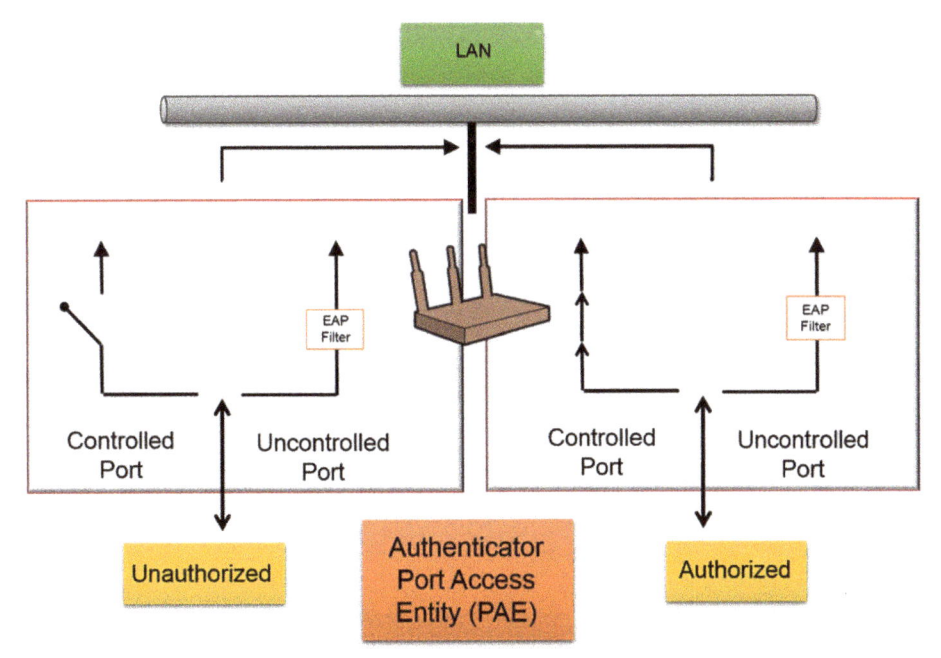

802.1X Architecture Visualized

> EAP over LAN (EAPoL) packets are used across the medium between the wireless client STAs and the AP/controller. Encapsulated EAP over RADIUS is used between the AP/controller and the authentication server (RADIUS).

In addition to improved encryption and integrity algorithms, the 802.11 standard specifies the use of 802.1X port-based access control and Extensible Authentication Protocol (EAP) to provide user authentication and dynamic key distribution. EAP is a Layer 2 authentication protocol used by both 802.3 Ethernet and 802.11 wireless networks as a flexible replacement for PAP and CHAP under PPP.

IEEE 802.1X restricts access to the network until a station has been authenticated by an authentication authority, usually residing within the wired network segment. The following factors should be considered when planning for and implementing 802.1X authentication solutions:

- The same 802.1X framework used for secure WLAN can also service wired 802.1X-secured environments.

- Access to the network is managed through the use of controlled and uncontrolled ports, which are logical entities on the same physical connection. Prior to successful authentication, the client station may only communicate over the uncontrolled port and the use of this communication is authentication only.

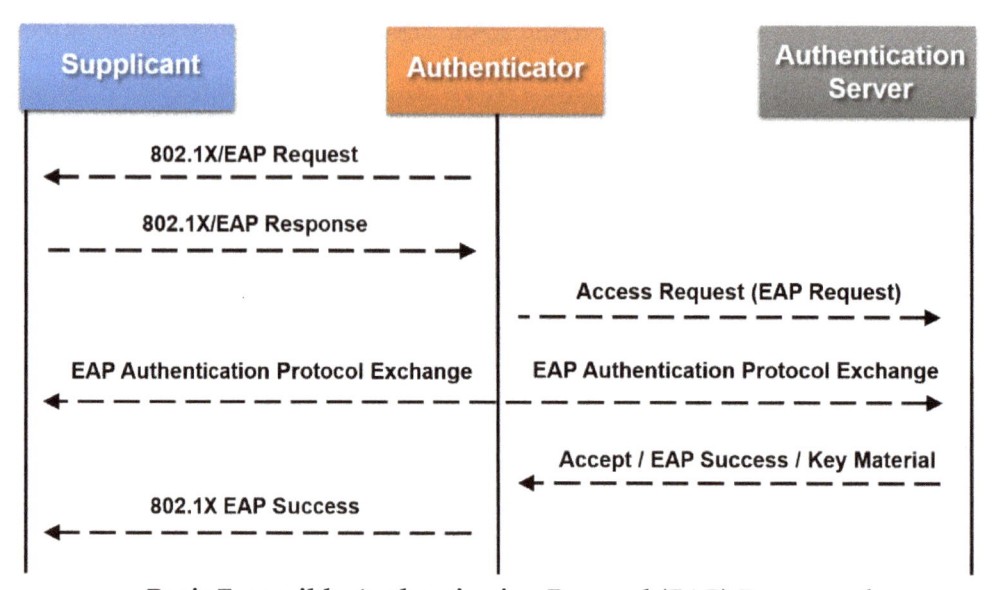

Basic Extensible Authentication Protocol (EAP) Framework

The third step in creating a RSNA is for the two STAs to have a matching pairwise master key (PMK). The PMK will be used to generate the pairwise transient key (PTK) for encryption purposes. Gaining the shared PMK is accomplished in one of two ways:

- Out-of-band
    - This method uses a preshared key (PSK) that is entered on both STAs either directly or created from a passphrase. Used with WPA/WPA2 Personal.

- In-band
    - This method uses an 802.1X/EAP with RADIUS infrastructure where the 802.1X/EAP mechanism creates the PMK.

Recall that the final step in creating a RSNA is the 4-way handshake, which results in the availability of the unicast and broadcast/multicast encryption keys on both the supplicant and the authenticator. At the conclusion of the handshake, each STA will have derived the same PTK. This PTK is used to secure unicast traffic, and it is used to exchange a group temporal key (GTK) to secure broadcast and multicast traffic. In an 802.1X/EAP enterprise scenario and upon successful completion of the 4-Way Handshake, the authenticator and supplicant have authenticated each other and the 802.1X controlled port is unblocked allowing encrypted data traffic to flow.

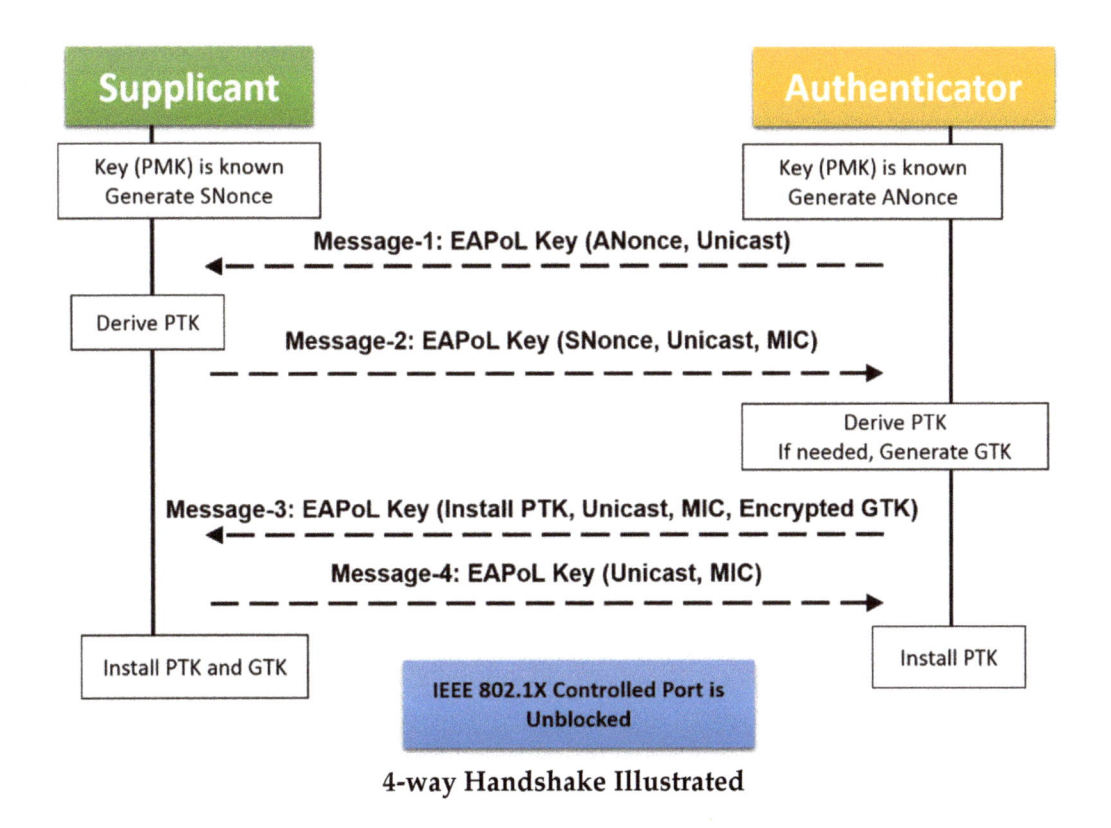

Supplicant | **Authenticator**

Key (PMK) is known
Generate SNonce

Key (PMK) is known
Generate ANonce

Message-1: EAPoL Key (ANonce, Unicast)

Derive PTK

Message-2: EAPoL Key (SNonce, Unicast, MIC)

Derive PTK
If needed, Generate GTK

Message-3: EAPoL Key (Install PTK, Unicast, MIC, Encrypted GTK)

Message-4: EAPoL Key (Unicast, MIC)

Install PTK and GTK | Install PTK

IEEE 802.1X Controlled Port is Unblocked

4-way Handshake Illustrated

> The order of communication is ANonce, SNonce with MIC, the transmission of the GTK and a final message to verify installation of the PTK sent from the supplicant to the authenticator.

You'll notice that we talk about concepts like 802.1X and the 4-Way handshake as distinct topics, and we put it all together in other discussions. There is a complexity here to appreciate and to attack from different angles. Do your best to keep it all straight at both the macro and micro levels, not just to get through the CWSP exam, but also to be a better wireless security professional.

EAP Type Comparison

There are many EAP types, each with their own advantages and disadvantages. Choosing the right EAP type is essential in a WLAN security deployment.

Remember that EAP is an authentication framework and does not advert to a specific authentication method. For that reason, multiple EAP types are available to choose from. The following comparison chart shows the features of many common EAP types.

	EAP-MD5	LEAP	EAP-TLS	EAP-TTLS	PEAP	EAP-Fast
Mutual Authentication	No	Yes	Yes	Yes	Yes	No
Certificates Required	No	No	Client/Server	Server Only	Server Only	No
Dynamic Key Generation	No	Yes	Yes	Yes	Yes	Yes
Costs and Management Overhead	Low	Low	High	Low/Medium	Low/Medium	Low
Industry Support	Low	High	High	High	High	Medium
Credential Security	Weak	Weak	Strong	Strong	Strong	Strong

Common EAP Types and their Features

Hopefully, we've impressed by now that EAP-MD5 should never be used on a wireless network because of several weaknesses, including the fact that it requires no digital certificates at all. EAP-MD5 does not provide mutual authentication, and it does not use tunneled authentication. As an authentication protocol, EAP-MD5 sends too much information across the network without encryption and is simply not a good protocol. As stated previously, it may be used for testing the links in the authentication chain, but it should not be trusted for any real-world authentication scenarios.

The following list shows EAP types that are more commonly used with wireless networks:

- EAP-TLS – client and server certificates required
- TTLS (EAP-MSCHAP-v2) – only server certificates required
- PEAPv0 (EAP-MSCHAP-v2) – only server certificates required
- PEAPv0 (EAP-TLS) – client and server certificates required

- PEAPv1 (EAP-GTC) – used with token card and directory-based authentication systems and only server certificates required

- EAP-SIM - EAP for GSM Subscriber Identity Module - mobile communicators

- EAP-AKA - for use with the UMTS Subscriber Identity Module - mobile communications

The EAP type used really depends on the organizational needs and capabilities of the network that is in place. Some EAP types such as EAP-TLS and PEAP (EAP-TLS) require the use of digital certificates on both the authentication server and the clients and therefore require a full private key infrastructure (PKI) which can become an expensive endeavor with extra management overhead.

> PEAP supports three different common internal methods on WLANs: MSCHAPv2, EAP-TLS, and EAP-GTC. EAP-GTC, specifically, is used with PEAPv1.

One common EAP method used is PEAP (EAP-MSCHAP-v2). It is included in the Microsoft Windows operating system, Android, Mac OS X, and Apple iOS, making it exceedingly popular. It can employ a username/password credential set and is widely available on several Linux variants. It generally strikes a popular balance between security and ease of use.

PEAP

Protected EAP is a common EAP implementation and is often referred to as EAP-in-EAP because it uses a second EAP type for client authentication after the server is authenticated.

All PEAP versions require server-side certificates, which are used to establish a TLS tunnel for the client authentication exchange. Establishment of the TLS tunnel is often referred to as Phase 1. Client authentication happens in Phase 2 inside the TLS tunnel and is specific to the particular PEAP implementation. Client authentication may include the use of a username and hashed passphrase (EAP-MSCHAPv2), a client certificate (EAP-TLS), or a token card (EAP-GTC), among others (such as POTP).

The following image shows the basic PEAP authentication flow for review.

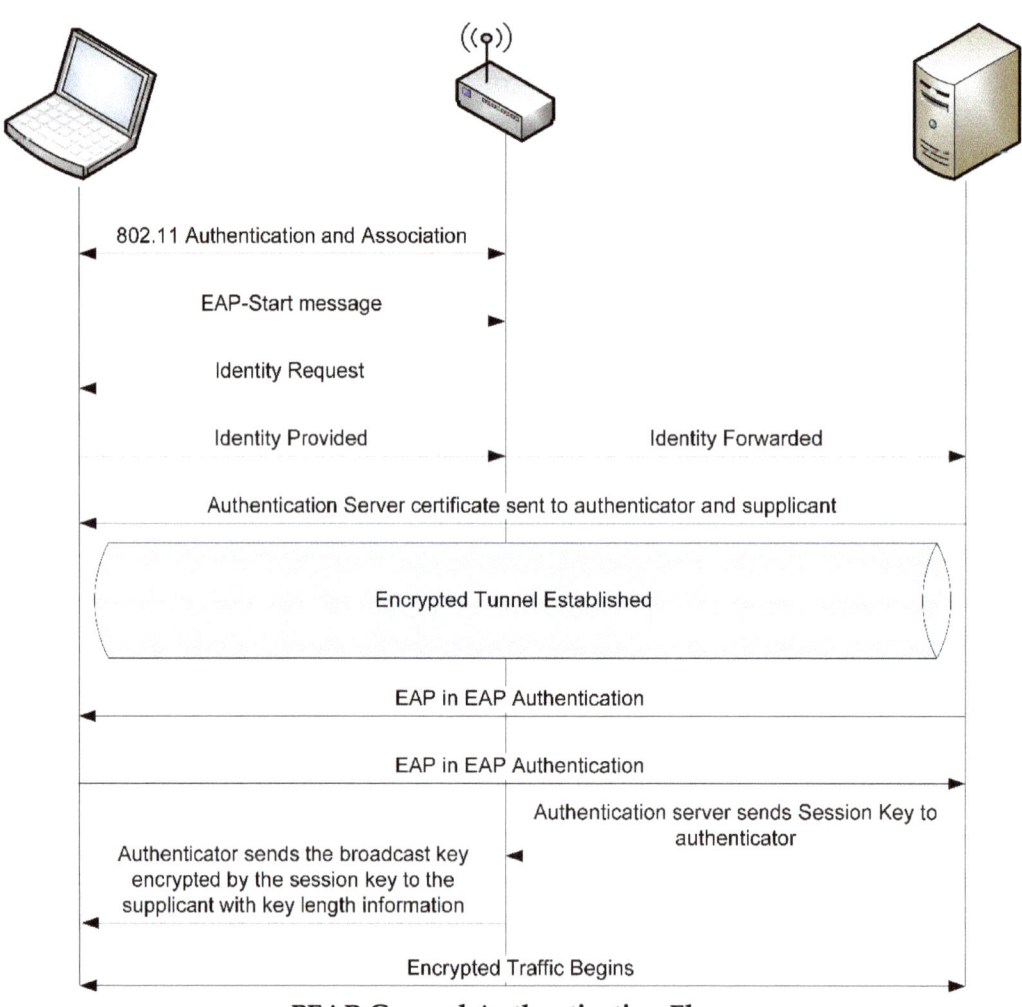

PEAP General Authentication Flow — sequence:
- 802.11 Authentication and Association
- EAP-Start message
- Identity Request
- Identity Provided
- Identity Forwarded
- Authentication Server certificate sent to authenticator and supplicant
- Encrypted Tunnel Established
- EAP in EAP Authentication
- EAP in EAP Authentication
- Authentication server sends Session Key to authenticator
- Authenticator sends the broadcast key encrypted by the session key to the supplicant with key length information
- Encrypted Traffic Begins

PEAP General Authentication Flow

PEAPv0/EAP-MSCHAPv2

PEAP (EAP-MSCHAPv2) is commonly implemented because it only requires username/password credentials from the client and because it is supported by many authentication servers. However, client-side certificates may be used instead of username/password credentials.

Phase 1 contains the establishment of the encrypted TLS tunnel and server authentication, while Phase 2 contains the client authentication and derivation of the session keys. PEAP requires EAP-in-EAP (e.g., PEAP/EAP-MSCHAPv2).

Note that PEAPv0 may also be used with tunneled EAP-TLS. This is called PEAPv0/EAP-TLS and is modeled after the EAP-TLS protocol, but performs client authentication inside a TLS tunnel. The frame flow is remarkably similar to that shown for PEAPv0/EAP-MSCHAPv2.

As stated, this EAP type is extremely popular because it has broad support on a wide range of client devices and operating systems.

Cisco LEAP

Long considered obsolete, LEAP provides mutual authentication, data encryption, per-user/per-session keys, dynamic key rotation at intervals, and a strong MIC. However, it has known weaknesses that prevent its utilization today.

LEAP requires only username/password credentials for authentication. The username is passed across the wireless medium as clear-text, and an MD4 hash is used as part of MS-CHAP authentication.

- Capturing user credentials is simple when strong passwords are not used
- 99% of all passwords that users select can be broken through the use of comprehensive dictionary files
- ASLEAP is an example of an offline dictionary attack utility
- Originally used for active and passive attacks against LEAP, ASLEAP can also be used to recover passwords contained in PPTP
- ASLEAP has been updated to support large file sizes
- Creating large dictionary files is simple and relatively fast
- High-capacity, portable hard drives (currently up to 2TB) are inexpensive and allow the use of terabyte-sized dictionary files with brute force exploits

LEAP Authentication Flow

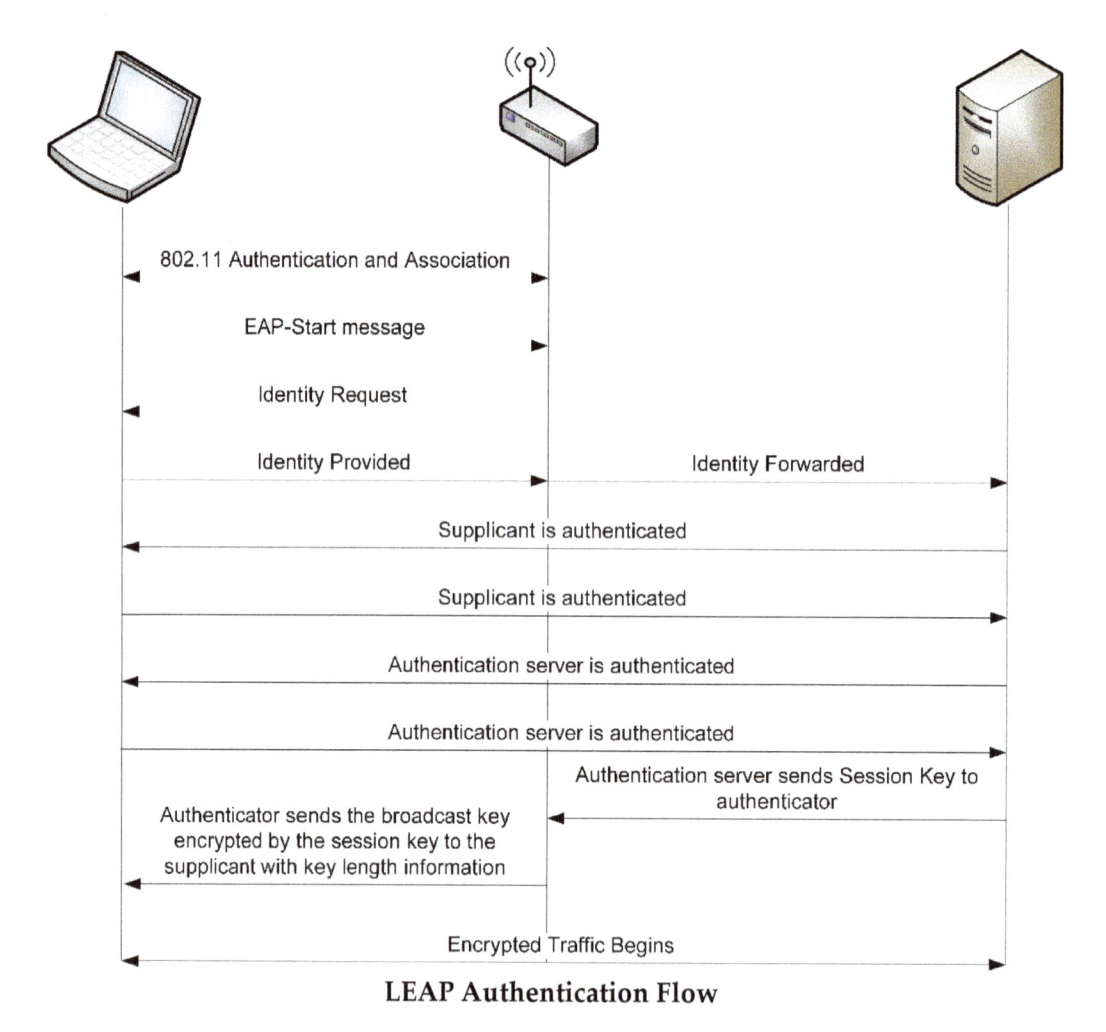

LEAP was cracked in early 2004 and should be avoided. Any installations that still use LEAP should work on a migration path to eliminate it from the network as quickly as possible.

EAP-FAST

EAP-FAST is Cisco's response to the vulnerabilities found in LEAP. EAP-FAST consists of three phases (0-2). Phase 0 is for provisioning Protected Access Credentials (PACs), which can occur either manually or through MS-CHAPv2. Phase 1 is for building a TLS tunnel for encrypting the client credentials when sent to the authentication server. Phase 2 is when the supplicant authenticates to the authentication server.

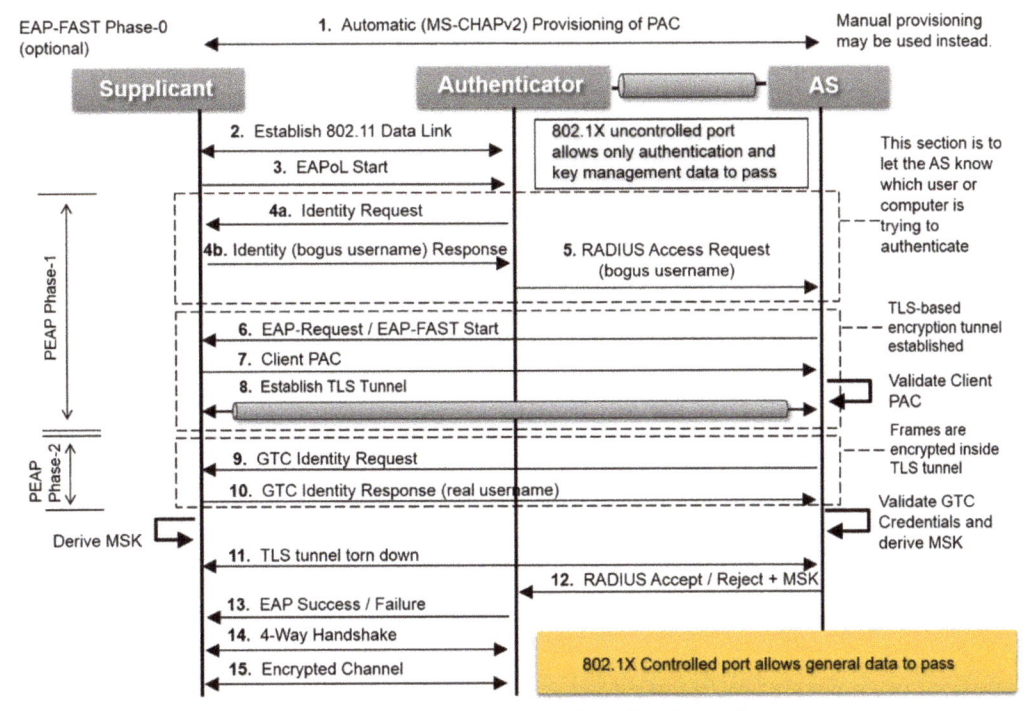

EAP-FAST Detailed Authentication Flow

According to Cisco (www.cisco.com/c/en/us/products/collateral/wireless/aironet-1300-series/prod_qas09186a00802030dc.html):

EAP-FAST uses symmetric key algorithms to achieve a tunneled authentication process. The tunnel establishment relies on a Protected Access Credential (PAC) that can be provisioned and managed dynamically by EAP-FAST through the authentication, authorization, and accounting (AAA) server (such as the Cisco Secure Access Control Server [ACS] v. 3.2.3). With a mutually authenticated tunnel, EAP-FAST offers protection from dictionary attacks and man-in-the-middle vulnerabilities:

- *Phase 1-Establish mutually authenticated tunnel-Client and AAA server use PAC to authenticate each other and establish a secure tunnel.*

- *Phase 2-Perform client authentication in the established tunnel-Client sends username and password to authenticate and establish client authorization policy.*

- *Optionally, Phase 0-This phase is used infrequently to enable the client to be dynamically provisioned with a PAC. During this phase, a per-user access credential is generated securely between the user and the network. This per-user credential, known as the PAC, is used in Phase 1 of EAP-FAST authentication.*

EAP-FAST supports 802.11, clause 8 key management. EAP-GTC is the only EAP type presently supported inside the EAP-FAST TLS tunnel. EAP-FAST was an acceptable interim solution for Cisco networks after LEAP was cracked and before newer EAP types became popular.

EAP-TLS

EAP-TLS requires the supplicant and authentication server to have their own x.509 certificates installed. A TLS tunnel is constructed to secure the key generation process. In fact, EAP-TLS has two modes: normal and tunneled. Since secure server certificates are used for server and client authentication, EAP-TLS implementations often forego the tunnel.

EAP-TLS supports mutual authentication and encryption key generation either through proprietary mechanisms or through the 802.11 4-way handshake. EAP-TLS is an excellent choice in environments distributing all usable network hardware due to required client certificates. When the organization only allows self-distributed hardware to be used on the network, they can more easily control the distribution of client certificates. In environments with requirements for BYOD and heavy secure guest access, PEAP or EAP-TTLS are more likely to be utilized.

EAP-TTLS/EAP-MSCHAPv2

EAP-TTLS supports the 802.11 4-way handshake, uses a TLS tunnel for encrypted user credential exchange, and supports various legacy authentication protocols (i.e., MD5, PAP, CHAP, MS-CHAP, and MS-CHAPv2), inside the TLS tunnel. EAP may also be tunneled (e.g., EAP-TTLS/EAP-MSCHAPv2). EAP-TTLS, developed by Funk Software and Certicom, has gained a loyal customer base in the industry because it supports authentication protocols compatible with legacy user databases.

You will notice several similarities between this EAP type (EAP-TTLS) and PEAP (EAP-MSCHAPv2). They were released within a short time of one another and are basically competitors. One major disadvantage to EAP-TTLS is that it requires installation of third-party supplicant software on devices running the Microsoft Windows operating system.

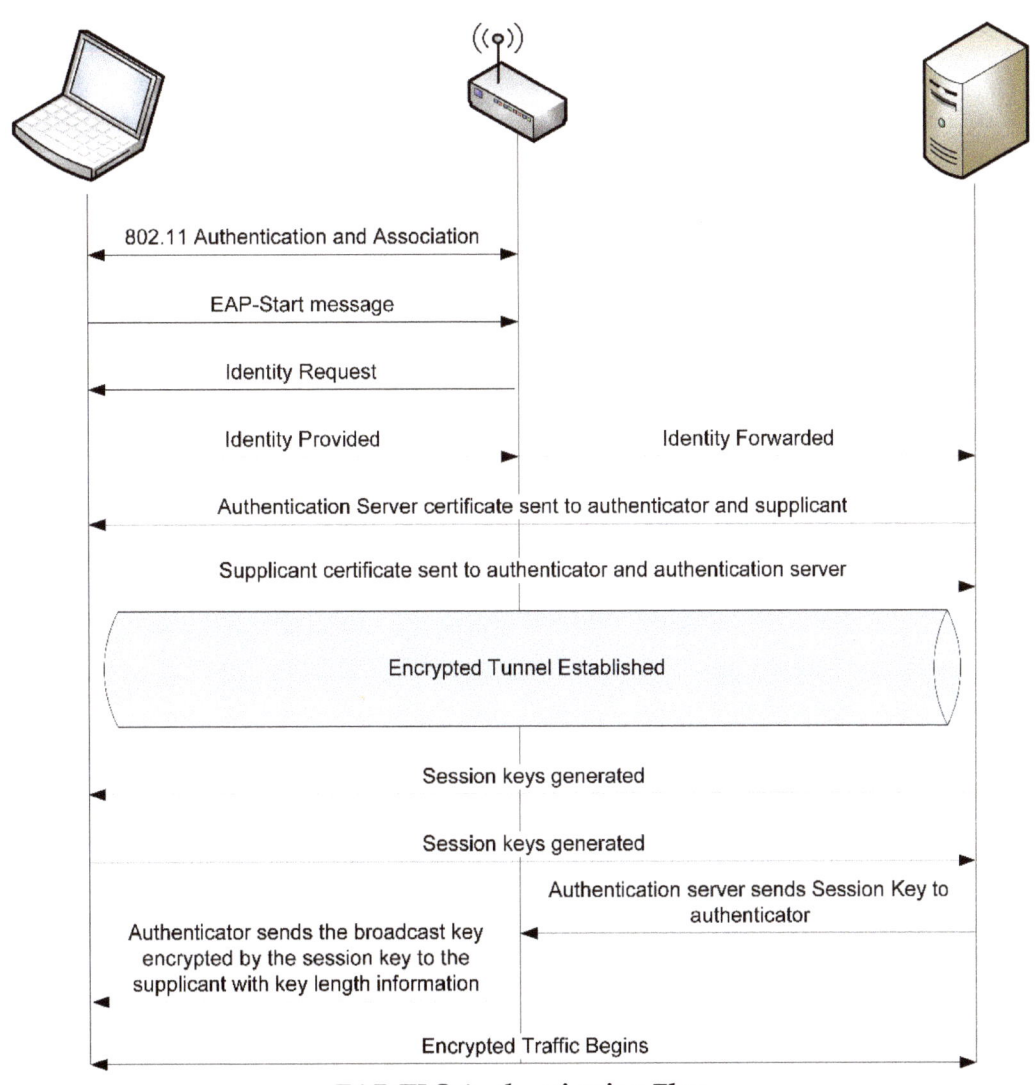

EAP-TLS Authentication Flow

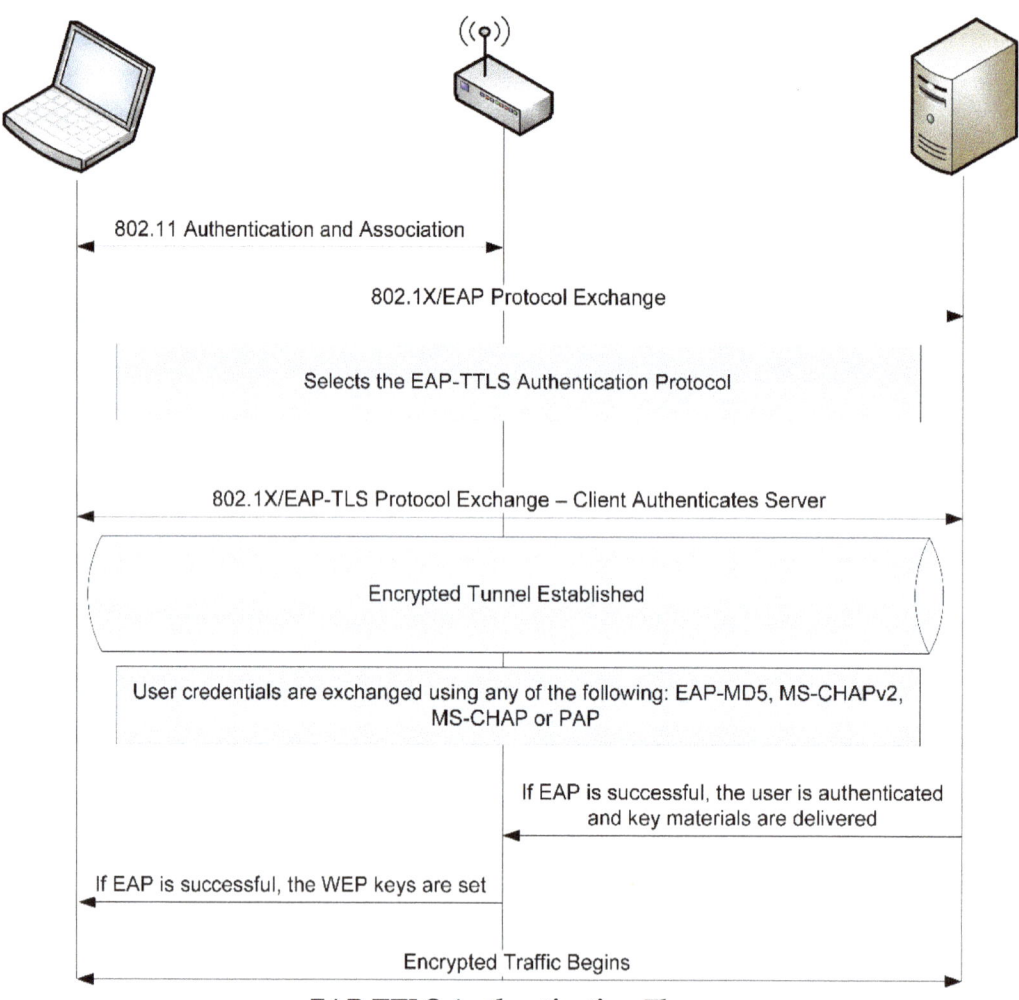

EAP-TTLS Authentication Flow

EAP-MD5

As previously mentioned, EAPMD5 is only an authentication protocol. It does not handle encryption keys of any type. After authentication, all messages are transmitted in clear text, and the authentication server is never authenticated by the client. EAP-MD5 requires only light processing but has no use in the production WLAN market.

Chapter Summary

In this chapter, you learned about the importance of authentication and the different authentication methods available in 802.11 WLANs. You explored the 802.1X/EAP enterprise authentication solutions and the different EAP types. This information is essential for the CWSP exam, but it's also important to know when making effective design security decisions in the real world. You were also reminded that WPA2 is current and a must-know in all its details, but that WPA3 is what comes next (as described in Chapter 10).

Review Questions

1. Which of the following packets are used across the medium between the wireless client stations and the AP or controller?
    a. EAP over RADIUS
    b. EAPoL
    c. EAP over PEAP
    d. LANEAP

2. You are discussing possible credential types with a new WLAN customer. Which of these isn't a valid WLAN credential type?
    a. Birth Certificate
    b. Digital Certificate
    c. Biometrics
    d. Tokens

3. Verification of a user's identity is the simple definition of _____.
    a. Authorization
    b. Accounting
    c. Enumeration
    d. Authentication

4. Where does the Master Session Key (MSK) originate on a PSK wireless network?
    a. RADIUS Server
    b. Nonce credential
    c. Passphrase
    d. GTK

5. You have a client that is interested in alternatives to standards-based PSK wireless security, but she doesn't want the administrative overhead of keeping up a RADIUS server. Which of these might work for her?
    a. PPSK
    b. DPSK
    c. MS-CHAP
    d. 802.1X
6. Which of the following keys is used to secure broadcast and multicast traffic, but not unicast?
    a. PTK
    b. GTK
    c. MTK
    d. RSNK

7. EAP is a Layer _____ authentication protocol?
    a. 7
    b. 3
    c. 2
    d. 1

8. During the 4-way handshake, which method of gaining the pairwise master keys is considered "in-band?
    a. 802.1X with RADIUS
    b. Preshared key
    c. Preshared RADIUS
    d. Token exchange

9. One available EAP type should never be used for wireless security because it only offers authentication; not authentication and encryption. What EAP type is this?
    a. EAP-TLS
    b. EAP-CLEAR
    c. EAP-FAST
    d. EAP-MD5

10. During the 802.1X process, which virtual port is used for authentication only?
    a. EAP port
    b. Controlled port
    c. Uncontrolled port
    d. PAE port

11. Where is RADIUS server configuration information located on a WLAN infrastructure device (i.e. a controller)?
    a. Device registry
    b. WLAN profile
    c. Access control list
    d. Server repository

12. While talking with a client's firewall administrator about a new secure WiFi rollout, she asks if you need UDP ports 1812 and 1813 open. What are these ports used for?
    a. 1812 is RADIUS accounting, 1813 is RADIUS authentication
    b. 1812 is EAP verification, 1813 is EAP accounting
    c. 1812 is 802.1X authorization, 1813 is 802.1X accounting
    d. 1812 is RADIUS authentication, 1813 is RADIUS accounting

13. Which of the following is not a valid example of a credential used in multifactor authentication?
    a. Something you know
    b. Something you can write
    c. Something you have
    d. Something you are

14. A new 802.1X-secured network has been implemented in your company headquarters, and you are explaining the key-related details to your WLAN technicians. Which key is used to secure unicast traffic?
    a. PTK
    b. GTK
    c. PMK
    d. APK

15. Which of the following is not a fundamental element of 802.1X?
    a. Supplicant
    b. Applicant
    c. Authenticator
    d. Authentication Server

Review Answers

1. **B.** EAPoL packets are transmitted between the supplicant (client) and the AP/controller.

2. **A.** There are many valid credential types you need to be aware of for secure WLAN. The birth certificate is not one of them.

3. **D.** Authentication concerns verifying a user's identity (or a host's).

4. **C.** On PSK networks, the MSK is derived from the passphrase.

5. **A.** PPSK is proprietary and does not come from any standard.

6. **B.** The Groupwise Temporal Key (GTK) is used to secure broadcast and multicast traffic.

7. **C.** Remember that all EAP and 802.11 security functions happen at Layer 2.

8. **A.** For the 4-way handshake, 802.1X use is considered in-band, while PSK is considered out-of-band.

9. **D.** MD5 is not an EAP type that should ever be used in WLAN security.

10. **C.** The uncontrolled port is used for authentication only.

11. **B.** The WLAN profile is the configuration set that includes RADIUS server configuration.

12. **A.** 1812 and 1813 ports are widely known RADIUS authentication and accounting ports.

13. **B.** The multifactor paradigm includes something you are/have/know.

14. **A.** The PTK secures unicast traffic.

15. **B.** The fundamental elements of RADIUS are the supplicant, authenticator, and authentication server.

Chapter 5: Authentication and Key Management

Objectives Covered

3.1 Select the appropriate security solution for a given implementation and ensure it is installed and configured according to policy requirements

3.2 Implement or recommend appropriate wired security configurations to support the WLAN

As you learned in Chapter 4, authentication is essential to WLAN security. When you are required to troubleshoot authentication issues, you must first understand the authentication process in-depth and how it eventually leads to the creation and distribution of encryption keys. Encryption itself, in greater detail, will be addressed in Chapter 6. Here in this chapter, you will explore authentication and key management (AKM), which addresses both authentication and the generation and distribution of encryption keys. We'll talk about operations through WPA2, and will address WPA3 in Chapter 10. The first step in understanding authentication and key management is properly grasping the terminology.

Terminology

The following definitions will provide a foundation for learning about AKM in wireless networks:

- **RSN** - Robust Security Network. A security network that allows only the creation of robust security network associations (RSNAs). An RSN can be identified by the indication in the RSN information element (IE) of Beacon frames that the group cipher suite specified is NOT Wired Equivalent Privacy (WEP).

- **RSNA** - Robust Security Network Association. The type of association used by a pair of stations (STAs) if the procedure to establish authentication or association between them includes the 4-Way Handshake. Note that the existence of an RSNA by a pair of devices does not of itself provide robust security. Robust security is provided when ALL devices in the network use RSNAs.

- **Pre-RSNA** - Pre-robust security network association. The type of association used by a pair of stations (STAs) if the procedure for establishing authentication or association between them did not include the 4-Way Handshake.

- **TSN** - Transition Security network. A security network that allows the creation of pre-robust security network associations (pre-RSNAs) as well as RSNAs. A TSN can be identified by the indication in the robust security network (RSN) information element of Beacon frames that the group cipher suite in use is Wired Equivalent Privacy (WEP).

- **MSK** - Master Session Key. Keying material that is derived between the Extensible Authentication Protocol (least EAP) peer and exported by the EAP method to the Authentication Server (AS). This key is at 64 octets in length.

- **PMK** - Pairwise Master Key. The highest order key used within this standard. The PMK may be derived from a key generated by an Extensible Authentication Protocol (EAP) method, or it may be obtained directly from a preshared key (PSK).

- **PTK** - Pairwise Transient Key. A value that is derived from the pairwise master key (PMK), Authenticator address (AA), Supplicant address (SPA), Authenticator nonce (ANonce), and Supplicant nonce (SNonce) using the pseudorandom function (PRF)- and that is split up into as many as five keys; i.e., temporal encryption key, two temporal message integrity code (MIC) keys, EAPOL-Key encryption key (KEK), EAPOL-Key confirmation key (KCK).

- **GMK** - Group Master Key. An auxiliary key that may be used to derive a group temporal key (GTK).

- **GTK** - Group Temporal Key. A random value, assigned by the broadcast/multicast source, used to protect broadcast/multicast medium access control (MAC) protocol data units (MPDUs) from that source. The GTK may be derived from a group master key (GMK).

- **KCK** - EAPOL-Key confirmation key (KCK). A key used to integrity-check an EAPOL-Key frame.

- **KEK** - EAPOL-Key encryption key. A key used to encrypt the Key Data field in an EAPOL-Key frame.

- **PMKSA** - Pairwise Master Key Security Association. The context resulting from a successful IEEE 802.1X authentication exchange between the peer and Authentication Server (AS) or from a preshared key (PSK).

- **PMKID** - Pairwise Master Key Identifier. The PMK is an identifier of a security association.

- PMKID = HMAC-SHA1-128(PMK, "PMK Name" || AA || SPA)

    - **PTKSA** - Pairwise Transient Key Security Association. The context resulting from a successful 4-Way Handshake exchange between the peer and Authenticator.

    - **GTKSA** - Group Temporal Key Security Association. The context resulting from a successful group temporal key (GTK) distribution exchange via either a Group Key Handshake or a 4-Way Handshake.

This set of defined terms may seem like a lot to digest, but it's also the language of wireless security when it comes to authentication and key management. It should become very familiar to you as you prepare for the CWSP exam, and as you work with secure wireless networks. While the list may seem daunting at first, you will quickly remember it through repeated exposure, which comes with designing and supporting networks that make use of these security mechanisms.

Pre-Robust Security Networks

You learned in Chapter 1 that the 802.11i amendment to the 802.11 standard defined two different types or classes of WLAN security. These are:

- Pre-Robust Security Network Association (pre-RSNA)
- Robust Security Network (RSN)

In this section, you will learn more about pre-RSNA. Then we will go on and discuss the RSN.

A wireless network that is classified as a pre-RSNA network consists of WEP and 802.11 entity authentication methods. Hopefully, that raises red flags in your mind, knowing that WEP has long been proven to be flawed. With the exception of the 802.11 Open System authentication method, all pre-RSNA security mechanisms have been deprecated due to their failure to meet their original security goals. The reality of the situation is that early attempts at WLAN security weren't good enough to be sustainable, as we've learned earlier in the text. New 802.11 standards-based implementations only support pre-RSNA methods to aid in the migration toward RSN security where legacy clients are in play and as part of a Transitional Security Network (TSN).

It is important to understand the concept of a TSN with respect to wireless networking. A TSN is typically a security network that allows a transition from one security solution to another, more secure solution. In relation to 802.11 wireless networking, the TSN includes the creation of pre-RSNAs, which use WEP, as well as RSNAs. So, we say WEP is bad, and must not be used. At the same time, it's not always easy to drop WEP without significant planning and budget implications.

A TSN can be identified by the inclusion of WEP in the group cipher suite in the RSN information element of Beacon frames. By contrast, pre-RSNA APs/STAs will generate Beacon and Probe Response frames without an RSN information element, and will also ignore the RSN information element in TSNs because it is unknown to them. This allows an RSNA STA to identify the pre-RSNA STAs from which it has received Beacon and Probe Response frames.

With all of that explained, let's clarify the current penetration level of TSN. In reality, most WLAN administrators should no longer have to deal with the TSN paradigm. Little excuse for TSN exists from the client-side, as very few clients remain in service that do not support at least WPA. An organization's infrastructure should be designed to support only required clients, and should not continue to support weak security solutions that are no longer required by any past generations of clients. There may still be small pockets of legacy specialty devices in use that require WEP, but eventually, they should all age out. It's for these circumstances that TSN is intended.

When working in an environment that still runs a TSN, find out if any client still requires WEP. If they don't, move quickly to migrate the environment to a robust security network. If some clients still require WEP, create a plan for transitioning from those clients, so you can move to a robust security network. If for whatever reason, specialty WEP clients cannot be upgraded, give serious consideration to how they can be isolated to minimize network exploitation through them.

Robust Security Networks (RSN)

As a reminder, an RSN is a wireless network that allows only the creation of robust security network associations (RSNAs). An RSN can be identified by the indication in the RSN Information Element (IE) of Beacon frames that the group cipher suite specified is not WEP. Stated plainly, if WEP is allowed, it is not considered an RSN, and RSN's do not allow WEP.

An RSNA is the type of association used by a pair of STAs if the procedure to establish authentication or association between them concludes with the 4-way handshake. It is important to know that the existence of an RSNA by a pair of devices does not provide robust security. Robust security is provided only when all devices in the network (such as those that support CCMP/AES exclusively) use RSNAs. One device allowed on the network (BSS) using WEP prevents the network from being an RSN, as the weakest security in use anywhere in the BSS defines the entire BSS.

The security afforded by the 802.11-2012 standard (carried into 802.11-2016) meets the following requirements:

- Protects user data
- Replaces legacy wireless security options
- Can meet governmental requirements for security, if implemented properly
- Is Comprised of two levels
    - One level for historical compatibility
    - One level for future security
- Both levels provide:
    - Continuously-changing encryption keys
    - A choice between two levels of user authentication
    - Replay protection
    - Removal of weak IVs
    - Better integrity protection than legacy ICV
    - Dynamic key management

RSN Information Element (IE)
The RSN IE is a set of frame fields included in certain WLAN management frames that are part of an RSN. The RSN IE defines the cipher suites used and authentication key

management (AKM) suites required and supported in the RSN. It also defines additional capabilities, such as preauthentication support.

The RSN IE is contained within the following 802.11 WLAN management frames:

- Beacon
- Probe Response
- Association Request
- Reassociation Request

> The RSN IE is important as it defines the security parameters of the BSS. It is included in beacon, probe response, association request, and reassociation request frames.

The following graphic shows a packet trace with an expanded view of the RSN IE. Two main fields here are the "PairwiseKey Cipher Suite List" field which shows the supported 802.11 cipher suites (encryption methods) for the BSS (in this case CCMP/AES), and the "AuthKey Management Suite List" field which shows if the service set is configured for preshared key (PSK) personal mode or 802.1X/EAP enterprise mode.

It is important to note that in order to be considered an RSN, the service set must support CCMP. However, TKIP is also allowed as an optional cipher suite for backward compatibility (though it should really not be used if there is a choice).

Recall that if WEP is supported, then the service set does not qualify as an RSN. One way to identify this is in the RSN IE "Group Key Cipher Suite Type" field. If this field did show support for WEP, then this BSS would not qualify as an RSN.

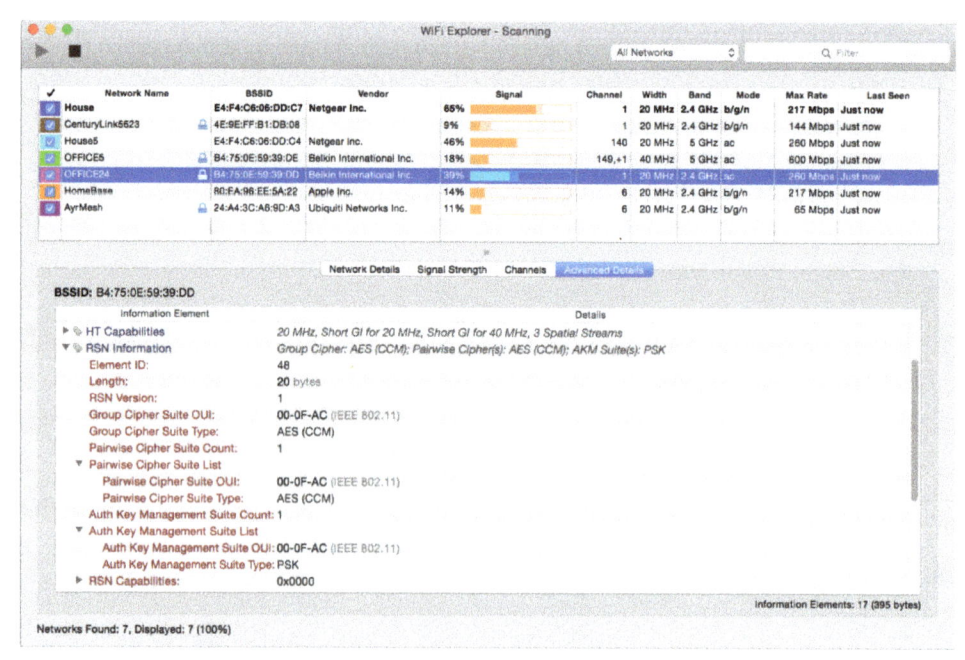

RSN IE Shown in WiFi Explorer on Mac OS

In the next image, notice the "PairwiseKey Cipher List" shows both TKIP and CCMP cipher suites are enabled for this service set; therefore, it does qualify as an RSN since CCMP is supported and WEP is not. However, devices (STAs) that do not support CCMP, but do support TKIP, would still be able to associate with the AP because of the supported TKIP cipher suite (encryption method). It is worth noting that TKIP is also aging out of the bigger WLAN landscape, as ever more devices are developed with support for CCMP and newer functionality.

In the next image, notice the "Authenticated Key Management Suite List" field, which shows a value of **00-0f-ac:02** for the "Authenticated Key Management Suite OUI." The fact that this ends in **02** indicates that this service set is configured as preshared key (PSK) and not 802.1X/EAP. The AKM suite list will be either 01 (802.1X) or 02 (PSK).

> The AKM Suite List field defines whether PSK or Enterprise (802.1X) authentication and key management are used. Stated differently, it defines whether Personal or Enterprise WPA or WPA2 is in use.

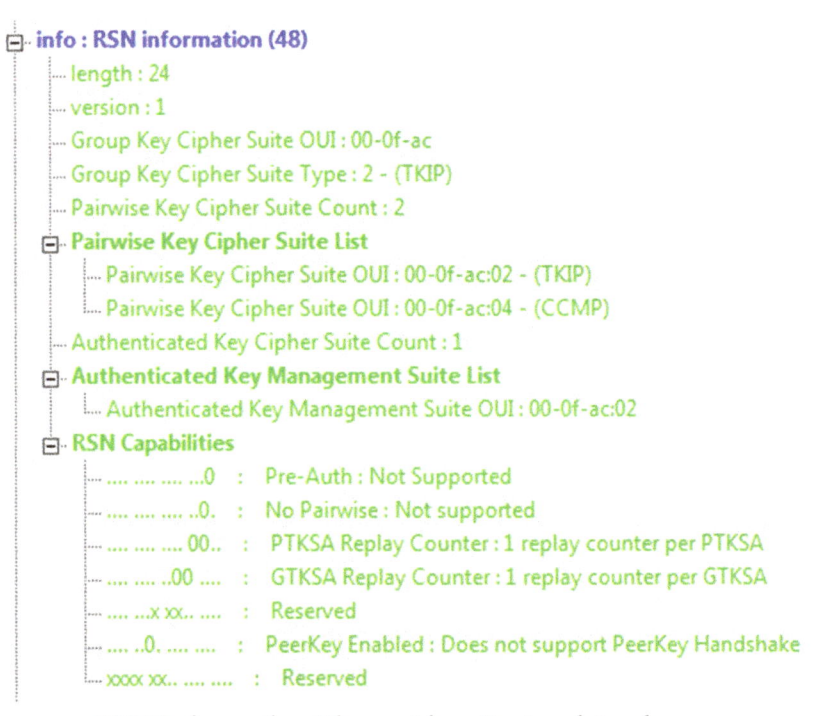

```
⊟ info : RSN information (48)
    ├── length : 24
    ├── version : 1
    ├── Group Key Cipher Suite OUI : 00-0f-ac
    ├── Group Key Cipher Suite Type : 2 - (TKIP)
    ├── Pairwise Key Cipher Suite Count : 2
    ⊟ Pairwise Key Cipher Suite List
    │   ├── Pairwise Key Cipher Suite OUI : 00-0f-ac:02 - (TKIP)
    │   └── Pairwise Key Cipher Suite OUI : 00-0f-ac:04 - (CCMP)
    ├── Authenticated Key Cipher Suite Count : 1
    ⊟ Authenticated Key Management Suite List
    │   └── Authenticated Key Management Suite OUI : 00-0f-ac:02
    ⊟ RSN Capabilities
        ├── .... .... .... ...0    :  Pre-Auth : Not Supported
        ├── .... .... .... ..0.    :  No Pairwise : Not supported
        ├── .... .... .... 00..    :  PTKSA Replay Counter : 1 replay counter per PTKSA
        ├── .... .... ..00 ....    :  GTKSA Replay Counter : 1 replay counter per GTKSA
        ├── .... ...x xx.. ....    :  Reserved
        ├── .... ..0. .... ....    :  PeerKey Enabled : Does not support PeerKey Handshake
        └── xxxx xx.. .... ....    :  Reserved
```

RSN Information Element in a Protocol Analyzer

The Cipher Suite Type or Cipher Suite OUI always starts with 00-0F-AC and is followed by a number indicating the actual suite. Suite values include:

- 00 – Use the Group Suite
- 01 – WEP-40
- 02 – TKIP
- 04 – CCMP
- 05 – WEP-104

It is important that you know these values for the CWSP exam, your work in wireless security, and for the Certified Wireless Analysis Professional (CWAP) course, should you eventually take that on. Note that 00-0F-AC-03 is reserved as of the 802.11-2016 version of the standard and the current draft of 802.11ax makes no changes to this either.

The RSN Capabilities fields will show additional supported features available for the service set that is configured as an RSN.

The *group cipher type* identifies what cipher type (encryption) is used for group traffic that traverses the wireless medium for the service set. This group cipher is identified in the "Group Cipher Type" field. When a service set is configured only for CCMP and not TKIP, it makes sense that CCMP would be used for group traffic encryption as well as for unicast traffic, because it is the only cipher suite available. The group cipher will always be the lowest possible encryption type that is used in the service set. For example, if you have a service set that is configured for both CCMP and TKIP, the lowest common type is TKIP, so the group cipher used to encrypt group traffic is TKIP. Any device capable of CCMP would also be able to understand TKIP; therefore, there are no compatibility issues with the group traffic.

Another important field contained within the RSN IE is the "PMKID Count" field. PMKID stands for pairwise master key identifier. The PMKID is a unique identifier created for each pairwise master key security association (PMKSA) that's been established between the AP and the client, (STA) when an RSNA is created. This is only used when fast, secure transition (roaming) features are enabled on the service set. The PMKID field is visible only in Association Request and Reassociation Request management frames. Fast, secure transition/roaming is fairly important for certain applications used over the WLAN, and so Chapter 8 is dedicated to the topic.

802.11 Association

The first step in creating an RSNA between two STAs (for example, the client device and AP) in an infrastructure BSS is to become 802.11 authenticated and associated. Anytime a STA (client device) connects to an infrastructure BSS or transitions to a different AP, the authentication and association process must occur. This allows for the STA to be connected to the AP. At this specific point in time, no wireless security measures are in place. In most public wireless hotspot networks, simple authentication and association are all that is required because these networks are open. For secure wireless networks, additional security methods must be put in place in order to protect the transmissions.

The next step in creating an RSNA is to continue the process of securing the wireless transmissions between STAs- again, generally the AP and the client device. This is the stage where each STA will receive the other's RSN information element describing their respective capabilities and requirements using the appropriate management frames. Notice in the following image, the Probe Response from the AP and the Association Request from the client STA contain the RSN IE information. If the capabilities match, the STA will be able to successfully associate to the AP.

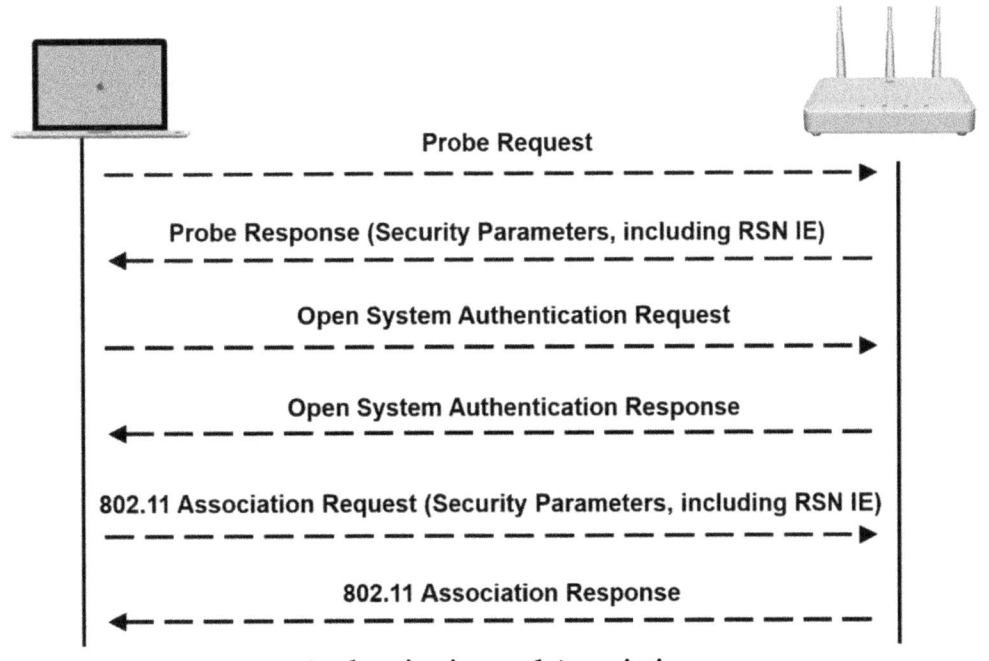

Authentication and Association

Key Hierarchy

One of the most confusing parts of security in 802.11 WLANs is the key hierarchy. This is because a single key is not generated and then used for encryption, as one unfamiliar with the topic might assume. A simple paradigm like that would not work in the complex communications environments of 802.11 networks. Think back to your CWNA studies; you will remember that sending a single file across the WLAN involved many frames. Management frames, control frames, and data frames all must traverse the network. Some frames are sent to every STA connected to the BSS while others are passed only between the AP and a single STA. The notion of several different keys being used for wireless security has to fit within this framework. To keep order, a key hierarchy is required. If you can grasp this key hierarchy, you will better accept the complexity introduced in the following pages.

The 802.11-2016 standard specifies an RSN key hierarchy for authentication and dynamic encryption keys. This is often referred to as authentication and key management (AKM). 802.11 AKM has several parts, but the overall scheme is illustrated in the pyramid structure below. The process works from the top down, starting with either a passphrase, PSK, or Master Session Key (MSK). This pyramid illustrates the key derivation process for both pre-shared key and 802.1X capable wireless networks (excluding WPA3).

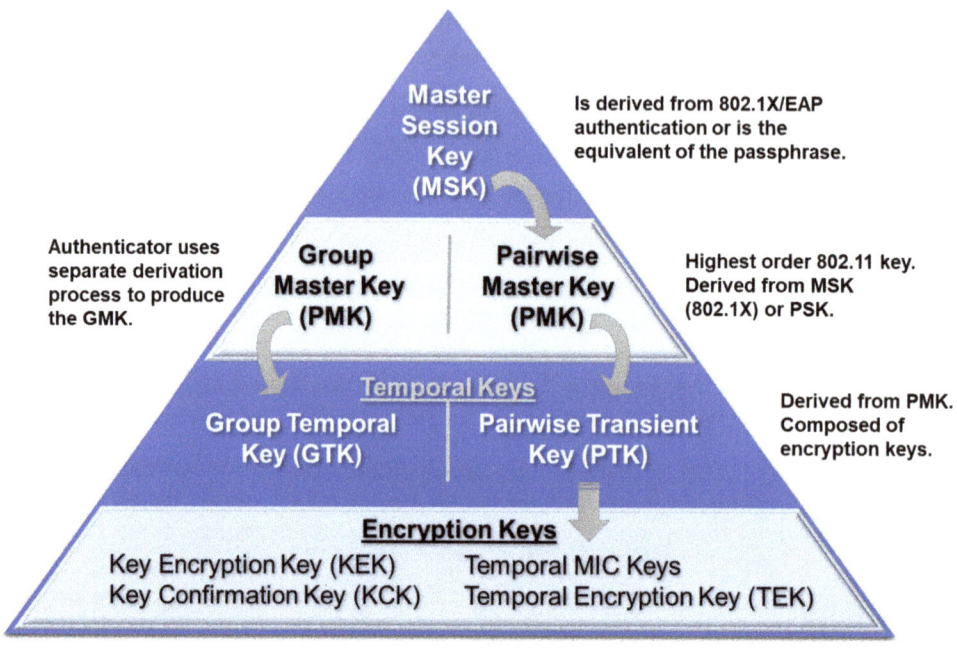

802.11 AKM Key Hierarchy

Authentication and key management should first be understood as it is used in PSK mode. When using PSK mode and if passphrases are used, the passphrase will be used to create a 256-bit PSK which is equivalent to the master session key (MSK). The MSK is then used to generate the PMK for the session. Therefore, when a passphrase is entered into the wireless client utility and the configuration of the wireless AP, wireless controller or cloud-managed WLAN device, the passphrase is used to create the PSK. This PSK is the functional equivalent of the MSK used in 802.1X AKM. It will be used to derive the other required keys in the hierarchy.

Alternatively, in some cases, the PMK can be entered directly as the PSK without the use of a passphrase. When the PMK is present on both STAs (AP and client device, for example), the 4-way handshake is the next step in the AKM process. The 4-way handshake uses "nonces" (arbitrary values that are only used once in the process) and other inputs with the PMK to create temporal unicast encryption keys which include the Pairwise Transient Key (PTK). The temporal keys are comprised of encryption and MIC keys. The 4-way handshake will be discussed in more detail later in this chapter.

In enterprise security settings where user-based 802.1X/EAP authentication is used instead of the pre-shared key method, authentication and key management is a somewhat different process. The MSK is now derived as part of the 802.1X authentication workflow. In this case, a series of frames are exchanged between STAs which in 802.1X terminology are the supplicant and the authenticator. These frames are used to derive the MSK with the aid of an authentication/RADIUS server. The derivation process continues as it does for the pre-shared key, and the PMK is generated. The 4-way handshake would then follow to create the PTK as it did with the pre-shared key model.

It is important to note that the authenticator (usually the AP or the wireless controller) is what derives a Group Master Key (GMK) using a separate derivation process and it uses the GMK to generate the Group Temporal Key (GTK). The GTK is then used to secure group traffic, which is both broadcast and multicast traffic.

PSK Key Hierarchy
WPA and WPA2 personal mode can secure wireless communications without the need for a RADIUS infrastructure by using a PSK security mechanism. The security input entered into all STAs, that will be part of the service set, can be a passphrase instead of directly entering a 256-bit key. The objective for using the passphrase is to lessen the chance of errors when a user manually inputs a key into their device settings. Since most users are familiar with using passwords, entering a passphrase is fairly straightforward and therefore will decrease the possibility of typographical errors. The passphrase entered uses a published algorithm, used to generate a 256-bit PSK, using a passphrase to PSK mapping. Using PSK results in a far simpler AKM key hierarchy.

The trade-off to the ease of using PSK is that the way passphrases are used may pose somewhat of a security risk depending on the length and complexity of the passphrase. After the 802.11 authentication and association completes, the 4-way handshake will occur. At the start of the 4-way handshake, the PMK is known by both STAs that are part of the frame exchange because it was entered into both devices. If an intruder was to capture the 4-way handshake with a protocol analyzer and use the right dictionary attack software, she might be able to extract the passphrase that can then be used to access the network and to possibly see encrypted traffic. If a weak passphrase is used, this attack is not difficult to accomplish. Therefore, passphrases should be strong and of sufficient length to lessen the chance of dictionary attacks. This is one of the weaknesses addressed by WPA3, which you'll read about in Chapter 10. The 802.11-2016 standard specifies that a key generated from a passphrase of less than about 20 characters is unlikely to deter attacks. Therefore long, complex passphrases are recommended. Interestingly, the Wi-Fi Alliance suggests an 8-character non-word passphrase on their website. CWNP recommends the longer 20-character passphrase instead.

With WPA2 personal mode, both TKIP/RC4 and CCMP/AES can be used with passphrases. Key derivation and distribution are identical with both models. WPA2 personal mode uses strong CCMP/AES for encryption but has the same authentication weakness as WPA Personal (i.e., if the passphrase is too short, it can be recovered by a dictionary attack).

It is important to distinguish the difference between unicast and multicast/broadcast encryption keys because unicast is encrypted differently from broadcast. Unicast information is either directed or a "one-to-one" exchange between STAs. User Data and directed management frames, such as Authentication or Association Request frames, are considered unicast communications. Broadcast communications are intended for all STAs that are part of the BSS. This means any STA within radio range should be able to hear broadcast traffic. Multicast traffic is similar to broadcast traffic in the sense that it is not one-to-one but is intended for a subset of STAs that are part of a multicast group, instead of all STAs.

Passphrase or PSK

Passphrase-to-PSK mapping

PMK

4-Way Handshake

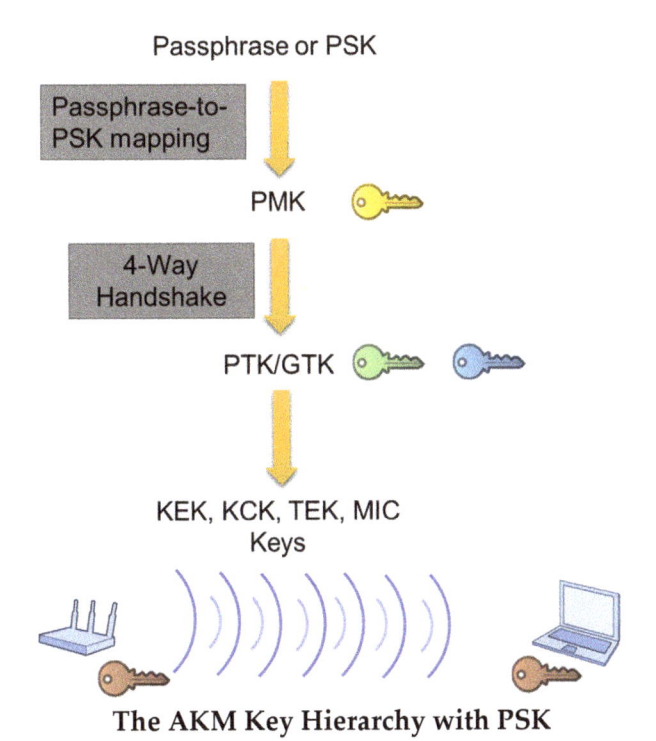

PTK/GTK

KEK, KCK, TEK, MIC Keys

The AKM Key Hierarchy with PSK

Unicast Encryption - The pairwise transient key (PTK) is a single key, unique to each STA association, and is used to encrypt all unicast traffic. The PTK is created during the 4-way handshake process using the PMK as the seed. Therefore, the PTK is unique between each pair of stations. An STA pair can be a wireless client device and an AP, or two client devices connected together in ad-hoc mode. Although all devices in a PSK service set share the same PMK, the PTK will be different for each pair of associated devices, therefore providing secure communications for all transmissions. You need to know the roles of the PMK and PTK here.

Broadcast / Multicast Encryption - A different key is used to secure broadcast and multicast traffic, as compared to unicast traffic. This key is the group temporal key (GTK). Recall, earlier in this chapter; it was stated that the GTK is derived from the GMK. This is a separate process than the PTK derivation process for unicast traffic described previously. The GMK is derived by the authenticator-- which is an AP, wireless controller or cloud-managed device. The GMK works as a seed to then derive the GTK. As with the PTK, the installation of the GTK occurs during the 4-way handshake process. In order for all STAs to share and understand group traffic, the

GTK will be common for all devices that are part of the same service set. Again, you need to fully understand the role of the GTK.

It is important to understand that in both WPA- and WPA2-Personal networks, each member of the network knows the PMK (or passphrase from which a PMK is derived). Since the other inputs to the PTK can be collected by observing the 4-way handshake, unicast encryption keys can be recreated by other members of the BSS. This is a potential security vulnerability if all members of a BSS cannot be trusted. In other words, in a PSK network, the network is only as secure as the users that know the PSK or passphrase.

From a network security management perspective, PSK networks are considered limited in scalability. This is because all STAs or devices that are part of the service set must have the same passphrase or PSK entered into the device. In a home or small network, this is easily handled. However, in enterprise networks, managing a PSK WLAN can be challenging. It is time-consuming to enter the keys when the network is first implemented and to periodically change the keys at responsible intervals. For the most part, PSK administration is a manual process; however, there are some proprietary and automated solutions available to help ease the burden of the passphrase or key management. PSK use and management (including limits) should be defined in the corporate security policy.

802.1X/EAP Key Hierarchy

When using legacy WPA or WPA2 enterprise, the 802.11 authentication and key management (AKM) scheme uses 802.1X port-based access control with EAP user-based authentication. In this scheme, the MSK is sometimes referred to as the authentication, authorization, and accounting (AAA) key. The MSK/AAA is exported out of the EAP process, and the first 256 bits of the MSK is considered to be the PMK from which encryption keys (PTKs and others) are generated during a 4-Way Handshake. This process is similar to that used with a PSK network discussed earlier. The main difference when 802.1X/EAP is used is that the session key is derived from a series of exchanged EAP messages.

Recall that WPA Enterprise uses 802.1X/EAP user-based authentications with the TKIP cipher suite for encryption. With Enterprise Mode the 802.1X/EAP method allows each new association between a pair of STAs to have its own unique key sets of both the PMK and PTK. Remember, that in a PSK network, all STAs that are part of the same service set share a common PMK. This distinction between the two types of networks is important.

WPA2 Enterprise uses 802.1X/EAP user-based authentication with support for the mandatory CCMP and optional TKIP cipher suites for encryption. Like WPA enterprise, this method also allows each new association between a pair of STAs to have its own unique key sets of both PMK and PTK, making this potentially a very secure solution. You can probably see why WPA2 Enterprise has so strongly been recommended as a security solution throughout the years. Despite its strengths, WPA2 Enterprise is no longer the high bar it once was, as you'll learn in Chapter 10 which covers WPA3. Given that it will take some years before WPA3 fully supplants WPA2, it's still very relevant to understand WPA2 deeply.

An important stage in 802.11 AKM is the "exporting" of dynamic encryption keys, which are created during the 802.1X/EAP authentication process. These keys are derived on both the authentication server (AS) and the supplicant (client STA). The AS must securely distribute this information to the authenticator (wireless AP or controller) before the 4-way handshake can make use of these keys to generate actual encryption keys used in protecting transmitted data. Creation of keying material is EAP method-specific. Similarly, the EAP framework does not define how the AS securely distributes the keys to the authenticator. This is left up to the developers of the individual EAP types, allowing for greater flexibility in the framework.

Keying material must be exported by the specific EAP type for RSN compliance. Mutual authentication is required for the creation of dynamic keying material and ensures the security of subsequent cipher suite encryption. Not all EAP methods included in RFC 3748 provide key generation. Those that don't provide key generation include EAP-MD5, EAP-OTP, and EAP-GTC. By contrast, EAP methods that do export dynamic keying material include PEAP (including PEAPv1), EAP-TLS, EAP-TTLS, EAP-LEAP, and EAP-FAST, among others.

In the 802.1X/EAP authentication model, the method used for generation of dynamic encryption keys is also EAP type-specific. The key generation, when offered, will be defined in the documentation that specifies the EAP method. After mutual authentication has occurred, both the authentication server and the supplicant will export an MSK, from which the PMK will be derived. Go back and look at the *802.11 AKM Key Hierarchy* graphic if needed.

The basic requirements common to all strong EAP types are:

- Mutual authentication is required for management of dynamic encryption keys.
- The dynamic keys must be securely distributed from the AS to the client. The distribution is beyond the scope of the EAP framework and will be defined by the specific EAP type.

The 802.11 standard defines mutual authentication as a dependency of an RSNA.

From IEEE 802.11-2012 (All sections carried into 802.11-2016):

4.3.4.3 Robust security network association (RSNA)

"An RSNA depends upon the use of an EAP method that supports mutual authentication of the AS and the STA, such as those that meet the requirements in IETF RFC 4017."

RFC 3748, Section 7.10 (Emphasis Added)

"In order to provide keying material for use in a subsequently negotiated cipher suite, an EAP method supporting key derivation MUST export a Master Session Key (MSK) of at least 64 octets, and an Extended Master Session Key (EMSK) of at least 64 octets. EAP Methods deriving keys MUST provide for mutual authentication between the EAP peer and the EAP Server."

RFC 3748, Section 7.2.1

"Mutual authentication -- This refers to an EAP method in which, within an interlocked exchange, the authenticator authenticates the peer and the peer authenticates the authenticator. Two independent one-way methods, running in opposite directions do not provide mutual authentication as defined here."

From IEEE 802.11:

"When IEEE 802.1X authentication is used, the specific EAP method used performs mutual authentication. This assumption is intrinsic to the design of RSN in IEEE 802.11 LANs and cannot be removed without exposing both the STAs to man-in-the-middle attacks. EAP-MD5 is an example of an EAP method that does not meet this constraint (see IETF RFC 3748 [B26]). Furthermore, the use of EAP authentication methods where server and client credentials cannot

be differentiated reduces the security of the method to that of a PSK due to the fact that malicious insiders can masquerade as servers and establish a man-in-the-middle attack.

"In particular, the mutual authentication requirement implies an unspecified prior enrollment process (e.g., a long-lived authentication key or establishment of trust through a third party such as a certification authority), as the STA must be able to identify the ESS or IBSS as a trustworthy entity and vice versa. The STA shares authentication credentials with the AS utilized by the selected AP or, in the case of PSK, the selected AP. The SSID provides an unprotected indication that the selected AP's authentication entity shares credentials with the STA. Only the successful completion of the IEEE 802.1X EAP or PSK authentication, after association, can validate any such indication that the AP is connected to an authorized network or service provider."

Stated plainly, for effective and secure distribution of dynamic encryption keys, mutual authentication is both intrinsically and explicitly required.

After the PMK is generated, the 4-way handshake will generate the actual PTK used for encryption. The 802.11 standard defines the 4-way handshake explicitly.

From IEEE 802.11:

4-Way Handshake

"RSNA defines a protocol using IEEE 802.1X EAPOL-Key frames called the 4-Way Handshake. The handshake completes the IEEE 802.1X authentication process. The information flow of the 4-Way Handshake is as follows:

Message 1: Authenticator → Supplicant: EAPOL-Key(0,0,1,0,P,0,0,ANonce,0,DataKD_M1) where DataKD_M1 = 0 or PMKID for PTK generation, or PMKID KDE (for sending SMKID) for STK generation

Message 2: Supplicant → Authenticator: EAPOL-Key(0,1,0,0,P,0,0,SNonce,MIC,DataKD_M2) where DataKD_M2 = RSNIE for creating PTK generation or peer RSNIE, Lifetime KDE, SMKID KDE (for sending SMKID) for STK generation

Message 3: Authenticator → Supplicant: EAPOL-Key(1,1,1,1,P,0,KeyRSC,ANonce,MIC,DataKD_M3) where DataKD_M3 = RSNIE,GTK[N] for creating PTK generation or initiator RSNIE, Lifetime KDE for STK generation

Message 4: Supplicant → Authenticator: EAPOL-Key(1,1,0,0,P,0,0,0,MIC,DataKD_M4) where DataKD_M4 = 0."

4-Way Handshake Frames

Four EAP over LAN (EAPoL) Key frames are used, with each acknowledged in a 4-way handshake, hence the name. The process of exchanging these four frames will allow for the creation of the PTK. The PTK is used to encrypt unicast traffic between STAs.

Remember that in a pre-shared key network/service set all STAs will share a common PSK that is entered into the STA as a 256-bit key, or created from a common passphrase that is entered on all STAs. The 4-way handshake occurs immediately following the 802.11 Open System authentication and association process. This is an indicator that the packet trace in the following image was likely taken from a PSK network since there are no 802.1X/EAP frames shown. However, it is not 100% guaranteed with the information that is shown here. The possibility exists that this could be an 802.1X/EAP session that is using fast, secure transition. The only way to be certain is to expand a frame that contains the RSN IE and then view the authentication key management suite list field, which will identify either PSK or 802.1X/EAP.

4-way Handshake with no EAP Authentication

It is important to consider that if the 4-way handshake is captured with a packet analyzer and PSK is used, software programs are available that will allow the collected information to be challenged against a dictionary to possibly determine the passphrase. Therefore, using strong passphrases will lessen the chances of an intruder discovering it. There is a possibility that this same process can be used for legitimate reasons such as troubleshooting or auditing. Many packet analyzer programs have built-in functionality allowing a network administrator to see encrypted traffic if the passphrase is known. The passphrase and SSID are entered into the analyzer, and once the 4-way handshake is captured, the administrator would then be able to view the encrypted information (which may be necessary for various support reasons).

On the topic of using a packet analyzer- it can be leveraged to reassemble VoIP communications, which is a common troubleshooting process. In a WLAN with a VoWi-Fi implementation, assuming encryption is in use, only a PSK deployment of VoWi-Fi would allow for the decryption of the voice stream in a packet analyzer. This is because the actual PMK is unknown in an enterprise deployment using 802.1X/EAP and is not exposed in any interface in a way that it can be retrieved. Therefore, even with the 4-way handshake data, the PTK cannot be derived.

A packet trace captured from a network configured for 802.1X/EAP will show a series of EAP messages after the 802.11 Open System authentication, ensuring that the association process completes before the 4-way handshake starts. This is a key indicator that 802.1X/EAP is in use on the network. If you were to view the authentication key management suite list field in a frame that carries the RSN IE, it would be identified as such.

Note again that the 4-way handshake being shown does not necessarily mean this connection is part of an RSN BSS, as the service set may be using TKIP/RC4, which alone is not considered an RSN. Remember, CCMP is mandatory to have an RSN, and TKIP is optional, though now deprecated in the standard. The only way to be certain whether an RSN is in use is to expand a frame that contains the RSN IE and view the pairwise key cipher list which would show CCMP for an RSNA.

Group Key Handshake

The Group Key Handshake is a group key management protocol defined by the 802.11 standard. It is used only to issue a new group temporal key (GTK) to peers with whom the local STA has already formed security associations.

Once the initial PTK and GTK are in place via the 4-way handshake, the group key (GTK), which is used to encrypt broadcast and multicast data traffic, may be changed by the authenticator (AP) for a number of reasons. Updating the stations with the new GTK is performed through a simple two-step Group Key Handshake, as shown in the following image.

```
┌─────────────┐                                      ┌──────────────┐
│ Supplicant  │                                      │ Authenticator│
└──────┬──────┘                                      └──────┬───────┘
       │                                                    │
       │                                              ┌─────┴──────────┐
       │                                              │ Generate GTK   │
       │                                              │ Encrypt GTK    │
       │                                              │ with PTK       │
       │                                              └─────┬──────────┘
       │   Message-1: EAPoL Key (Encrypted GTK, Group, MIC) │
       │ ◄─────────────────────────────────────────────────│
┌──────┴──────┐                                             │
│ Decrypt and │                                             │
│ install GTK │                                             │
└──────┬──────┘                                             │
       │         Message-2: EAPoL Key (Group, MIC)          │
       │ ──────────────────────────────────────────────────►│
```

Group Key Handshake Protocol

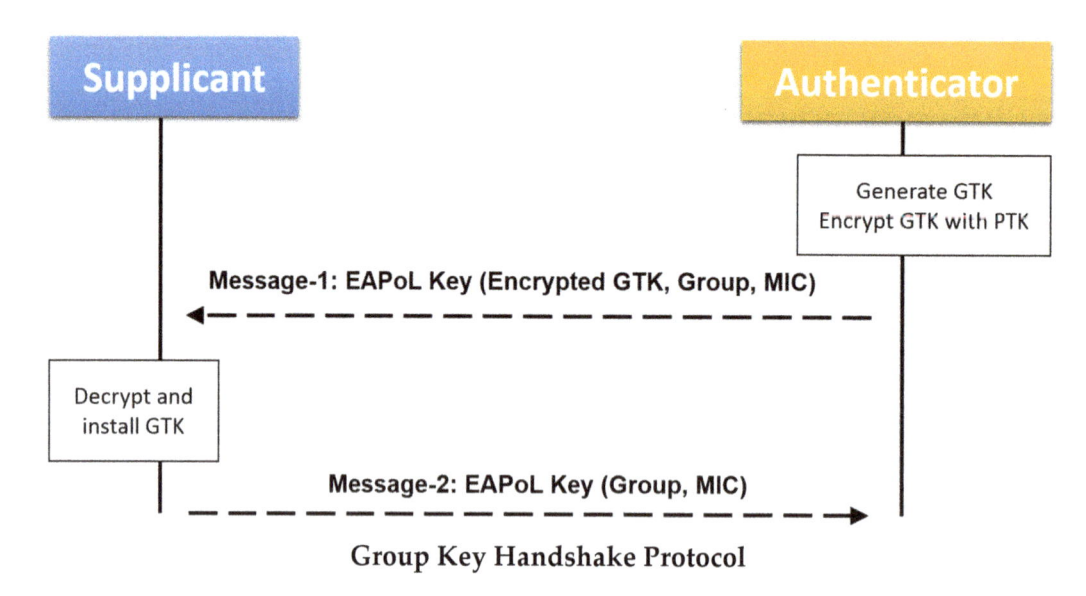

The authenticator may initiate a group key handshake when an STA is disassociated or deauthenticated to protect the integrity of BSS multicast or broadcast communications.

Chapter Summary

In this chapter, you learned about the key derivation processes based on the authentication system in use. This is known as authentication and key management (AKM). You learned how to identify AKM information in the RSN information element (IE) and the meaning of many terms, such as PMK, PTK, GTK, and more, all related to AKM. You were reminded that WEP is obsolete and TKIP has been deprecated, yet both may still be found on occasion. You also learned that WPA2 is strong, but has been superseded by WPA3. It will take some time for WPA2 to be pushed aside, so is still considered current. WPA3 is covered in Chapter 10

Review Questions

1. When will WEP be specified in the RSN IE of Beacon Frames on Robust Security Networks?
   a. When RSNAs are in use
   b. After 4-way handshakes
   c. When TSNs fail
   d. Never

2. Which of the following is not a valid term in Authentication and Key Management?
   a. TSN
   b. MSK
   c. GSK
   d. GMK

3. The security parameters of a given BSS are defined in which Information Element (IE)?
   a. RSN IE
   b. SSL IE
   c. MSK IE
   d. PMK IE

4. Which key is derived from the 802.1X process or from the PSK in pre-share networks?
   a. RSK
   b. PMK
   c. GMK
   d. PTK

5. Of the following, which is not a valid Cipher Suite type value?
    a. 00-0F-AC-00
    b. 00-0F-AC-01
    c. 00-0F-AC-03
    d. 00-0F-AC-05

6. The _____ key secures group traffic, while the _____ secures unicast traffic.
    a. GTK, PTK
    b. PTK, GTK
    c. GTK, MSK
    d. MSK, PTK

7. With WPA2 Personal mode, which of the following can be used?
    a. CCMP/AES
    b. TKIP/RC4
    c. Neither
    d. Both

8. Only one of these EAP types provides key generation. Select the one that does:
    a. EAP-GTC
    b. EAP-MD5
    c. EAP-FAST
    d. EAP-OTP

9. When is the PTK used to encrypt unicast traffic generated?
    a. During the 4-way handshake
    b. During Open System authentication
    c. During the Group Key handshake
    d. After the EAP is created

10. Which of the following is required for secure distribution of encryption keys?
    a. Bounded medium
    b. 802.1X
    c. Unicast encryption
    d. Mutual Authentication

11. For WPA and WPA2 Enterprise, the MSK is exported out of what?
    a. The EAP process
    b. The GMK
    c. The PTK
    d. The PSK

12. When a Cipher Suite Type of 00-0F-AC-05 is in use, what cipher suite is present?
    a. WEP-40
    b. TKIP
    c. WEP-104
    d. CCMP

13. Which of these represents the Authenticated Key Management Suite List value for 802.1X?
    a. 00-0f-ac-00
    b. 00-0f-ac-03
    c. 00-0f-ac-01
    d. 00-0f-ac-02

14. The RSN IE defines _____ and _____ in certain WLAN management frames.
    a. Cipher suites, AKM suites
    b. Cipher suites, SSID suites
    c. RSNA suites, Cipher suites
    d. AKM suites, RSNA suites

15. Select the option that represents the proper AKM key hierarchy.
    a. MSK, GTK, PMK, KCK
    b. KCK, PMK, MSK, GTK
    c. MSK, PMK, GTK, KCK
    d. MSK, PMK, GTK, TEK

Review Answers

1. **D.** WEP is not allowed in the RSN framework.
2. **C.** There is no shortage of acronyms in wireless networking. GSK does not exist as a valid WLAN security term.
3. **A.** The RSN IE is the only valid choice, and it includes security parameters of a given BSS.
4. **D.** It's important that you fully learn about the PMK.
5. **C.** -03 is not valid.
6. **A.** The Group Temporal Key secures multicast and broadcast traffic, while the Pairwise Transient Key secures unicast traffic.
7. **D.** Both options are valid with WPA2 Personal.
8. **C.** Of those listed, only EAP-FAST provides keying material used in key generation.
9. **A.** The PTK is generated during the 4-way handshake.
10. **B.** 802.1X is used for secure distribution of encryption keys, among its other discrete functions.
11. **A.** MSK is exported out of the EAP process.
12. **C.** You'll need to memorize the Cipher Suite types, an -05 is used for WEP-104.
13. **C.** -01 is the AKM Suite List value for 802.1X.
14. **A.** Cipher suites and AKM suites are defined in the RSN IE of certain management frames.
15. **D.** Understanding key hierarchy is critical to the CWSP exam and understanding WLAN security.

Chapter 6: Encryption

Objectives Covered

3.1　Select the appropriate security solution for a given implementation and ensure it is installed and configured according to policy requirements

3.2　Implement or recommend appropriate wired security configurations to support the WLAN

Encryption is defined as the process of modifying information (data) with an algorithm called a *cipher* that results in unreadable or meaningless data to those without the key used in the algorithm. Encryption is very important in wireless communications because, as you well know, wireless networks use RF to carry data through the air. CWSP must understand the basics of encryption as it relates to WLANs because it is central to a variety of wireless security approaches. First, let's explore the foundational terms and definitions related to encryption that we'll need for this chapter. New encryption paradigms related to WPA3 will be discussed in Chapter 10.

Terminology

The following terms should be understood by the CWSP:

- **Encryption algorithm** - Encryption algorithms are mathematical procedures used to obscure information, so it appears as seemingly meaningless data to an unintended recipient without a key. AES RC4, RC5, and RC6 are examples of encryption algorithms.

- **Hash function or hashing algorithm** - A cryptographic hash function is a deterministic procedure that takes an arbitrary block of data and returns a fixed-size bit string, the (cryptographic) hash value, such that an accidental or intentional change to the data will change the hash value.

- **Cipher suite** - A cipher suite is a named combination of authentication, encryption, and message authentication code (MAC) algorithms used to negotiate the security settings for a network connection.

- **Stream cipher** - A stream cipher is a symmetric key cipher where plaintext bits are combined with a pseudorandom cipher bit stream (keystream), typically by an exclusive-or (XOR) operation. In a stream cipher, the plaintext digits are encrypted one at a time, and the transformation of successive digits varies during the encryption.

- **Block cipher** - In cryptography, a block cipher is a symmetric key cipher operating on fixed-length groups of bits, called blocks, with an unvarying transformation.

- **Symmetric key encryption** - Symmetric-key algorithms are a class of algorithms for cryptography that use trivially related, often identical, cryptographic keys for both decryption and encryption.

- **Asymmetric key encryption** - Asymmetric-key algorithms are a class of algorithms for cryptography that uses separate key pairs for encryption and decryption. Key pairs are typically deployed as shared public and secret private keys.

Symmetric Key Encryption

Two common encryption techniques used with electronic information are symmetric key encryption and asymmetric key encryption. Let's start with symmetric key encryption.

Symmetric key encryption is the primary standardized frame encryption technology used with IEEE 802.11 standards-based wireless networks today. With symmetric-key encryption, matching keying material is passed to both parties over an encrypted link that has been created. With static key implementations (such as long-since deprecated WEP), the keying material is used as a direct input to the encryption process as it is entered manually on all devices that belong to the wireless service set. With dynamic key implementations (such as TKIP and CCMP using 802.1X/EAP), this initial keying material is used to generate subsequent encryption keys which are then used for encryption and are created during the 4-way handshake process as discussed in the previous chapter.

In this process, the actual encryption keys are never transmitted across the wireless medium. Instead, some of the required key inputs from the authenticator and supplicant, including nonces and other required information are transmitted. Then each participating device derives (generates) the keys. This adds a layer of security to the key creation process.

The image below illustrates that the dynamic encryption key generation process begins with symmetric keys. The 4-way handshake is used to derive mutual frame encryption keys that will be used to encrypt information that traverses the wireless medium. These frame encryption keys are symmetric encryption keys.

Symmetric encryption is less processor intensive than asymmetric encryption. For this reason, many file encryption systems will use a symmetric key and algorithm to encrypt data (Microsoft calls this the File Encryption Key- FEK) and an asymmetric encryption key and algorithm to encrypt the data encryption key. The asymmetric encryption key is typically a digital user certificate, and the symmetric data encryption key is usually unique to a file or encryption process. The main takeaway is that the symmetric encryption process is less computationally intensive regardless of where it is utilized, and so is typically better for large-scale encryption tasks.

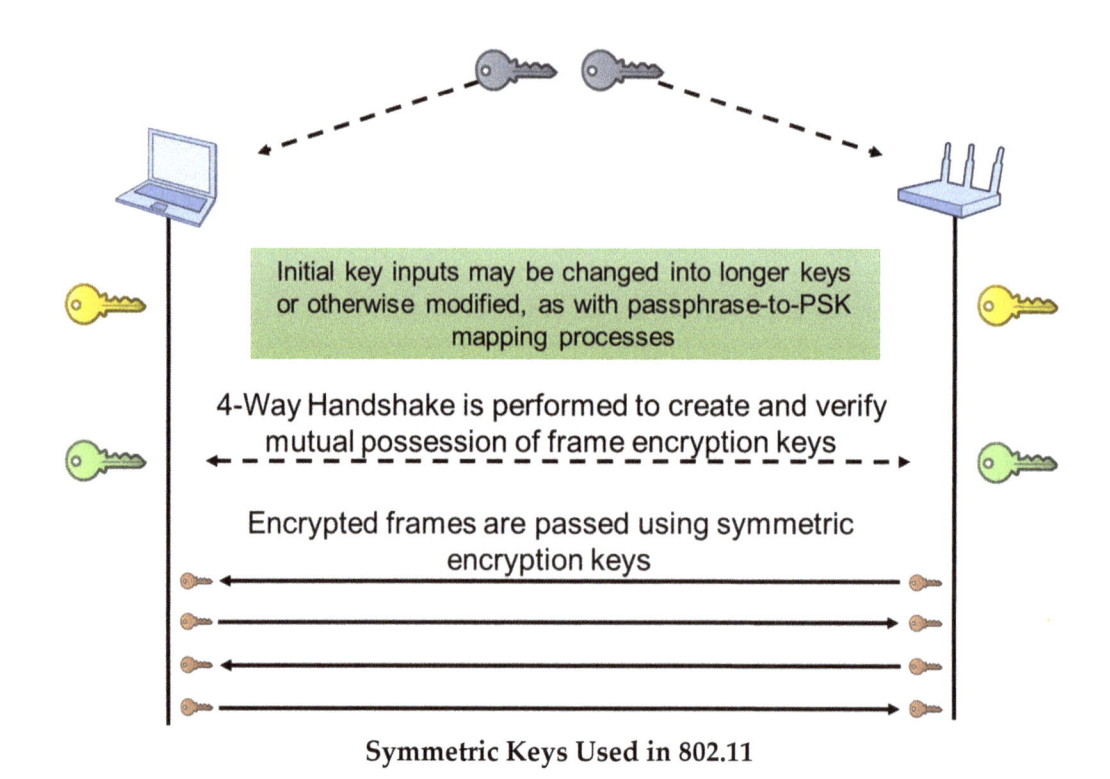

Symmetric Keys Used in 802.11

Asymmetric Key Encryption

Many different types of asymmetric key encryption exist, and each of them works a bit differently. Asymmetric key encryption is also called public-key cryptography.

In the simplest terms, asymmetric key encryption uses both a public and a private encryption key. An initial entity (often a server, which we will call Entity A) possesses a matched private and public key (the asymmetric keys). The public key is passed to any and all entities (Entity B) with whom a secure connection is desired, but the private key is never shared with other entities. When a frame is encrypted with the public key by Entity B, it can only be decrypted by the private key. Thus, only Entity A can successfully receive this information because Entity A is the only one with the private key. Others who possess the public key cannot decrypt a frame that is encrypted with the public key. Hence, asymmetric encryption.

The following image illustrates just one use of asymmetric encryption.

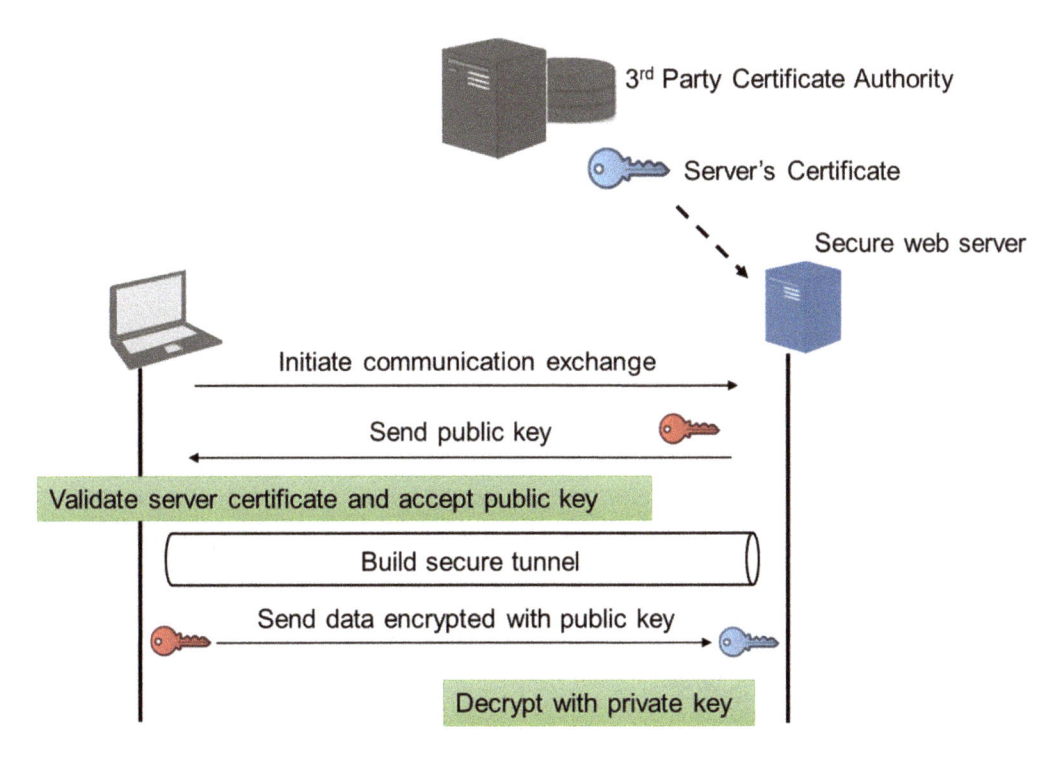

Use of Asymmetric Encryption

This type of cryptography generally requires two sets of keys. Each communicating entity has a private/public key pair for secure information exchange. The public key is distributed to others, but the private key is kept by only one entity.

In some cases, asymmetric keys are used initially so that symmetric encryption keying can then be established. Symmetric keying material can be transmitted inside an asymmetrically encrypted frame. The benefit of asymmetric key encryption is that initial keying material must only be distributed to one device at the beginning of the process.

Outside of the WLAN space, we see other examples of comparable encryption. For instance, when using a public hotspot and accessing HTTPS-based web sites, you are using asymmetric encryption based on the HTTPS server certificate (which contains the public key of the server). This is a perfect example because the server sends the session encryption key (symmetric key) or materials used to generate the key to the client using asymmetric encryption processes. This example also makes the point that a fair amount of what you will learn as a CWSP is applicable far beyond the Wi-Fi realm.

Stream Ciphers

We've mentioned ciphers a few times so far in this chapter, but let's dig in deeper on the concept. A cipher is a mechanism that will allow for the encryption and decryption of information to occur. Stream ciphers take an initial plaintext input data stream and encrypt this stream one bit at a time. The plaintext input is typically processed using exclusive OR (XOR) math against a keystream and, bit-by-bit, the outcome of an encrypted text is created. 802.11 WLAN technology uses the RC4 stream cipher with both legacy WEP and TKIP.

The following image describes how the stream cipher process works. You can see that the plaintext information, in this case, WLAN computer data, is combined with a created keystream. The logical combination of the plaintext data and the keystream using an XOR process will result in ciphertext (encrypted information) that will be sent across the wireless medium.

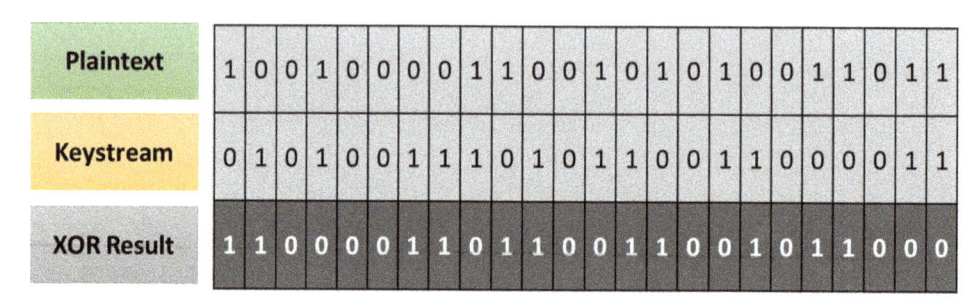

Plaintext	1	0	0	1	0	0	0	0	1	1	0	0	1	0	1	0	1	0	0	1	1	0	1	1
Keystream	0	1	0	1	0	0	1	1	1	0	1	0	1	1	0	0	1	1	0	0	0	0	1	1
XOR Result	1	1	0	0	0	0	1	1	0	1	1	0	0	1	1	0	0	1	0	1	1	0	0	0

Stream Cipher Process (Showing the importance of protecting the key and, therefore, the keystream.)

Stream ciphers can be fast, and by comparison are often computationally faster than block ciphers, which we will discuss next. If not implemented correctly, though, stream ciphers can allow for security weaknesses. This is one of the reasons why WEP had its share of vulnerabilities and was cracked early in its tenure, but not because RC4 is necessarily "bad." WEP's use of a plaintext initialization vector (IV) that was sent across the wireless medium was certainly problematic. This IV was also used as a seed with the WEP key that was entered into the client software to create the needed keystream.

Stream ciphers also lack integrity protection. When using WEP (or TKIP), the integrity check value (ICV) is added to the process. The ICV as it was implemented in WEP was also vulnerable to a *bit flip attack* (used to alter transmitted data), which constitutes yet another weakness of WEP.

Remember that RC4 is also used with TKIP. However, changes were made in TKIP to help combat the problems associated with RC4 that were part of the WEP process. Remember, WEP wasn't problematic because of RC4, but rather the way it was implemented and that TKIP greatly improved upon that. We'll get into TKIP deeper, later in the chapter.

Block Ciphers

There are several different types of block ciphers. In contrast to stream ciphers, block ciphers encrypt plaintext data in blocks, or chunks of bits, instead of sequentially one bit at a time. Block ciphers specify the size of the block to be encrypted, and CCMP/AES uses a 128-bit block. Other block ciphers use different block sizes.

The CBC block cipher mode is used with CCMP, and as shown in the following image, and uses a chaining process whereby each encrypted block is used as an input to the encryption of the next block. This type of encryption adds strength to the cipher because it builds upon the strength of the previously encrypted blocks.

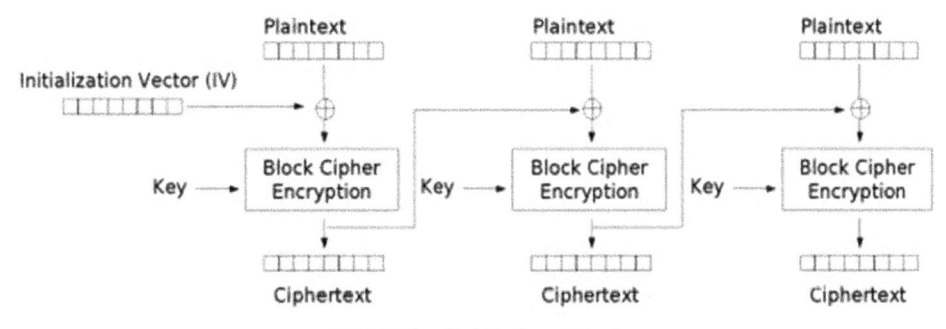

CBC Block Cipher Mode

Like stream ciphers, block ciphers also use an XOR process to encode plaintext information. Block ciphers can be slower than stream ciphers, and to operate correctly and efficiently, may require more powerful hardware. This requirement is why many older APs could not be upgraded to support WPA2 through firmware, but, rather, required hardware replacement. Thankfully, wireless hardware devices have progressed to the point where the last few generations' worth have all been built with the resources to handle stronger encryption.

Frame Encryption

It is important to understand what protection is afforded by any specific type of encryption. We often describe an encryption scheme in accordance with the OSI layer at which it is applied. For example, 802.11 standardized encryption methods, such as TKIP and CCMP, use Layer 2 encryption mechanisms. This description simply means that they are applied at Layer 2 of the OSI model, protecting only the higher-layer data, and not the MAC sublayer data (headers).

If you recall from CWNA and previous networking studies, the OSI model has 7 layers: Physical, Data Link, Internet, Transport, Session, Presentation, and Application, listed from Layer 1 to Layer 7. The Physical (Layer 1) and Data Link (Layer 2) layers are subdivided into two sublayers. The Data Link Layer sublayers are the MAC and the LLC. The Physical Layer sublayers are the PLCP and the PMD.

Since WLAN encryption is applied at the MAC Layer, it protects the LLC sublayer and the higher-layer contents, but it does not protect MAC sublayer data. MAC sublayer information such as MAC addresses must remain unencrypted in order to be correctly transmitted and received. In many cases, the application data is what we are trying to protect, but it is also helpful to obscure IP-layer information as well as other pieces of the networking puzzle. The following image illustrates the encrypted portion of the 802.11 frames.

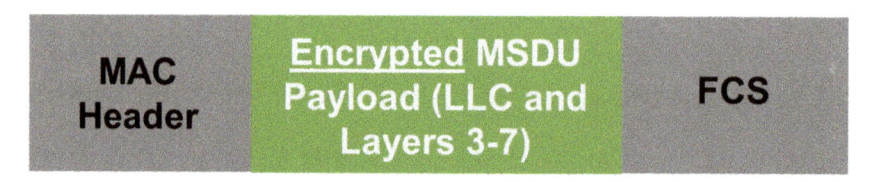

| MAC Header | Encrypted MSDU Payload (LLC and Layers 3-7) | FCS |

Illustrating 802.11 Frame Encryption

Capturing encrypted WLAN frames with a packet analyzer will show the MAC Layer information that is transmitted in cleartext. However, the frames that carry Application Layer (Layer 7) data payload will be obfuscated and not viewable. These nuances are critical to both CWSP study and in supporting secure wireless networks.

Encryption and Decryption

Encryption and decryption can either be centralized or distributed. When centralized, a WLAN controller performs the encryption and decryption processes. When distributed, these processes occur in the AP. Most deployments today use distributed data forwarding and, therefore, perform distributed decryption as the encryption is only employed for the benefit of the wireless link. At the same time, encryption all the way to the controller may be marketed as a differentiator.

Older model APs did not adequately secure secret keys in their configuration settings, which posed a security threat if the AP was stolen. Attackers could potentially recover network passwords/secrets. For that reason, it was advisable to move the keys to a centralized WLAN controller. The centralized model of encryption/decryption protects the encrypted data all the way from the client (edge), across the WLAN and switching along the way, to the WLAN controller (core). Thankfully, modern APs in distributed security environments do a better job of protecting keys.

Conversely, distributed data forwarding models require key storage and encryption/decryption processes locally on the AP. This allows for more expedient processing and forwarding of data frames directly to a destination from the edge, instead of going through the WLAN controller and then on to the destination. Because the unbounded wireless medium poses the greatest security vulnerability (attackers can easily intercept traffic with simple eavesdropping techniques), encryption operations at the AP provide sufficient security in most cases. "The wire" is generally regarded as secure for most environments.

As you can see, neither method is necessarily better all the time. Business and security requirements dictate the type of encryption/decryption model to be used. Government networks often see the centralized model as an advantage, as some wired privacy is also provided. However, distributed forwarding and entirely distributed WLAN architectures are pushing encryption and decryption to the edge in most deployments today.

Part of the encryption method selection decision may also be dependent upon the other network security features supported, such as firewalls. If the firewall can only be applied in the WLAN controller, then it makes sense to also store keys there. If, on the other hand, the AP can perform policy and firewall services, the keys must be stored there. Understanding the various paradigms will make you better at your WLAN duties and will help you guide clients to the best solutions for their circumstances.

Encryption Algorithms

Encryption algorithms are mathematical procedures used to obscure information, so it appears as seemingly meaningless data to an unintended recipient without the correct key.

These are some of the more common algorithms:

- Rivest Cipher 4 or "Ron's Code 4" (RC4)
- Advanced Encryption Standard (AES)
- Data Encryption Standard (DES)
- Triple Data Encryption Algorithm (3DES)
- Rivest Cipher 5 (RC5)
- Rivest Cipher 6 (RC6)

For CWSP purposes, we will focus only on RC4 and AES, as these are commonly used with 802.11 standards-based WLAN technologies.

RC4 was developed by Ron Rivest of RSA Security in 1987. In 802.11 WLAN technology, RC4 is used in conjunction with legacy WEP and TKIP. As you know from earlier lessons, WEP was cracked early on. The main problem with WEP was not with RC4 encryption algorithm, and it's important to understand the nuance here. WEP was weak because of how RC4 was used within the WEP framework, and with WEP's plain text Initialization Vector (IV). TKIP eventually provided a fix for some of the weaknesses in WEP as an interim solution until the 802.11i amendment was ratified, which we learned earlier provided much stronger security. Interestingly, the use of TKIP lasted much longer than the intended interim period, and it still is regularly found in many environments today – particularly in older PSK deployments within small businesses. The notion of backward-compatibility and the fairly fragmented nature of the client device space in regard to feature sets keeps many older devices and security protocols around longer than many of us would like. TKIP also uses RC4 but with several enhancements to provide stronger security than WEP did. Though TKIP doesn't suffer the same deep disregard as WEP, it too should be avoided in favor of AES (and WPA3 options) when possible.

In 802.11 standards-based WLAN technology, AES is used in conjunction with CCMP. AES uses the Rijndael algorithm and is a block cipher that was established by the U.S. National Institute of Standards and Technology (NIST) in 2001 to replace the older 1970s DES encryption algorithm. AES has a block size of 128 bits and can use three different key lengths, 128-bit, 192-bit and 256-bits. AES is considered to be quite secure within the framework of today's available technology. It would take a large amount of computing power and many years to be able to crack AES. (Bear in mind, even AES will one day be considered obsolete. No security protocols have an indefinite shelf-life.)

WEP (Pre-RSNA)

By now, you're well-versed that WEP was the initial security solution in the original 802.11 standard, and that WEP is no longer appropriate for securing any wireless environment. The stated goal of WEP was to provide an equivalent level of security to that which is normally found in a wired LAN, but it fell short. WEP clients are getting fewer and farther between, and as a CWSP, part of your charge is to help transition any remaining WEP-only clients away. Let's examine WEP yet again, from the encryption angle, to better understand the specifics of how poorly-implemented encryption can bite us.

The best practice is to state in your wireless use policy that WPA2 (and looking forward, WPA3) must be used in place of WEP and even WPA if possible. Where WEP must be used for whatever reason, the policy should require testing to ensure that the signal from a WEP SSID does not reach outside of your building. Of course, in the real world, it is nearly impossible to say that you are certain WEP frames are not reaching outside of your building. Remember, that the signal may not appear with standard client devices, but if an attacker has a wireless NIC with a very good receive sensitivity rating and a high gain antenna, he or she may be able to pick up the signal from surprising distances.

We need to start our discussion with yet another nod to the OSI model. The MAC service data unit (MSDU) contains upper-layer information that is present at the Data Link Layer (Layer 2), and which has passed down the OSI model from the Application layer. The MSDU has the appropriate layer-specific information added as it traverses the upper layers on its way "down the stack." As the name implies, this is a data unit that will be serviced by the MAC sublayer of the Data Link Layer. Once the MAC sublayer header is added, this data unit becomes what is known as the MAC protocol data unit (MPDU).

WEP encapsulates the MPDU data payload with a 4 octet IV (initialization vector) and a 4 octet ICV and extends the length of the MPDU by a total of 8 octets. This is what is known as *frame expansion*. (Prior to the 802.11n amendment to the standard, the maximum frame body size was 2304 bytes. With the additional 8 octets used for WEP, the frame size would increase to 2312 bytes. It is important to note that newer WLAN technology such as 802.11n/ac allows for larger frame body sizes.)

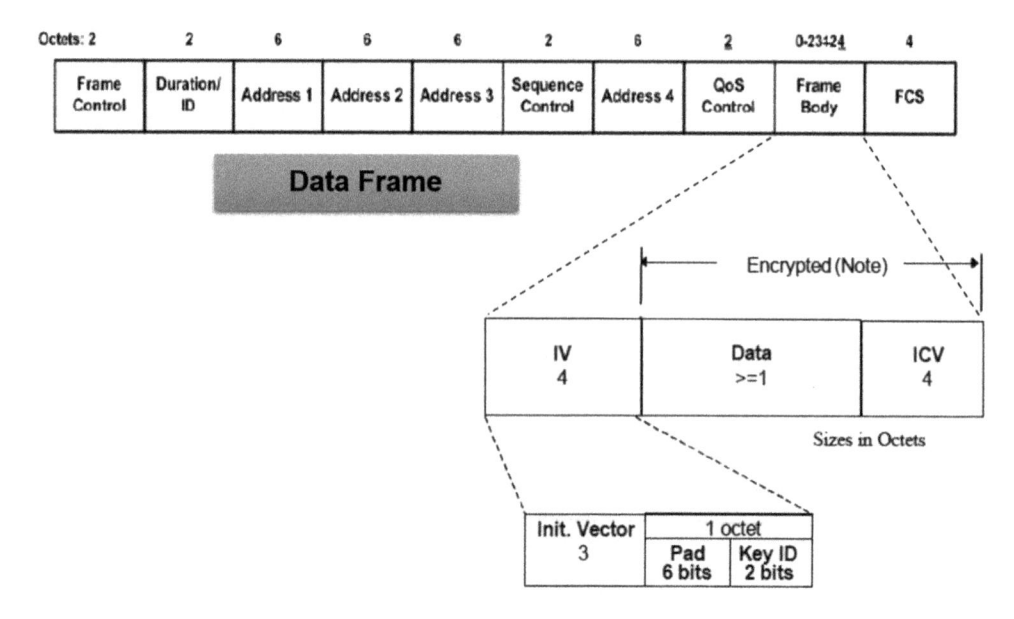

Octets: 2	2	6	6	6	2	6	2	0-2324	4
Frame Control	Duration/ ID	Address 1	Address 2	Address 3	Sequence Control	Address 4	QoS Control	Frame Body	FCS

Data Frame

Encrypted (Note)

IV 4	Data >=1	ICV 4

Sizes in Octets

Init. Vector 3	1 octet
	Pad 6 bits / Key ID 2 bits

Expanded WEP MPDU

WEP Weaknesses

WEP was never intended to provide impenetrable security, but it was supposed to protect against casual eavesdropping. With the rapid increase in processor speeds, cracking WEP became an easy task. WEP has long since been considered as sufficient protection for any wireless network, because of its vulnerability to any and all of these:

- Brute Force Attacks
- Dictionary Attacks
- Weak IV Attacks
- Re-Injection Attacks
- Storage Attacks

In late 2000 and early 2001, the security weaknesses of WEP became public knowledge. After that, many attack methods were developed, and tools created that made cracking WEP simple to implement for even entry-level technical individuals. As the weakness of WEP was exposed, wireless professionals were introduced to a range of new concerns. Sticking with WEP as a discussion focus for a bit, let's examine the attacks that have now become part of our lexicon.

A *brute force attack* is a key-guessing methodology that attempts every possible key in order to crack encryption. Each failed attempt is followed by another automated attempt with a new key until the attack is successful. With 104-bit WEP, brute-force was really not a feasible attack method for most attacks. However, 40-bit WEP could usually be cracked in one or two days with brute force attacks using more than 20 distributed computers using the technology of the early 2000s. The short timeframe was accomplished using a distributed cracking tool like jc-wepcrack. The jc-wepcrack utility was actually two tools: the client and the server. You would first start the tool on the server and configure it for the WEP key size you thought was in use on the wireless LAN that you were cracking. Then you'd provide it with a pcap file (a capture of encrypted frames) from that network. Next, you launched the client program and configured it to connect to the server. The client program would request a portion of the keys to be guessed and would attempt to access the encrypted frames with those keys.

With the modern addition of Field Programmable Gate Arrays (FPGAs) - which were add-on boards for hardware acceleration - the time to crack could be reduced by more than a factor of 30x. In fairness, the 20 computers would have to be Pentium P4 machines or better, which were fairly hard to come by in the early half of the last decade (2001-2010). If you chose to go the FPGA route, you would have been spending a lot of money to crack that WEP key in those days. If you could take even a modern PC back in time to those days, you'd be cracking WEP in just a fraction of the time it used to take.

Dictionary attack methodology relies on the fact that humans often use words or common strings as passwords. When using dictionary attacks against WEP, the pivotal step was to use a dictionary cracking tool that understood the conversion algorithm used by a given hardware vendor to convert the typed password into the WEP key. This algorithm was not specified in 802.11 and so was implemented differently by each client device vendor. Many vendors allowed the user to type a passphrase that was then converted to the WEP key using the Neesus Datacom or MD5 WEP key generation algorithms. The Neesus Datacom algorithm was notoriously insecure and had resulted in what was sometimes called the Newsham-21-bit attack because it reduced the usable WEP key pool to 21-bits instead of 40 when using a 40-bit WEP key. This smaller pool could be exhausted in about 6-7 seconds on a P4 3.6 GHz single machine using cracking tools against a pcap file. Even MD5-based conversion algorithms were far too weak and should not have been considered secure because they were still used to implement WEP (which was insecure due to weak IVs as well). Hindsight is 20-20, as they say. One silver lining that came of the turbulent WEP years is an overall raised awareness of the need for strong security in general, and also how various attacks work.

Weak IV attacks were based on the faulty implementation of RC4 in the WEP protocols. The IV was prepended to the static WEP key to form the full WEP encryption key used by the RC4 algorithm. This meant than an attacker already knew the first 24 bits of the encryption key since that IV was sent in cleartext as part of the frame header. Additionally, researchers Fluhrer, Mantin and Shamir identified "weak" IVs in a paper released in 2001. These weak IVs resulted in certain values becoming more statistically probable than others and made it easier to crack the static WEP key. The 802.11 frames that used these weak IVs had come to be known as interesting frames. With enough interesting frames collected, you could crack the WEP key in a matter of seconds. This reduced the total attack time down to less than 5-6 minutes on a busy wireless LAN.

HISTORIC NOTE: The weak IVs discovered by Fluhrer, Mantin, and Shamir are now among a larger pool of known weak IVs. After 2001, many more classes of weak IVs were discovered by David Hulton (h1kari) and KoreK. Thankfully, wireless security flaws are made public, so they can then be fixed.

What if the WEP-enabled network under attack was not busy and you could not capture enough interesting frames in a short window of time? After all, there were fewer Wi-Fi clients back in the day, so the WLAN environment may have been far less busy. The answer was a *re-injection attack*. This kind of attack usually re-injected ARP packets onto the wireless LAN. The program Aireplay could detect ARP packets based on their unique size and did not need to decrypt the packet. By re-injecting the ARP packets back onto the wireless LAN, it would force the other clients to reply and cause the creation of large amounts of wireless LAN traffic very quickly. For 40-bit WEP cracking, you usually wanted around 300,000 total frames to get enough interesting frames and for 104-bit WEP cracking you wanted about 1,000,000 frames. Though these seem like high counts, automated attacks made short work of getting more frames in the air to exploit.

Storage attacks are those methods used to recover WEP or WPA keys from their storage locations. On Windows computers, for example, WEP keys were often stored in an encrypted form in the registry. An older version of this attack method was the Lucent Registry. An application named Wzcook could retrieve the stored WEP keys used by Windows' Wireless Zero Configuration. This application recovered WEP or WPA-PSK keys (since they were effectively the same, WPA just improves the way the key is managed and implemented) and was part of the Aircrack-ng suite of tools used for cracking the keys. The application only worked if you had administrator access to the local machine, but in an environment with poor physical security and poor user training, it was not difficult to find a machine that was logged on and using Wi-Fi for carrying out the storage attack.

WEP makes up the core of pre-RSNA security in 802.11 networks and is mostly relegated to the past. We study it as we study significant historical developments in any discipline, and it still occasionally surfaces in corner cases. By thoroughly understanding the attack vectors used against WEP, the CWSP is better equipped to understand various types of attacks as they are developed and to educate customers on the topic.

Dynamic WEP

Dynamic WEP is an interesting footnote in the history of WLAN security. It was a non-standard interim solution introduced prior to the 802.11i amendment. Using the 802.1X/EAP framework to produce dynamic keys, manufacturers began supporting a proprietary WEP solution that used these keys dynamically. Many of the same weaknesses are present in dynamic WEP, and if enough frames are transmitted with dynamic WEP keys, the key can be recovered. As with 802.11's WEP, this proprietary WLAN security solution is long since considered obsolete. Dynamic WEP never really saw wide-scale adoption.

TKIP (WPA)

Working down the evolutionary timeline of WLAN security, TKIP was introduced to resolve the weaknesses of WEP. TKIP was part of the draft 802.11i amendment when the Wi-Fi Alliance chose to certify it as WPA so vendors could begin implementing it in firmware for existing devices, and out-of-the-box with new devices. TKIP/WPA added four new algorithms to WEP to address its security weaknesses:

- Michael - Message Integrity Check (MIC) to prevent forgery attacks (yes, it is really referred to as "Michael" for whatever reason)
- 48-bit Initialization Vector and IV sequence counter to prevent replay attacks
    - MPDUs received out-of-order are dropped by the receiver
- Per-packet key mixing of the IV to de-correlate IVs from weak keys
    - 48-bit IV called TKIP Sequence Counter (TSC)
    - TSC updated each packet
    - 2^{48} frames allowed per single temporal key, would require 100 years to exhaust key space
- Dynamic re-keying mechanism to change encryption and integrity keys

- Temporal key, transmitter address, and TSC combined into per-packet key
- Split into 104-bit RC4 key and 24-bit IV for WEP compatibility

Earlier you saw that WEP encapsulated the MAC protocol data unit (MPDU) payload with a 4 octet IV and a 4 octet ICV, for a total of 8 octets. When TKIP is implemented, it adds the additional overhead of an extended IV of 4 octets (32 bits) and an additional MIC of 8 octets inside of WEP's encapsulation, which is a total of 12 additional octets. The total encryption overhead becomes 20 octets per frame vs. 12 octets for WEP, so the maximum frame body becomes a total of 2324 octets.

Michael is the name of the integrity algorithm, which enhances the legacy ICV mechanism used with TKIP. Michael was meant to improve integrity protection while remaining backward compatible with (at the time TKIP came out) millions of limited-feature legacy radios since it was required to operate within a very small computing budget.

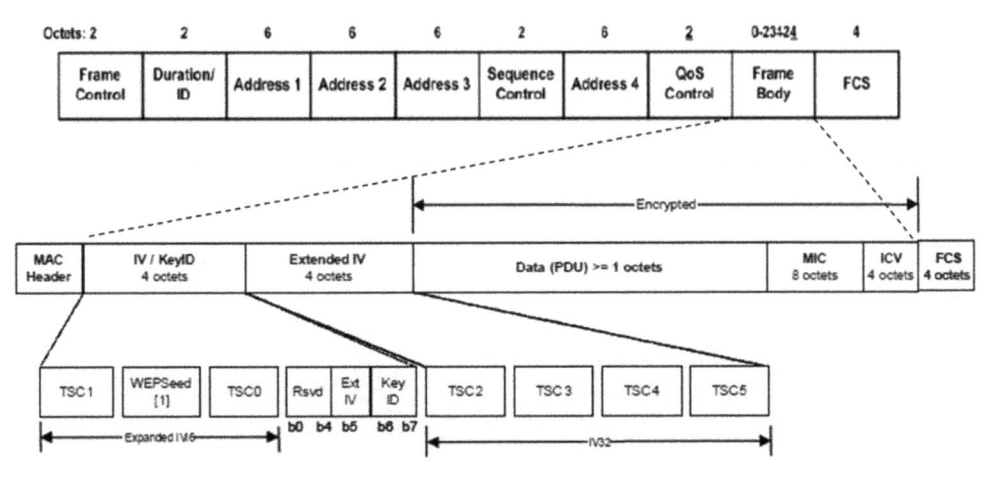

TKIP MPDU Expansion

The Michael MIC contains only 20 bits of effective security strength, and so is vulnerable to brute force attacks. For further protection, Michael is capable of implementing countermeasures if it detects an attack. Using these countermeasures, STAs or APs that detect two MIC failures within 60 seconds of each other, must disable all TKIP receptions for 60 seconds.

In theory, these MIC failures should be logged for follow-up by a security administrator. It should be noted that the Michael countermeasure mechanism could be used as a DoS exploit, although there are much easier DoS attacks that could be used.

The TKIP-mandated 60-sec disablement period can also cause problems in healthy WLANs when a client has a defective driver or similar that triggers the countermeasure. Nearby clients are essentially disabled for Wi-Fi during the penalty period, and the effect can feel like a malfunctioning network. It's not uncommon for network administrators to override the 60-second non-transmit period to avoid false alarms related to Michael, but this also negates an important security feature of TKIP.

The Michael MIC was designed to help prevent forgery attacks, which are a vulnerability of WEP.

In a frame decode that uses TKIP, but does not support CCMP, you will not see an RSN IE. This is because, in order to qualify as an RSN, you'll recall that the service set must support CCMP. Instead of the RSN IE, you will see a manufacturer-specific information element, commonly "WPA Information" or "WPA IE (221)", with either containing most of the same information as an RSN IE.

One thing to note is that if support for both CCMP and TKIP is enabled on the wireless infrastructure device (which includes APs, WLAN controllers or cloud-managed devices), in the appropriate management frame decode, you will see both the RSN IE and the WPA IE. This makes sense as a TKIP-only device would not be able to interpret the RSN information contained within the management frame.

The following image shows the 802.11 TKIP Data field in a protocol decode.

```
802.11 TKIP Data
    IV:                 0x002001 [26-28]
        RC4Key[0]:      0x00 [26]
        RC4Key[1]:      0x20 [27]
        RC4Key[2]:      0x01 [28]
    Key Index:          0x20 [29]
        Key ID:         %00  Key ID=1 [29 Mask 0xC0]
        Ext IV:         %1 [29 Mask 0x20]
        Reserved:       %00000 [29 Mask 0x1F]
    Extended IV:        0x00000000 [30-33]
    TKIP Data:          (107 bytes) [34-140]
    MIC:                0x08D0894652D19F0B [141-148]
    ICV:                0xD31775A9 [149-152 Mask 0x0000FFFF]
FCS - Frame Check Sequence
    FCS:                0xC483228F  Calculated
```
TKIP in an 802.11 Frame

WPA was a huge step forward from WEP; particularly regarding the recognition of the importance of well-designed encryption. As WLAN use gained in popularity, wireless networks became a bigger target for the bad guys. The game of cat-and-mouse continued with the development of WPA2 and will continue with WPA3 and beyond.

CCMP (WPA2)

CCMP is based on the CCM of the AES encryption algorithm. CCM combines CTR (counter) mode for data confidentiality and CBC-MAC for authentication and integrity. CCM protects the integrity of both the MPDU Data field and selected portions of the 802.11 MPDU header. The AES algorithm is defined in FIPS PUB 197-2001. All AES processing used within CCMP uses AES with a 128-bit key and a 128-bit block size.

Note that WPA2:

- Replaces RC4 with the Advanced Encryption Standard (AES) (Rijndael algorithm) in Counter mode (for data privacy) with Cipher Block Chaining-Message Authentication Code (CBC-MAC) for data authenticity – CCMP/AES
    o AES is a symmetric, iterated block cipher

- Uses a 128-bit encryption key size, and encrypts in 128-bit fixed-length blocks
- Has a 48-bit IV (called Packet Number or PN) derived from the AES Key
- Does encryption and MIC calculation in parallel
- Renders per-packet keys unnecessary due to the strength of the AES cipher
- Features an 8-byte MIC which is considered much stronger than Michael
- Uses a separate chip to perform computation-intensive AES ciphering

Earlier you saw that prior to 802.11n, the largest frame body size was 2304 bytes. This was without any encryption methods used. When encryption is used, the frame body is expanded.

WEP added 8 octets of overhead, which increased the frame body to a maximum of 2312 bytes. TKIP added an additional 12 octets of overhead (which is in addition to the 8 octets for WEP) and increased the maximum frame body size to 2324 bytes.

Since CCMP is much more efficient than both WEP and TKIP, and some of the CCMP encryption processing is handled by improved hardware technology, less overhead is required in the frame body. Therefore, CCMP adds only an additional 16 bytes of overhead to the frame body; 8 octets for the CCMP header and another 8 octets for the MIC. The maximum frame body size that uses CCMP will be 2320 bytes in pre-802.11n deployments.

> The CWSP exam no longer tests on frame overhead knowledge when considering WEP, TKIP, and CCMP. This was removed from the exam because organizations should only be using CCMP moving forward. It is provided here for current operational knowledge only.

Although CCMP is much stronger and more secure than WEP and TKIP, it will not require additional overhead in the frame body.

> Frame body sizes vary greatly with 802.11n, 802.11ac and aggregation features. You will not be tested on frame sizes on the CWSP exam at all, but a general knowledge will serve you in your wireless duties.

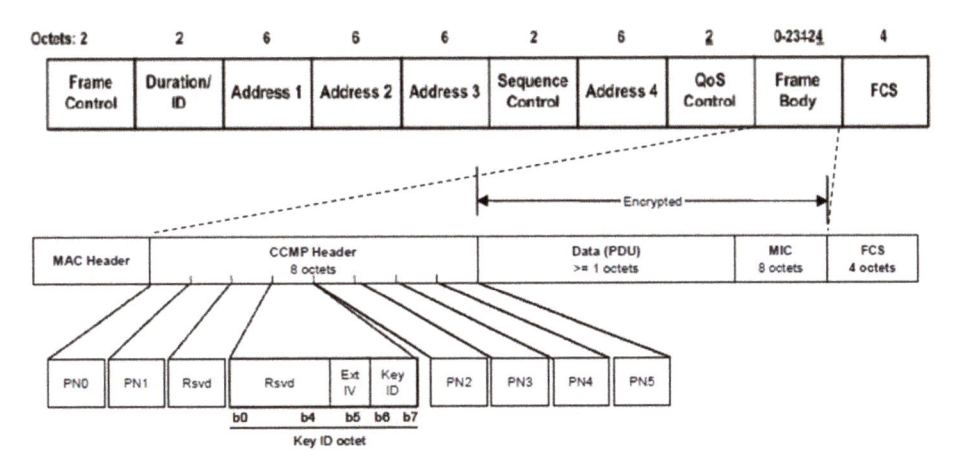

Octets: 2	2	6	6	6	2	6	2	0-2312	4
Frame Control	Duration/ID	Address 1	Address 2	Address 3	Sequence Control	Address 4	QoS Control	Frame Body	FCS

MAC Header | CCMP Header 8 octets | Data (PDU) >= 1 octets | MIC 8 octets | FCS 4 octets

PN0 | PN1 | Rsvd | Rsvd | Ext IV | Key ID | PN2 | PN3 | PN4 | PN5

b0 b4 b5 b6 b7
Key ID octet

CCMP Frame Expansion

It is important to note that newer WLAN standards such as 802.11n and 802.11ac allow for TKIP use, but with a tremendous performance penalty. If TKIP is used instead of CCMP and the 802.11 standard is followed, no 802.11n or 802.11ac MCS rates will be available. Therefore, with TKIP, the highest achievable data rate is only 802.11a/g-era 54 Mbps! This is yet one more reason that TKIP should no longer be used on enterprise networks desiring the data rate advantages offered by 802.11n and 802.11ac. In reality, if you use TKIP in an 802.11n network, you're really running an 802.11g or 802.11a network, because 802.11n and forward no longer allowed for the use of any cipher suite but CCMP.

According to 802.11-2012 (carried forward in 802.11-2016), "The use of TKIP is deprecated. The TKIP algorithm is unsuitable for the purposes of this standard." (Section 11.1.1) Further, it says explicitly in 11.1.6, "An HT STA shall not use either of the pairwise cipher suite selectors. Use "group cipher suite" or TKIP to communicate with another HT STA." This, of course, applies forward to the 802.11 VHT PHY as well (and will also apply to 802.11ax).

Chapter Summary

In this chapter, you studied encryption and its uses in 802.11 networks, excluding WPA3. You began by learning or reviewing definitions of several key terms related to encryption, and then explored the various encryption solutions from the 802.11 historical timelines. We re-enforced that WEP is not a suitable solution for any modern network and that TKIP enhanced the RC4 implementation so that the weaknesses of WEP were overcome. However, you also learned that the IEEE no longer recommends the use of TKIP despite its enhancements over WEP and that it is only included in the standard in a deprecated manner. All new wireless networks should be implemented with CCMP (WPA2) and eventually WPA3. We also highlighted that implementing 802.11n or later equipment with TKIP (if the configuration interfaces even allowed it) would result in only basic 802.11a/g data rates. We reviewed a number of encryption-targeting attacks, and how wireless security has improved, to combat these attacks. We've also noted that WLAN security is evolutionary out of necessity.

Review Questions

1. Which two encryption algorithms are commonly used with 802.11 standards?
    a. AES and 3DES
    b. AES and RC4
    c. RC4 and 3DES
    d. RC5 and AES

2. A procedure that takes an arbitrary block of data and returns a fixed-size bit string is _____?
    a. Hashing algorithm
    b. Cipher suite
    c. Encryption algorithm
    d. Keying suite

3. Two valid dynamic key implementations related to secure 802.11 networks are:
    a. RC5 and TKIP
    b. 3DES and CCMP
    c. TKIP and CCMP
    d. RC4 and CCMP

4. You are using a public Wi-Fi hotspot and are accessing a web site using HTTPS. Your encryption is _____?
    a. Unequal
    b. Deprecated
    c. Symmetric
    d. Asymmetric

5. Which of the following is a valid statement comparing asymmetric encryption to symmetric encryption?
    a. Symmetric is more processor intensive
    b. Asymmetric is more processor intensive
    c. Symmetric is better
    d. Asymmetric is better

6. In 802.11 networks, where, in the OSI model, are standardized encryption methods applied?
    a. Layers 1 and 2
    b. Layer 1
    c. Layer 2
    d. It depends on the encryption type

7. CCMP/AES uses what size block to be encrypted?
    a. 40-bit
    b. 128-bit
    c. 256-bit
    d. A variable-size bit that changes with packet sizes

8. As implemented in WEP, what specific attack was the ICV vulnerable to?
    a. Bit-blasting
    b. RF DOS
    c. MTM
    d. Bit-flipping

9. Layer 1 is made up of two sub-layers. What are they?
    a. PLCP and LLDP
    b. PMD and PLCP
    c. LDP and PMD
    d. PPTP and PMD

10. With centralized mode of encryption in use on 802.11 networks, encryption protects data from _____ to _____.
    a. Client to access point
    b. Access point to core
    c. Client to Internet
    d. Client to core

11. Which method uses public and private encryption keys?
    a. Public/Private
    b. Asymmetric
    c. Symmetric
    d. Reflexive

12. The weak IV attacks used against WEP were based on faulty implementation of what?
    a. RC4
    b. AES
    c. TKIP
    d. CCMP

13. What sort of attack would be used to recover WEP or WPA keys from the registry of the Windows operating system?
    a. Key probe attack
    b. Stored value attack
    c. Storage attack
    d. Bit-flip attack

14. Michael was a _____ associated with which encryption type?
    a. Message Integrity Check, WEP
    b. Error Correction Protocol, WPA
    c. Message Integrity Check, WPA
    d. Error Correction Protocol, CCMP

15. What makes up the core of pre-RSNA security in 802.11 networks?
    a. TKIP
    b. RC4
    c. AES
    d. WEP

Review Answers

1. **B**. AES and RC4 are common encryption algorithms used with today's 802.11 secure networks

2. **A**. This is the definition of a hashing algorithm

3. **C**. TKIP and CCMP are dynamic key mechanisms- don't confuse them with encryption algorithms

4. **C**. HTTPS uses Asymmetric encryption

5. **B**. Asymmetric encryption is more processor-intensive than symmetric encryption

6. **C**. 802.11 encryption takes place at Layer 2

7. **B**. CCMP/AES uses a 128-bit block cipher

8. **D**. Bit-flipping took advantage of the weak WEP IV

9. **B**. Layer 1 is made up of PLCP and PMD sub-layers

10. **D**. Client to core encryption is the hallmark of centralized WLAN encryption

11. **B**. Asymmetric encryption uses public and private keys

12. **A**. RC4 itself wasn't the problem with WEP; it was the way it was implemented

13. **C**. Storage attacks are used to find WEP keys in Windows' registries

14. **C**. Michael is the oddly named Message Integrity Check used in WPA

15. **D**. WEP is the main element of pre-RSNA security

Chapter 7: Security Design Scenarios

Objectives Covered

3.1   Select the appropriate security solution for a given implementation and ensure it is installed and configured according to policy requirements

3.2   Implement or recommend appropriate wired security configurations to support the WLAN

3.3   Implement authentication and security services

Although different network and business scenarios require specific security configurations tailored to each situation, there is a finite number of approaches from which to choose. As a wireless security professional, it's important to understand what methodologies are available and to know when each is a good choice to implement, to meet the requirements set forth by security policy. In this chapter, we will examine several of the primary security options available to us for different situations and discuss when each makes sense. This chapter summarizes a number of non-802.11-specific security measures that are commonplace across the business WLAN landscape.

Virtual Private Networking Basics

A virtual private network (VPN) provides the capability to create private network communication over a public network infrastructure, such as the Internet. VPN technology is used in many different networking scenarios, including 802.11 wireless networking. In fact, VPN is one of the most pervasive security mechanisms in use across LAN, WLAN, and WAN alike. VPNs are Internet Protocol (IP) based, so they commonly operate at the Network Layer (Layer 3) of the OSI model. Some VPN protocols will also operate at other layers, or even over multiple layers. VPN technology can consist of different configurations, including client-to-server or site-to-site (gateway-gateway), and use a range of protocols, including:

- Point-to-point tunneling protocol (PPTP)

- Layer 2 tunneling protocol (L2TP) with Internet Protocol Security (IPSec) - L2TP/IPSec

- Internet Protocol Security (IPSec)

- Transport Layer Security (TLS), Secure Sockets Layer (SSL) - SSL/TLS

- Secure Shell (SSH)

- Datagram Transport Layer Security (DTLS)

VPN use was very common in enterprise network deployments prior to the ratification of the .11i amendment to the 802.11 standard and is still a very common remote access security solution. Due to advancements in WLAN security protocols and Wi-Fi Alliance (WPA, WPA2, and WPA3) certifications, Data Link Layer (Layer 2) security solutions have become stronger to the point where VPN is not as widely used within enterprise LANs for client access as it once was. However, VPN still remains a powerful security solution for remote access in both wired and wireless networking and does see limited use within the borders of WLAN environments when particular security needs are in play.

The VPN general construct consists of two parts, a tunneling component, and an encryption component. A standalone VPN tunnel does not necessarily provide data encryption in and of itself, and VPN tunnels are created across Internet Protocol (IP) networks. In a very basic sense, VPNs use encapsulation methods where one IP frame is encapsulated within a second IP frame. The encryption of VPNs is performed as a separate, stand-alone function in many implementations.

A VPN consists of *endpoints*, which are the devices that create the tunneled architecture. VPN endpoints can consist of various infrastructure devices including:

- Computers
- Layer 3 routers
- Wireless LAN controllers
- Wireless APs
- Dedicated servers
- VPN concentrators
- Firewalls
- Edge security appliances from different vendors
- Network Management System (NMS) platforms

The most common uses of VPNs in the WLAN space are for remote APs, remote client access to network resources across the Internet, proprietary bridging, and vendor-defined proprietary communications between WLAN devices, like WLAN controllers or APs. Knowing these use cases is important for CWSPs.

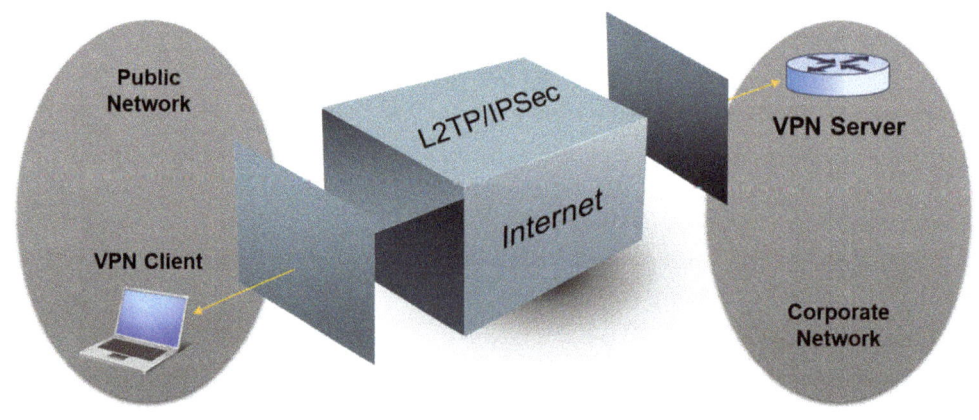

Client VPN Example

Common VPN Protocols

Two common types of VPN protocols are:

- Point-to-Point Tunneling Protocol (PPTP)
- Layer 2 Tunneling Protocol (L2TP)

We've talked a bit about PPTP's shortcomings back in the early chapters, but let's expand the discussion on this long-running VPN protocol. The Point-to-Point Tunneling Protocol was developed by a vendor consortium that included Microsoft. PPTP has been very popular because of its ease of configuration and has been included in all Microsoft Windows operating systems starting with Windows 95. PPTP uses Microsoft Point-to-Point Encryption (MPPE-128) Protocol for encryption. PPTP operates at Layer 2 and uses Generic Routing Encapsulation (GRE) tunneling to encapsulate point-to-point protocol (PPP) packets. You'll find GRE tunnels are fairly common in a number of wired and wireless applications in modern networking environments.) The PPTP VPN process provides both tunneling and encryption capabilities for the user's data. Vulnerabilities have been found in PPTP, in the implementation of MSCHAP authentication used along with MPPE encryption. Although PPTP is easy to configure and did provide necessary security at its initial implementation, the VPN protocol lost much ground after the introduction of Layer 2 Tunneling Protocol (L2TP) and the discovery of several vulnerabilities. You may still find PPTP out there in the wild, but know that through the lens of today's security metrics, PPTP will fail most audits.

It is important to note that with respect to wireless networking, the authentication process of PPTP has been cracked and, therefore, it should not be used with a wireless network. This statement applies when MS-CHAPv2 is used for the user authentication. If the authentication process (wireless frames) were captured using a wireless protocol analyzer and dictionary attack software, the user credentials could be discovered through analysis of the captured frames. This would allow the intruder to logon to the network with stolen credentials. You will recall from past chapters that a dictionary attack is performed by software that challenges the encrypted password (or pre-shared key, or whatever is being targeted) against common words or phrases in a text file that acts as the dictionary in this scenario. Text files used in these attacks are commonplace in the hacker community. This is very similar to the process that can be used to crack Cisco Systems LEAP and is considered a fairly low-skill cracking exercise. Therefore, using PPTP VPN on a wireless network with MS-CHAPv2 (or MS-CHAPv1) should be avoided.

By contrast, L2TP is the combination of two different tunneling protocols: Cisco's Layer 2 Forwarding (Layer 2F) and Microsoft's Point-to-Point Tunneling Protocol (PPTP). L2TP defines the tunneling process, which requires some level of encryption in order to function. With L2TP, a popular choice of encryption is Internet Protocol Security (IPSec), which provides authentication and encryption for each individual IP packet in a data stream. Since it was published in 1999 as a proposed standard, and because it is more secure than PPTP, L2TP has gained much popularity and is a recommended replacement for PPTP. Many systems no longer provide the choice to use the weaker PPTP option and only support L2TP, so L2TP/IPSec has become a very common VPN solution. You should note that L2TP should always be used instead of PPTP.

IPSec is a VPN protocol designed to authenticate and encrypt packets using the Layer 3 Internet Protocol. IPSec includes two possible implementations:

- **Authenticated Header (AH)** - Provides only authentication
- **Encapsulation Security Payload (ESP)** - Provides encryption for the data payload in addition to authentication and integrity verification

Furthermore, ESP operates in two modes:

- **Transport mode** - This mode with ESP would be appropriate for client-server or site-to-site communications. This excludes remote WLAN endpoint connectivity, where only tunneled mode should be used with ESP. With transport mode, endpoint devices will encrypt/decrypt the data between each endpoint

- **Tunneled mode** - This mode is able to communicate from one private IP address directly to another private IP address because the devices build a virtual tunnel. Again, remote WLAN endpoints should only use Tunneled Mode.

An additional VPN protocol that is gaining popularity in Microsoft environments is Secure Socket Tunneling Protocol (SSTP). SSTP implements HTTPS on TCP port 443 in order to allow passage through common firewall configurations. Interestingly, EAP-TLS is a common authentication protocol used with this VPN solution because it allows the passing of PPP traffic over the connection. SSTP is supported on all currently supported Microsoft operating systems, having started with Windows Vista and Server 2000.

Proprietary VPN protocols may also be used to secure communications between wireless bridge links or possibly infrastructure devices, such as WLAN controllers. This is an area where WLAN vendors very much do their own thing and is one reason why you can't typically use Vendor A's wireless controllers with Vendor B's in the same WLAN.

VPN Functionality

A client-server VPN solution consists of three components:

- Client-side endpoint
- Network infrastructure (public or private)
- Server-side endpoint

As mentioned earlier, the client-side and server side are known as VPN endpoints. The infrastructure in most cases is an unsecured public access network such as the Internet, though some may still use dedicated private leased lines from telecom providers. The client-side endpoint typically consists of software on a PC, tablet or smartphone, which allows it to be configured for the VPN connection. If it doesn't come native to the specific operating system, this software is available at a nominal cost from a variety of manufacturers. Microsoft Windows and Apple operating systems have included VPN client software for both PPTP and L2TP for several years. The VPN can terminate either at an AP or across the Internet to a VPN endpoint on the corporate network set up for terminating remote connections. As a CWSP, you may find yourself dealing with several different VPN scenarios within the same organizational environment.

Typically, there are three steps in creating a VPN, after the basic configurations have been accomplished:

1. Perform the required authentication
2. Build the virtual tunnel
3. Encrypt the data

VPN networks will encapsulate one IP packet into another IP packet. The packet that has been encapsulated will contain the data payload. This action will prevent unauthorized users from being able to see any data that is sent over the secure tunnel.

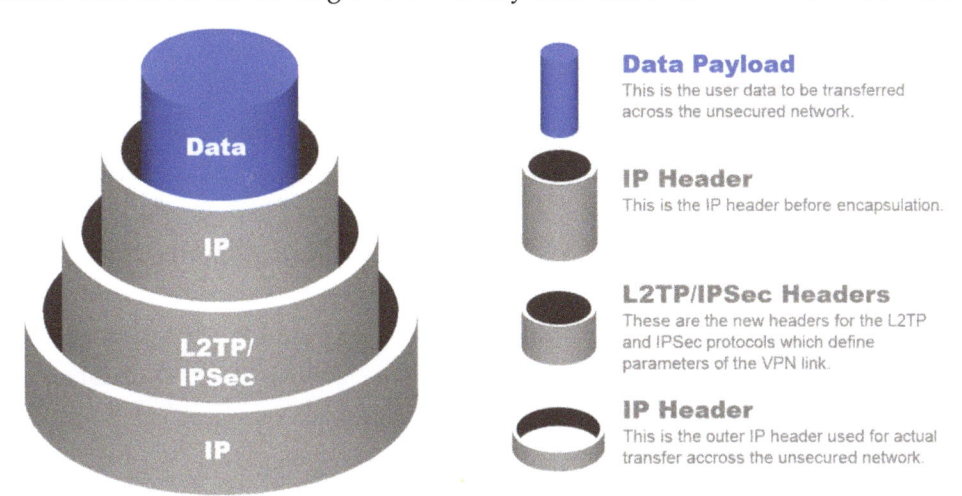

Data Payload
This is the user data to be transferred across the unsecured network.

IP Header
This is the IP header before encapsulation.

L2TP/IPSec Headers
These are the new headers for the L2TP and IPSec protocols which define parameters of the VPN link.

IP Header
This is the outer IP header used for actual transfer accross the unsecured network.

IPSec Tunnel Mode Illustrated

Common Wireless VPN Uses

Three common uses of wireless VPNs that CWSPS must know include:

- Remote APs

- Wireless bridging
- Remote client connectivity

Remote networking has become very common as wireless access has proliferated to the home and branch office networks. Some manufacturers have developed dedicated "remote APs" that automatically build a tunnel to a remote WLAN controller or VPN concentrator to extend the main network to the remote site. After establishing a secure tunnel, the AP receives a WLAN policy, which it then broadcasts locally for wireless access. This approach securely tunnels remote users to corporate networks for access to network resources and ensures that the remote network instance complies with organizational policy. It may also be used to protect users from unknowingly compromising their corporate resources on their computer.

Remote APs will allow a user to connect a wireless AP to remote Local Area Network with an active Internet connection. The remote AP will use the Internet to create a secure connection to the organization's corporate network. This will, in turn, provide secured wireless access from the remote location to the corporate network. This is a common scenario for remote offices or home office users and for those that work "on the road," such as sales or field service personnel.

A wireless bridge is a common tool for the wireless professional and will connect two or more local area networks (LANs) together. Wireless bridges can provide cost savings for the organization because it will not require installation of physical infrastructure between the end sites, and there will be no recurring monthly fees as with leased lines. Since many wireless bridging technologies use proprietary protocols and do not provide client connectivity, they often use proprietary VPN protocols for security. Securing a bridge is critical as the connection can span for long distances and, being wireless, the signal is not contained within a physical space.

Similar to the remote AP scenario described above, individual clients connecting to unsecured networks often use VPN technologies to secure their data traffic. This type of technology employs VPN software (instead of hardware, as with remote APs) that runs on the client computer. The software establishes an encrypted tunnel with a remote VPN terminator for access to network resources, or for protection from local threats associated with using open public wireless networks.

Let's look at each VPN use case in more detail.

Remote AP

We've established that remote APs provide secure access to traveling and home users by linking organizationally provided APs back to the corporate network. From a compatibility perspective, because they tend to use proprietary protocols, remote APs must usually be from the same wireless vendor as the main network. One of the great benefits of remote AP technology is that administrators maintain control of remote APs and can provision them in a way that reliably maintains the corporate security policies and prevents errant configurations. Similarly, the users' connectivity experience doesn't have to change when they're at home, on the road, or at the office. If configured for it, they connect to the same WLAN profiles as when on the main network, they retain mobility, and access to corporate resources is only limited by the organizational network to which they are connected—and not by their local infrastructure (beyond ISP capacity limits).

A remote AP is usually very easy to use. These devices are configured by the wireless network administrators of the organization unless otherwise coordinated. The remote user will then plug the AP into an available Ethernet port on the local network. This could be at a home office, hotel conference room, company branch office or anyplace with an active Internet connection from which the user wishes to connect.

Once the AP is connected locally, it will use the Internet to create a secure VPN tunnel from the remote location to the corporate office. The process is very similar to the client-server VPN model in which a client device will use VPN software to connect to the corporate office. The difference is that the remote AP has been configured to handle everything behind the scenes.

Remote Client Access

Secure remote client access is the most common use of a wireless network client-server VPN solution. With the continued growth of open public access wireless hotspot networks (free access in many cases), this VPN solution is routinely used every day by network users versed in the importance of network security. Remote client VPN using recognized strong protocols provides adequate security when connected to an open-access public hotspot and will allow for communications between a remote wireless client and a corporate network across the Internet.

Remote AP Illustrated

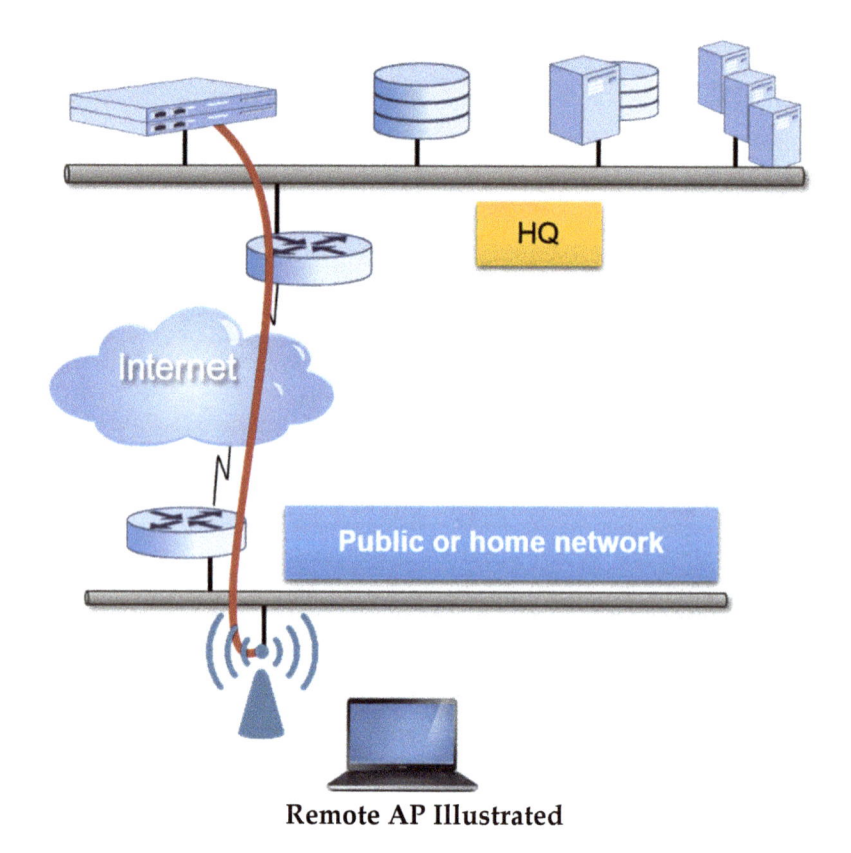

As wireless hotspots have become more common and wireless security vulnerabilities have received greater publicity, VPN implementations for remote clients have become increasingly popular. Properly maintained client VPN technologies along with client endpoint software or NAC solutions could offer strong protection for remote users connecting to open networks. However, there is still reason for concern. This type of VPN can only be applied after the user has associated with the open network. This often leaves users open to other vulnerabilities, such as hijacking or man-in-the-middle (MITM) attacks, and is one of the drivers behind OWE encryption that comes from the Wi-Fi Alliance's Enhanced Open certification that may eventually come to displace VPN as the recommended Wi-Fi hotspot security measure.

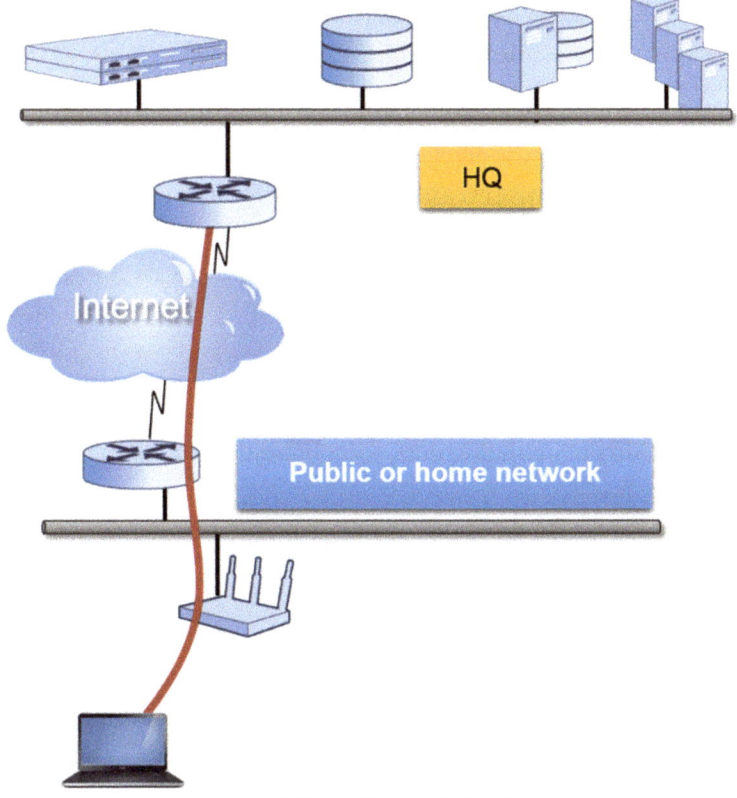

Remote Client Access Illustrated

It is common for less savvy users to assume that a VPN connection solves every potential security problem, but this is just not true. VPN by itself is only a secure connection between two endpoints and not a complete security environment. Basic security policies and solutions should be enforced for users operating from unsecured wireless networks. These include personal firewalls, up-to-date antivirus software, endpoint software/agents, and network access control (NAC) solutions all derived from the operational policy.

As discussed earlier in the text, the network security policy should define the requirements for remote client connectivity. Several considerations must be made here. First, enterprises should define the operating system rights/permissions of the end-users. Will they be capable of making configuration changes to client utilities? Will they be allowed access on open wireless networks? Is BYOD allowed for remote connectivity? Restricting the privileges of the wireless client is not always popular from the user satisfaction standpoint, but best practices for security demand tight control of corporate assets.

For the best security, client endpoint agents and MDM solutions should be used to manage devices on the wireless network to which a client has access, as well as to track usage and monitor network behavior and threats. Endpoint agent software can provide powerful, unique views from the client perspective. At the same time, distributing and up-keeping agents can be challenging.

If you desire to work with a free VPN server that works on Kali Linux, install and configure Openswan. A tutorial on the installation and configuration options is available here:

https://github.com/xelerance/Openswan/wiki

Tunneling and Split Tunneling

VPNs are examples of secure tunneling. As we just learned, tunneling is the process of encapsulating one IP packet within another IP packet.

- The original packet becomes the payload of the second packet.
- The source and destination IP addresses of the second packet typically point to the virtual IP address of the VPN client software (source) and the IP address of the VPN endpoint (destination).

Secure tunneling encrypts the original packet and obfuscates the original source and destination IP addresses. The VPN endpoint decapsulates (sometimes "de-encapsulates" is used in conversation) the tunneled packet onto the trusted network. This restores the original packet, with its original source and destination addresses intact. This process works very well and is widely accepted and supported as a network security solution.

There are many different tunneling protocols available. Two common examples that we've already touched on are the less-desirable PPTP and L2TP protocols. When configuring a VPN tunnel, both endpoints must understand the specific mechanism used for that tunnel and must agree on the shared configuration, which will include various settings such as encryption type and any other required parameters. Data transferred between the endpoints and over the tunnel is typically sent using a protocol such as user datagram protocol (UDP). However, other protocols may be employed, and special VPN-related protocols are used to build and tear down the tunnel.

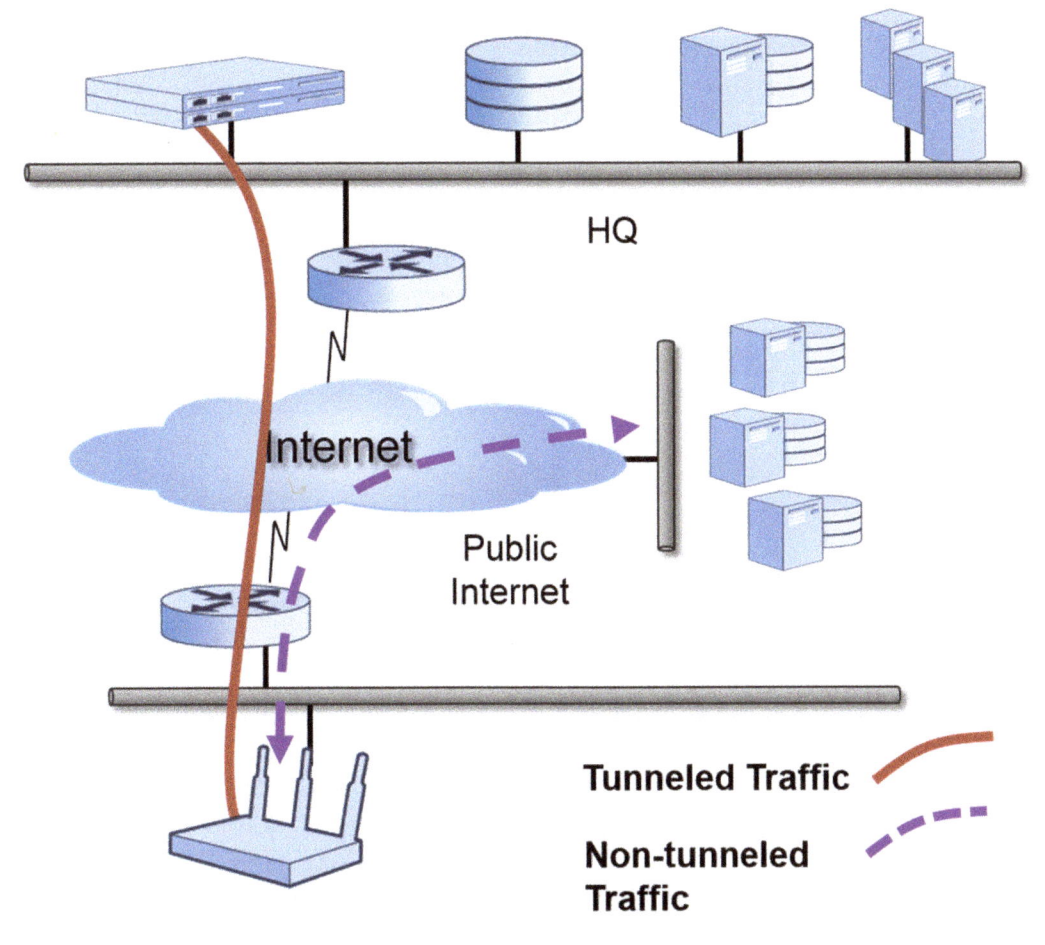

Split Tunneling Illustrated

The split tunnel paradigm was designed to reduce the processing overhead incurred by VPN usage and is very empowering in branch-office and remote client VPN situations. In a split tunnel scenario, the only traffic sent to and from the private network is organizational traffic that belongs on the main network. This organizational traffic is protected by VPN, but all other traffic, including local LAN activity and web-based activities, are not encapsulated within secure tunnels and uses the local Internet

259

connection to get in and out of the network. One positive aspect of split tunneling is keeping traffic that doesn't "need" to be in the tunnel out of it, but there are trade-offs. For example, split-tunneling can result in a vulnerability whereby a malicious intruder in the public WLAN space may be able to piggyback the secure connection through the unsecured local connection and inject a Trojan horse, malware, rootkit, backdoor, or virus into the corporate environment. It also allows intruders access to the wireless client's local resources. For this reason, CWNP's position is that only full-tunnel VPNs, which send all TCP/IP traffic through the VPN tunnel, should be allowed by remote access endpoints for the VPN implementation to be considered secure.

Public Access Networks

In review, public access wireless networks are commonly known as wireless hotspots and are increasingly available in many different locations and business types. This type of wireless network can be found at places including, but not limited to:

- Hotels
- Airports
- Coffee shops
- Restaurants
- Retail chain stores
- Public libraries
- Cruise ships
- Transportation - automobiles, airplanes, trains and other public transportation methods
- Sporting and convention venues (large public venues/LPV)

In some cases, these networks are available for free, while others are fee-based. Those that are offered at no cost are typically provided as a draw to bring customers into the business as a value-added service. Those that charge a fee use guest Wi-Fi for a revenue stream. Depending on how the public access WLAN was implemented, there could certainly be security risks for users as we'll discuss next.

Proper WLAN system configurations, including client-to-client blocking features, should be enabled to help lessen the chances of certain types of wireless intrusions. Other configs may block protocols or ports to help prevent spamming and other Internet-related attacks. In some cases, content filtering may be used to control access. Keep in mind that all of these practices may be controversial since the user will be limited in what they can do while connected. This is yet another example of the frequent tension between security and ease of use, which is a fact of life for CWSPs.

Public Access Risks

By now, you hopefully understand the many security concerns that arise when talking about public access networks, but let's review them again as we discuss VPN. These popular networks typically provide no authentication or encryption in the absence of new strategies like OWE, so users are vulnerable to a number of textbook attacks right away.

Public networks have always been attractive for intruders that want to gather information, and in some cases, hotspots can yield data that is ultimately used for financial gain or even identity theft. In many cases, the client device connected to the unsecured network itself is not properly secured and so may include the following vulnerabilities:

- Outdated or no anti-virus software
- Improperly configured or disabled firewalls
- Weak or missing passwords
- Unsecured file shares
- Missing operating system updates or service packs
- Saved password and account files in plain text

The service provider also has a number of concerns to address depending on their own policies, such as limiting their liability with a use/abuse policy, restricting network access in accordance with usage guidelines, and maintaining a captive portal for user pass-through.

Educating and informing end users of potential security risks is also beneficial. Many users on public wireless networks do not fully understand the risks and potential threats associated with open public networks and are easily taken in by the lure of free wireless access.

The providing host of public Wi-Fi needs to be aware of the pros and cons that come with making WLAN available to customers and visitors but often is not. If the host is a coffee shop or small restaurant that provides free Wi-Fi to its customers as a business draw, they may not have a technical frame of reference for making good network choices. The general approach is often simply plugging a consumer-grade AP into an Ethernet port that has access to the Internet or using a low-end wireless router as the business' entire single-VLAN network infrastructure.

In these simple do-it-yourself scenarios, security features like client-to-client blocking or protocol/port blocking features are likely not even options, and there is little that can be leveraged to comply with regulatory requirements like PCI. Some hotspot providers may turn a blind eye to risks because they might view security as an impediment to the free hotspot service that is used to attract customers. If you are consulting in this kind of situation, you should encourage the business to implement a solution that provides a balance of security and usability to both their organization and the customers.

Other host networks, especially those that are fee-based, tend to use more elaborate enterprise-grade wireless equipment and will often follow a specific policy that was written for public Wi-Fi. This network may be handled by IT resources within the corporation, or it could be outsourced to a third-party service provider. As a CWSP, you may find yourself in either role at some point in your career.

Public access users should use common sense and follow best practices when connecting to a public host network. Some of the threats mentioned earlier can be mitigated if a bit of caution is employed, and proper client-side security features are installed, enabled, and configured. These client-side security settings include:

- Anti-virus software installed and up-to-date
- Firewall enabled and correctly configured
- Operating system updates installed and configured; this is important for all operating systems as none are immune to attack.
- No open or unsecured file system shares

- Use strong login credentials

- Only install mobile apps from trusted sources

- Watch for broadcasting SSIDs that clearly don't belong in the public environment- like your home network. Never connect to these.

- Use VPN

- As OWE clients come to market, try to use it as often as possible (will require the public network to support it as well).

Taking these steps into consideration will help to lessen potential security threats and should provide the user with a secure connection on the host wireless network. Unfortunately, many users don't give proper attention to their own devices, and CWSPs often have to educate on the topic.

Captive Portals

Captive portals (sometimes referred to as captive web portals, or abbreviated as CP) can serve many different functions depending upon the network provider's goals. They are often used to usher users through an acceptable use agreement, which arguably offloads some amount of legal liability to the service provider by clearly spelling out the terms and conditions associated with using their network.

Other implementations may be provided as for-profit services, in which the ISP will want to collect money in exchange for network access. For those network hosts who want to restrict network access to paying customers only, a captive portal can be used for authentication or to provide verification of services, such as with a receipt or customer number.

A captive portal works by redirecting a user to an authentication source of some type before they are allowed wireless network access. This authentication source, in the form of a web page, will require a user to "authenticate" in some way and may include the following:

- Enter user credentials (username and password)

- Input payment information

- Agree to terms and conditions

When one or more of these methods is complete, a wireless device will then be able to access the network and use whatever resources to which they have permission. Many public access wireless networks will have some type of captive portal enabled. A captive portal may help to protect both the provider (host) and the user of the wireless network. Many organizations use captive portals minimally, so the user agrees to the wireless network terms and conditions. For some environments, some hard-to-spoof identifier like a cell phone number may be required as the captive portal login by organizational policy. Most enterprise-grade wireless APs, including cloud-based APs and WLAN controllers, have built-in captive portal capabilities that are fairly straightforward to implement. Captive portals are also finding increased use in wireless NAC systems for various functions.

Captive portals have their place but can complicate the user experience. From the client perspective, some mobile devices, such as smartphones or tablets, may experience problems while trying to connect to a network with a captive portal enabled. This could happen because the mobile operating system is in use or other application related issues. For example, some devices have custom web browsers installed on them, and such browsers may not function properly when connecting to a captive portal. One common problem is with apps trying to synchronize before the user actually opens a browser and goes through the portal.

Most captive portals work such that when the user connects to the WLAN, DNS requests are intercepted so that no matter what URL the user attempts to access, they are directed to the captive portal web page. Additional interception methods may also be utilized to capture direct IP connection attempts.

Captive Portal Configuration

Captive portal configuration options allow administrators to specify all the specific parameters they would like to apply to their network from those available on the specific portal platform. These parameters may include VLAN segmentation, requiring an acceptable use agreement, and bandwidth and time limitations for users. In many captive portal implementations, system configuration is a fairly intuitive and flexible process. The precise steps to create a captive portal will vary based on the specific infrastructure device or software used. Here are some common basic steps in bringing a portal to life on the WLAN:

1. Set up and customize the captive portal functional flow and graphical elements
2. Create a WLAN profile; this is commonly the guest wireless profile
3. Do not enable any security features; this should be 802.11 Open System authentication
4. Assign the profile to the captive portal functionality

Once a user connects to the captive portal enabled SSID, they will need to perform additional steps to gain access to wireless network resources. This restriction includes opening a web browser and attempting to access any web page. When web access is attempted, the user will be "captured," or redirected to the specific, configured web page for the wireless network. The user will need to meet specified requirements, such as accepting terms and conditions or entering provided credentials. After the requirements are validated, the user will then be able to perform any task or access resources for which she has been granted.

Although not as common, in some cases Data Link Layer security such as WPA2 Personal or Enterprise may be used in conjunction with captive portal implementations. This configuration would provide some level of security (through encryption) for users that have connected to the captive portal network.

It is very important for those who use wireless networks with captive portals to be properly educated about the potential security concerns regarding their use. In the setup steps above, it was stated that the WLAN profile used would often be configured for Open System authentication. This setup results in all data being sent and received to the client device in clear text unless other security measures, such as VPN or other secure protocols, are used.

If the captive portal page required any kind of authentication such as a personal identification number (PIN) or other credentials supplied by the host, it might give the user a false sense of security. Because they are required to enter credentials of some sort, they may get the impression that the wireless transmissions are secure. It is important that the terms of service spell out what security is, or is not, provided for the captive portal.

Some wireless infrastructure device manufacturers also provide built-in services for the management of billing plans (just one example of how WLAN systems are now going beyond providing simple access). Other implementations use dedicated gateway appliances for this function.

Network Segmentation

Segmentation is a common networking technique used to limit the resources to which a device has access and is used for both LAN and WLAN implementations. Segmentation commonly includes the use of virtual local area networks, (VLANs), access control lists (ACLs), and firewalls to filter and funnel users to specific resources. For a guest network, the accessed target resource is usually the Internet and no "internal" network resources. By isolating guest clients to the Internet, Internet service providers (ISPs) prevent guest users from accessing and/or exploiting corporate network resources, which are in place only for corporate users.

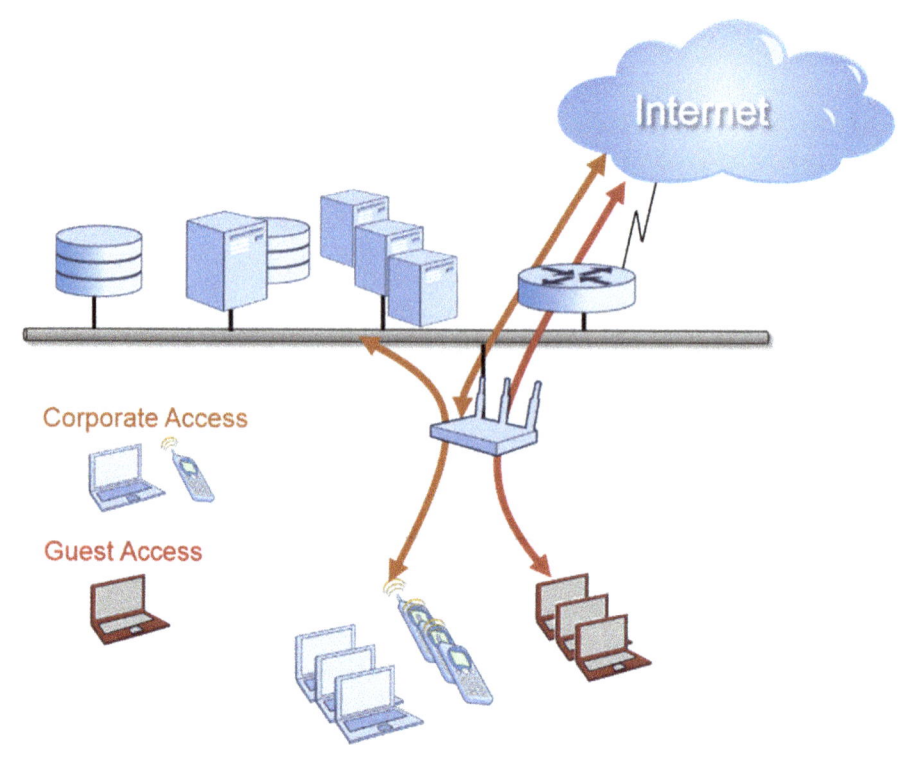

Network Segmentation

Most current network infrastructure devices have the capability to securely segment different types of network traffic. These infrastructure building blocks include devices that are used with wireless networking, including APs and WLAN controllers. If configured correctly and policy requires it, the corporate network traffic will remain completely separate for the guest network traffic. Segmentation of wireless traffic can be accomplished using role-based access control (RBAC) methods, or WLAN profiles with correct policies and access control.

As you can see in the preceding image (Network Segmentation), a single AP is used to provide secure wireless access to the corporate network using the corporate SSID and to the Internet using a separate guest SSID, with each usually tied to a specific VLAN. This configuration allows properly authenticated users to the corporate segment, to access both corporate network resources and to the Internet. Conversely, the guest SSID will only allow connected users access to the Internet, and guests will not be able to access secured corporate network resources. This is the prevailing model of typical guest access used for the exam, but CWSPs may be called on to be more creative with guest topologies in cases where guest privileges are more nuanced and may actually include specific internal resources.

In many cases, a captive portal is enabled on the guest wireless segment in order to restrict access to the guest SSID to only those who comply with whatever connection parameters are specified in the captive portal web page. Also, the captive portal might be configured to keep corporate users and devices that should be on the secure business WLAN, out of the guest topology. Portal platforms tend to be very flexible, so clear policy is required to know how to configure them for each specific situation.

VLANs provide an excellent management technique for wireless LANs. VLANs are supported on enterprise-class access points for the purpose of extending VLAN functionality all the way through network switching out to the mobile client device. 802.1q VLAN tagging is the only non-proprietary implementation available, so most WLAN devices use it.

While the full criteria for each wireless VLAN deployment are likely unique, some standard characteristics exist for most rollouts, including:

- Common applications used by all WLAN end users. The wireless network administrator should define:

- Wired network resources commonly accessed by WLAN users
- Quality of service (QoS) level required by each application
- Common devices used to access the WLAN. The wireless network administrator should define:
  - Security mechanisms (WPA2/WPA3/OWE/802.1X/EAP, VPN, etc.) supported by each device type
  - Wired network resources commonly accessed by wireless LAN device groups
  - QoS level needed by each device group

After specific wireless VLAN deployment criteria have been defined, the deployment strategy must be determined. Two standard deployment strategies are:

Segmentation by user groups: Segmentation of the WLAN user community and enforcement of specific access-security policies per user group. For example, three wired and wireless VLANs in an enterprise environment could be created for full-time employees, part-time employees, and guest access. In higher education, it may be a four-way division between students, faculty, staff, and guests as another example.

Segmentation by device types: Segmentation of the WLAN to allow different devices with independent access-security "levels" to access the wireless network. For example, it is not recommended to allow legacy handheld computers that support only 40/128-bit static-WEP to co-exist with other wireless LAN client devices using 802.1X with WPA2 in the same VLAN. In this scenario, devices should be grouped and isolated with different "levels" of access security into separate VLANs.

Implementation criteria such as those listed above are then defined to include:

- Use of policy filters to map wired policies to the wireless side
- Use of 802.1X to control user access to VLANs using either RADIUS-based VLAN assignment or RADIUS-based SSID access control
- Use of separate VLANs to implement different Classes of Service (CoS)

Wireless vendor Cisco Systems recommends several best practices for the wired infrastructure when 802.1q VLAN tagging is extended to access points and bridges.

These best practices are dependent upon the features being supported by the wired infrastructure to which the wireless devices are connected and reflect traditional networking versus Software-Defined Access topologies. The following are examples of what might be done (not hard and fast "rules" that apply to every scenario):

- Limit broadcast and multicast traffic to the access point and bridge by enabling VLAN filtering and Internet Group Management Protocol (IGMP) snooping on the switch ports. Where 802.1q trunks are used to extend VLANs to access points, filter to allow only active VLANs in the ESS. Enabling IGMP snooping prevents the switch from flooding all switch ports with Layer 3 multicast traffic.
- Map wireless security policies to the wired infrastructure with ACLs and other relevant mechanisms.
- The access point does not support Virtual Terminal Protocol (VTP) or Generic Attribute Registration Protocol VLAN Registration Protocol (GVRP) protocols for dynamic management of VLANs because the access point acts as a "stub" node. The wireless LAN administrator must use the wired infrastructure to maintain and manage the wired VLANs.
- Enforce network security policies via Layer 3 ACLs on both the "guest" and management VLANs.
    - The wireless LAN administrator could implement ACLs on the wired infrastructure to force all "guest" VLAN traffic to the Internet gateway.
    - The wireless LAN administrator should restrict user access to the native/default VLAN of the access points and bridges with the use of Layer 3 ACLs and policies on the wired infrastructure.

With wireless VLANs, each SSID is mapped to a default VLAN-ID on the wired side of the access point (or controller). The wireless LAN administrator may wish to impose RADIUS-based VLAN access control using 802.1X or MAC address authentication mechanisms. For example, if the wireless LAN is configured such that all VLANs use 802.1X/EAP and similar encryption mechanisms for wireless LAN user access, then a user can hop from one VLAN to another by simply changing their SSID and successfully authenticating to the access point (using 802.1X/EAP). This may not be preferred if the wireless LAN user is supposed to be confined to a particular VLAN. There are two different ways to implement RADIUS-based VLAN access control features:

RADIUS-based SSID access control: Upon successful 802.1X/EAP or MAC address authentication, the RADIUS server passes back the allowed SSID list for the wireless LAN user to the access point or bridge. If the client device used an SSID on the allowed SSID list, then it is allowed to associate to the wireless LAN. Otherwise, the user is disassociated from the access point or bridge.

RADIUS-based VLAN assignment: Upon successful 802.1X/EAP or MAC address authentication, the RADIUS server assigns the user to a predetermined VLAN-ID on the wired side. The SSID used for wireless LAN access doesn't matter because the user is always assigned to this predetermined VLAN-ID.

In order to have RADIUS return the appropriate attributes to the access point, the RADIUS server must implement the AP vendor's Vendor-Specific Attributes (VSA) that define the allowed SSIDs or static VLAN assignment. As you can see from the explanation above, wireless VLAN functionality gives the access point somewhat similar functions to wireless middleware (controllers) while maintaining infrastructure security at the network edge (the AP). The use of Vendor-Specific Attributes can be very powerful when complex WLAN security requirements are in play, and CWSPs should understand the concepts in play even if they are not using them in a given environment.

BYOD

The concept of bring your own device (BYOD) has taken deep root over the last decade or so. When a user works for an organization that allows BYOD, there will be a number of specific security concerns that must be dealt with. As you know, the core premise of BYOD is that a user will access the corporate wireless network in addition to home and other public access networks from a device that is the personal property of the employee. Personal devices will generally have access to both corporate and personal apps and data. Segmentation is often used to separate BYOD client devices (and certainly guest users) from the corporate users.

Security policy is very important when it comes to BYOD due to the potential security threats that come with this usage paradigm. All of the previously mentioned concerns associated with other types of client devices and public access networks also exist in BYOD situations, along with more risks that are specific to BYOD's unique framework. Concepts such as *data containers* should be considered. Tools are available, allowing network administrators to logically separate corporate data from personal data on the same device (containerization) according to well-defined security policy. Containerization will help to ensure the integrity of the corporate data as it lives on personal BYOD devices by portioning the BYOD device into at least two distinct operational environments. Tools also allow for remote access and control features such as remote lock, unlock and remote wipe features to help ensure the integrity or verifiable destruction of the data in the event a device is lost or stolen.

As a reminder, BYOD, in many cases, fosters personal satisfaction and potentially increases productivity because the users are allowed to use their own devices for organizational business. This concept provides a certain comfort level because BYOD users are already familiar with the hardware now being used with corporate apps. As a bonus to the users, they are also able to access their own personal data and apps from their own devices, when allowed, as specified in corporate security policy.

Onboarding is an important additional process related to BYOD and a topic that CWSPs must firmly grasp. Onboarding allows devices to join the network through either pre-registration or self-registration. With pre-registration, the devices are added by an administrator before the users can connect. With self-registration, the user goes through a process which allows them to gain access to the network in a secure manner. Self-registration is often done via a web form or a portal.

MDM

Mobile device management (MDM) solutions are used to manage devices that are connected to a corporate wireless network. Because of the increase in popularity of mobile devices in general, MDM solutions have matured into an industry of their own.

Some manufacturers of enterprise wireless infrastructure devices will integrate MDM solutions within their platforms, while others will partner with third-party companies that specialize in the technology. Many MDM solutions are available as both on-premise and cloud-based Software as a Service (SaaS) solutions. It is important to ensure that the one selected and used will meet the needs of the organization, especially when it comes to security.

MDM solutions typically provide a plethora of BYOD-oriented security strategies and features that will help to ensure corporate security policy is followed and maintained. Some of these security features include, but are not limited to:

- Compliance reporting

- Device registration (self-service portal)
- Location-based services
- Password control
- Policy enforcement
- Remote lock, unlock and wipe
- Secure communications (virtual private network)
- Location-based services
- Geofencing
- Geolocation
- App management, containerization

Choosing the best MDM solution to use in your environment is an important decision to make. Evaluation of these products and the security features that they provide is something that needs to be closely considered before the purchase and implementation of the solution, as not all solutions will have the same look, feel, and administrative capabilities.

If MDM does get rolled out, formal change-management is a must to minimize the pain of suddenly controlling what mobile devices can do. Change management will help with an orderly deployment, documentation of what devices were impacted by what security controls, and effectively communicate to management and end-users all details of the change for the sake of business continuity. (For that matter, the change control process is important to any new wireless security step being implemented or removed.)

Client Management Strategies

MDM is one client-side management strategy that can be used to manage both personal and corporate-owned mobile devices, but there are others to consider as well. These options offer a range of interesting functionality for client management. Although different solutions may contain common feature sets, it is still important to evaluate various solutions in order to choose the one that is best for the organization's specific needs. For example, MDM may be used in conjunction with other management options such as a WLAN management system (WNMS), to provide a complete wireless network management solution.

Client management strategies will depend on the type of client devices used on the network, the features and capabilities of those devices, and the corporate security policy that is in place. Like MDM solutions, manufacturers of WLAN infrastructure devices may offer integrated client-side management features. Most management products provided by manufacturers are created to work only with their own network components but will usually be able to manage a variety of different client devices. Others provide a more vendor-neutral approach that will work with any manufacturer's infrastructure devices. As WLAN systems get ever more feature-rich, multi-vendor WLAN management strategies become scarcer on the market.

Depending on the organization and the user population, different strategies can be used when it comes to managing connected devices. In some circumstances, the user may have complete access and full control over the wireless device, in which case they will be personally responsible for maintaining proper security control. In other situations, the user will have little or no administrative control, and the network manager is responsible for ensuring proper posture and security compliance.

How corporate security policy is written, and the specific security goals of the organization, will influence the management of client devices. This includes what hardware, applications, and behaviors are either permitted or disallowed. Options such as remote access, the use of removable devices, installation or removal of applications, and operating system permissions are a few examples that will vary according to corporate security policy.

Chapter Summary

In this chapter, you studied several specific design scenarios, including the use of VPNs to secure public access. Additionally, you learned about captive portals and their basic operation. You also explored client management strategies and the factors that impact the selection process related to client management solutions.

Review Questions

1. Choose the option that specifies two valid VPN modes used to secure communications on wireless networks.
    a. Tunneled and overlay
    b. Tunneled and transport
    c. Transport and split
    d. Split and tunneled

2. Which of the following is not a valid defense against attack for a wireless user on a public access wireless network?
    a. Use of VPN
    b. Personal firewall
    c. OWE
    d. Wireless controller

3. Of the following VPN protocols, which is known to have been cracked and is therefore typically not recommended for use in enterprise wireless environments?
    a. PPTP
    b. L2TP/IPSec
    c. SSH
    d. SSL/TLS

4. Which one of these is not a common use of wireless VPN?
    a. Site-to-site tunnels
    b. Remote APs
    c. Wireless bridging
    d. Remote client connectivity

5. Split-tunneling has a major disadvantage. What is it?
    a. Data is encrypted
    b. Increased processing overhead
    c. Exposure to the public network
    d. Encryption keys must be made public

6. Using open public wireless access networks exposes users to a number of risks. Which of the following is typically not one of those risks?
    a. Weak passwords
    b. Unsecured file shares
    c. Improperly configured firewalls
    d. Poorly constructed WPA2 pre-share keys

7. There are typically three steps to building a VPN connection. One of the following is not part of that process. Select the option that isn't part of building a VPN connection.
    a. Perform required authentication
    b. Build a virtual tunnel
    c. Traverse captive portal
    d. Encrypt the data

8. What WLAN security posture is typically present on SSIDs that make use of captive portals?
    a. WPA2 Pre-share
    b. WEP
    c. VPN
    d. Open System Authentication

9. Which of the following is used to segment guest traffic from corporate data?
    a. ACLs
    b. Firewalls
    c. VLANs
    d. All of the above

10. What standard is associated with VLANs?
    a. 802.1q
    b. 802.11q
    c. 801.2q
    d. 802.1X

11. What aspect of a captive portal implementation is leveraged to protect the service provider from liability?
    a. Walled garden
    b. Encryption
    c. Terms of service
    d. VPN statement

12. Which of these is not a fundamental component of the VPN?
    a. Server-side endpoint
    b. Network infrastructure
    c. Client-side endpoint
    d. Security tunnel broker

13. What should guide every security solution, including VPN, captive portals, VLANs, and MDM implementations?
    a. Security policy
    b. Client behavior
    c. Government regulations
    d. Client device capabilities

14. The ability for network administrators to remotely lock or wipe a lost or stolen device is a functional feature of _____.
    a. Security policy
    b. MDM
    c. Roaming admin platform
    d. Super user role

15. The ability to automatically place an authenticated user on a specific VLAN is enabled by which of these?
    a. VLAN attributes
    b. 802.1X attributes
    c. Policy attributes
    d. RADIUS attributes

Review Answers

1. **B.** Tunneled mode and transport mode are two common VPN modes

2. **D.** There are a number of recommended practices to protect individual wireless users, but wireless controllers are infrastructure devices

3. **A.** PPTP is considered a cracked VPN protocol at this point

4. **A.** Site-to-Site VPN tunneling is common, but not generally applicable to wireless VPN

5. **C.** Split tunneling exposes traffic directly to the Internet

6. **D.** WPA2 is usually not utilized on open public networks

7. **C.** Captive portals have nothing to do with VPN per se

8. **D.** Captive portals generally don't use encryption

9. **D.** ACLs, VLANs, and firewalls are all common segmentation methods

10. **A.** 802.1q is the VLAN standard for wired and wireless networks

11. **C.** Terms of Service are included in Captive Portal pages to protect the WLAN provider from liability

12. **D.** VPN requires server-side endpoint, client-side endpoint, and network infrastructure

13. **A.** The security policy is the overarching touchstone for any potential security solution and can't be left out of the planning process

14. **B.** MDM's remote wipe capability is a powerful security feature for lost or stolen devices

15. **D.** RADIUS attributes can steer users to the VLAN that policy says they should be on

Chapter 8: Secure Roaming

Objectives Covered

3.4   Implement secure transitioning (roaming) solutions

When 802.11 was new, the focus was on simply providing wireless access to laptop computers (which at the time were not exactly mainstream yet) and the occasional desktop PC. Since the adoption of 802.11a/g, the entire usage paradigm for Wi-Fi has changed. We've come a long way since basic client access, as an accessory to Ethernet was the only purpose of the WLAN. Modern business WLAN is far more complicated, and many critical and/or latency-sensitive applications are used over Wi-Fi. With the promised speeds of 802.11ax, it's reasonable to expect that wireless will further become the access method of choice in even more environments. Every facet of the WLAN has evolved, and as device and application sophistication increases, so does the complexity of WLAN configuration.

Today, WLANs are accessed by laptops, desktops, tablets, specialty devices, mobile phones, and VoIP handsets. While all wireless devices can benefit from properly designed roaming between Wi-Fi cells, VoIP phones require it in nearly all scenarios for the voice communications to be usable. The roaming from one AP to the next must be fast, within a certain metric, or call quality suffers, and it must allow for security throughout the roaming process to protect the privacy of the conversation.

The primary focus of this chapter is on available roaming techniques in current vendor solutions that are based on both standard and proprietary technologies. We'll also look briefly at the latest roaming capabilities introduced by 802.11r, which is now part of 802.11-2016. We will see why even though .11r is a valid and standardized option, many vendors still rely on earlier developed roaming techniques.

IEEE 802.11 Roaming Basics

In simplest terms, an 802.11 roam includes the movement of a client association from one AP to another. Of course, it is not as easy as simply "moving" the association. The new AP must also authenticate the roaming client, and dynamic encryption keys must be established along with temporal encryption keys (assuming the client is not on an open network). These processes take time, especially in WPA-Enterprise or WPA2-Enterprise networks. Therefore, special roaming procedures are required to facilitate fast roaming. The 802.11 standard refers to this as a *transition*; however, the term "roam" is more common across the WLAN industry.

When a wireless client device moves an association from one AP to another, the process can be straightforward, but it may also be quite complex depending on the specific scenario. You learned in earlier chapters that every time a WLAN client device connects to an AP, it must perform an 802.11 Open System authentication and then an association. This process is what provides the Physical Layer and Data Link Layer connections to the network.

If a client device roams from one AP to another, it will have to perform what is called a *reassociation*. Though 802.11 association and reassociation frames are almost identical, CWSPs must recognize the subtle differences. A client device can only be 802.11-associated to one AP at any one time; therefore, moving the association, or *reassociating*, is required when a device moves to another AP.

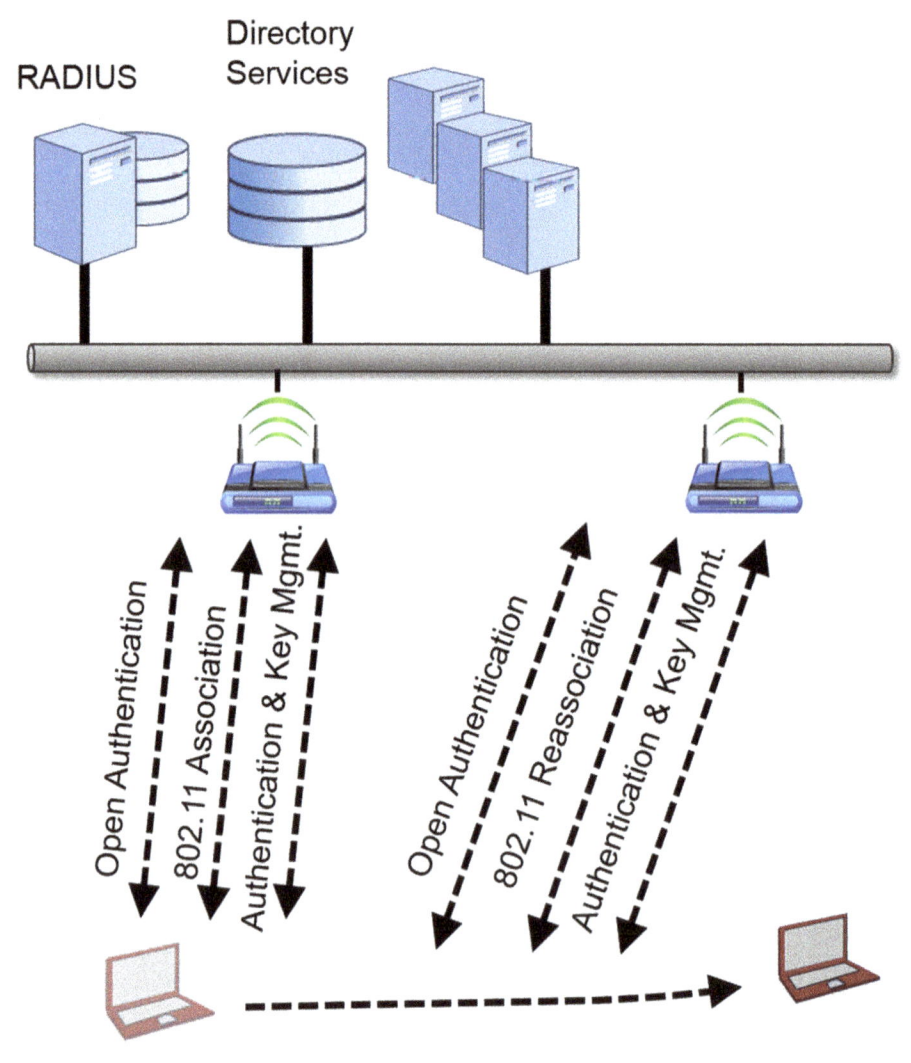

Basic Roaming Flow

Before the reassociation can happen, the client must also first perform an 802.11 Open System authentication. Reassociation happens as a device moves among the APs that make up the ESSID of the WLAN, after the initial association. As described so far, this process pertains to an open network connection without any Layer 2 security features. If any 802.11 standard security features are enabled, the reassociation process gets more complicated as you will see later in the chapter.

To assist your understanding of roaming, you should be aware of the following terms referenced in earlier chapters, but playing a larger role here:

- **PMKSA** - Pairwise Master Key Security Association. The context resulting from a successful 802.1X authentication exchange between the peer and Authentication Server (AS), or from a preshared key (PSK).

- **PMKID** - Pairwise Master Key Identifier. The PMKID is an identifier of a security association.
    - PMKID = HMAC-SHA1-128(PMK, "PMK Name" || AA || SPA)

- **PTKSA** - Pairwise Transient Key Security Association. The context resulting from a successful 4-Way Handshake exchange between the peer and Authenticator.

PSK Roaming (WPA/WPA2)

In previous chapters, you learned about 802.11i, now also incorporated into 802.11-2016, and WPA/WPA2 personal and enterprise modes. Remember that if an 802.11 wireless network is configured for WPA or WPA2 personal mode, the wireless client device will perform a 4-way handshake after the 802.11 Open System authentication and association has completed. Also, recall that the purpose of the 4-way handshake was to allow wireless devices that are connecting together (i.e. AP and client device) to exchange some keying material in order to create the keys required to encrypt unicast and broadcast/multicast traffic for that device.

The following image shows a capture of the 802.11 Open System authentication and association, in addition to the 4-way handshake. Notice that the relative time, which is the time it took from the first authentication frame to the final acknowledgment frame, was 44ms in this example. It's important to note that this time is not as operationally significant in an initial 802.11 authentication and association as it is when a device roams from one AP to another. The reason is this: at initial association, the client device is being on-boarded to the WLAN, with no application traffic yet passing. But during a roam, the client is passing data and fully engaged in network operations. If the time it takes to connect to a new AP and perform the necessary frame exchanges for wireless security is too long, chances are, the client device will have to perform an entirely new authentication process. For some wireless applications, such as Voice over IP, this can cause serious performance or connectivity issues. Simply put; faster roaming is better.

IEEE 802.11 wireless networks supporting WPA Personal (which hopefully is no longer in use much) and WPA2 Personal typically do not require any special enhancements for fast roaming between APs. A full authentication typically takes less than 50ms, so there is no need to improve upon this time. In this case, there should be no roaming issues.

Time for Authentication, Association, and 4-Way Handshake

But what happens when what is known as a *slow roam* occurs in a WLAN that is configured for 802.1X/EAP with no fast roaming features? In a standard roaming environment, also known as a slow/secure roaming environment (when you are dealing with a robust security network (RSN)), each reassociation requires a full 802.1X/EAP reauthentication. This requirement is especially true in autonomous AP environments but is often the case in controller-based and other environments as well. If there are no Fast BSS Transition (FT) protocols in place on the supplicant (client device) and authenticator (AP or WLAN controller), each reassociation may take a painfully slow 500ms or more, depending on a number of variables. At best, un-enhanced roaming, almost always, takes more than 200-250ms. This can be problematic.

A roam that takes 500ms is much too slow to maintain the integrity of the Wi-Fi connection. Many wireless networking best practices recommend roaming times to be less than 150ms maximum. Longer roam times will cause issues for the wireless client device that is making a move, leading to user confusion and trouble tickets.

These steps occur during the slow roam process:

1. Open System authentication

2. Association

3. 802.1X/EAP authentication

4. 4-way handshake

As you can see, the same basic processes occur as those in the preceding image (time for Authentication, Association, and the 4-Way Handshake), which required just under 50ms alone, but the added requirement of 802.1X/EAP authentication is significant in the time it adds. Keep in mind, the time delay between the AP and the client STA is no longer the only consideration. Now, the time delay between the AP or controller and the RADIUS server must also be considered, along with added time for interaction between the RADIUS server and LDAP server where individual user credentials are stored. In addition, processing delay must be considered, as the RADIUS server may be overtaxed at times, or the LDAP server may slow in high-usage periods. With all of these individual opportunities for processing delay, you can see why full, slow roams can take more than 500ms to complete.

VoIP clients are among the most susceptible to issues that come from long roam times. This is because real-time VoIP protocols have stringent requirements for providing toll-quality calls (calls as good as on the PSTN). Voice data packets come with demands that are not seen in traditional delay-tolerant data packets.

For example, if you are sending a file to a server via FTP, it doesn't matter if a few packets arrive out of order, or if a delay occurs between the arrival of one packet (that is greater than a particular threshold) and the next. If an extended delay occurs, it will certainly slow down the communication, but the data will eventually arrive at the destination with no ill effects and likely without the human user even noticing. However, voice traffic will not tolerate such occurrences.

If an extended delay occurs with VoIP, the call will be dropped, or its quality will suffer. With the PSTN quickly falling, in favor of cellular and VoIP, the need for fast, secure roaming is only increasing.

Humans are at both ends of a Voice over IP (VoIP) communications link. They will both talk and listen, and they have expectations that have been set by the analog telephone network. If they do not hear anything for some variable length of time, they assume the call has been dropped, or the person on the other end has disconnected. If the sound quality is inferior, particularly to the point where they cannot understand one another, they may give up on the conversation. Expectations of session quality and predictability of performance exist with voice over Wi-Fi. Those expectations must be met with VoIP data even though they have not traditionally been required of other data types.

Because VoIP packets are transmitted over the same physical network as traditional application data packets (such as e-mail, database access, file transfer, printing, etc.), you can say that voice is layered over the data network. The same network devices, cables, and software that are used for traditional data are also used to transfer voice data. This layering places a new demand on the network. The data network must be able to differentiate between packet types and give priority to voice packets in order to meet the quality expectations of the VoIP users. The technology that enables this differentiation is known as Quality of Service (QoS) and is beyond the direct scope of this study guide (though it is addressed, in detail, in the CWDP Official Study Guide). Additionally, fast and secure roaming is needed on the wireless network.

Because voice traffic must move rapidly and without excessive delay across the network, and because retransmissions of lost or corrupted voice packets would provide no functional value given the real-time nature of voice, UDP is used to send most VoIP data packets. UDP is a connectionless protocol, unlike TCP. TCP has far too much overhead to transmit voice packets as rapidly as they must be transmitted.

If you have not worked with VoIP, you may wonder why there is no benefit from resending corrupted or lost voice packets. The reason is simple, and let's look at a theoretical example to make the point. Think about how long it takes you to say the word "don't". If you're like most people, it will take you far less than a second. Now, imagine you're having a conversation on a VoIP phone, and you say the following sentences, "Don't push the button. Pull the lever." Next, further imagine that the word "don't" was lost in transmission and the system decided to resend it. Because of the sequencing problem, the user, on the other end, hears the following, "Push the button. Don't pull the lever." This reordering could theoretically happen because the phrase "push the button" made it through while the word "don't" didn't make it through. When the word "don't" was retransmitted, it was placed before the phrase "pull the lever." The result is the transmission of what amounts to the complete opposite of the intended message. Do you see why retransmitting lost audio packets would be useless and possibly damaging?

In reality, this exact scenario is not likely to occur, even with retransmissions. It would definitely not occur in real UDP-based VoIP implementations because they do not retransmit lost packets. Instead, the listener would likely just not hear the word "don't"; however, in reality, it all gets a bit more complex. More than likely the listener would hear something like, "D---t pu—the ---ton. Pu-- --- --ver." The dashes represent either sounds that are unintelligible or complete silences. The point is that the network doesn't usually drop exact words, but rather portions of audio much less than a complete word. This results in what we usually call a "bad connection."

So why does a CWSP need to understand Voice as a network application? We're working towards understanding non-disruptive, secure roaming with this discussion, and so this certainly falls within the scope of the wireless security professional. VoIP problems are rarely related to bandwidth or throughput; rather, they are related to a more precise term: *delay*. When the delay is excessive in a WLAN link, the call can suffer in quality, or simply be dropped. It is true that bandwidth and modulation can certainly impact delay because they limit the total amount of data that can be transmitted, but the network can typically be engineered to sustain voice traffic. With a limited number of allowed concurrent calls, a properly configured wireless network can provide call quality given even old 802.11b data rates. But when you introduce roaming-based delays, all of the other engineering benefits are lost as soon as users begin to move around the Wi-Fi environment while talking.

VoIP systems typically require less than 150ms of unidirectional delay. The ITU-T recommends in Recommendation G.114 that the round-trip time (RTT) or round-trip delay not exceed 300ms in a telephone network- 150ms in each direction.

So, what are the options for meeting the timing requirements for wireless VoIP? You can do one of three things to reduce delay, using the traffic on a highway as an analogy:

1. Install more generic lanes
2. Identify one or more lanes as high-occupancy lanes
3. Teach people to drive better

Installing more generic lanes will increase the lane-width (similar to network bandwidth) and should reduce your delay. However, this solution is a very expensive one and can slow down traffic while the extra lanes are being installed. If we apply this concept to the data network, we face the same negative impacts. If we simply install more APs with the same efficiency as current ones, we may reduce delay in the WLAN while greatly increasing the cost of doing so (or we may create new problems if the additional APs create co-channel interference).

The second option of identifying a lane exclusively for high-occupancy vehicles (HOV), will not have the same overall impact as adding more lanes, but it will be far less expensive and should not result in as many delays during initial implementation. Installing HOV lanes is analogous to implementing QoS on a network or implementing fast, secure roaming techniques in a WLAN. QoS and roaming methods cannot increase the available data rates as defined by the 802.11 standards, but it can indicate that specific traffic types have a higher priority than general traffic.

The third and final option for managing priority traffic is probably the most difficult in real life. "Teaching people to drive better" amounts to enticing people to work together so that all drivers arrive at their destination in a window of time that is relatively acceptable to all. Put another way; one driver is not gaining an advantage while he or she delays other drivers. Ultimately, we're talking about implementing better collaborative driving algorithms. Thankfully, while this task is very difficult with human drivers on the highway, it is fairly simple to accomplish with data that is transmitted across our wireless networks.

 Although they may be using the same 40 MHz of frequency bandwidth, remember that 802.11ac provides a higher throughput and data rate than 802.11n. You could say that both 802.11ac and 802.11n use 40 lanes for data transfer, but 802.11ac potentially gets more data through. How is this difference in throughput possible? The answer is that 802.11ac drives better, with more efficiency. And… 802.11ax drives even better than 802.11ac.

Basic Roaming Review
Basic roaming works in one of three primary ways, depending on the specifics of each Wi-Fi environment:

- Layer 2 roaming across APs within a single controller, or without a controller
- Layer 2 roaming across APs connected to separate controllers
- Layer 3 roaming

When Layer 2 roaming occurs, the IP configuration on the device in motion is NOT lost. With the same IP address and roaming times of typically less than 40 milliseconds, Layer 2 roaming on non-secured open WLANs can support streaming technologies like VoIP. Layer 2 roaming across APs within a single controller is called *intracontroller roaming*. Layer 2 roaming across APs connected to separate controllers is called *intercontroller* roaming. Vendors handle the actions that take place within or between the controllers according to their proprietary algorithms (this can be frustrating for WLAN professionals who frequently lament that roaming ought to be standardized for the greater good, but, at this point, it is what it is). The 802.11 standard defines only WHAT should take place as a client STA roams from one AP to another, and they do not specify exactly HOW the communications must occur within the infrastructure. This flexibility allows the vendors to provide competitive features in this area while working against the mixing of different vendors' network hardware in the same WLAN. With the demand for VoIP support, some WLAN solutions are eliminated from consideration if the vendor's roaming solutions are inefficient. The good news is that the major WLAN vendors all have both intracontroller and intercontroller Layer 2 roaming solutions that can accomplish the roaming speeds required for wireless voice over IP. It's also worth noting that vendors are supplying ever-larger controller options which may help keep more roaming actions intracontroller.

Layer 3 roaming occurs when the client STA roams to an AP that cannot provide the same IP configuration because that AP is located on a different wired network (or uses a different L2 space for the same SSID). In this roaming scenario, the new IP address must be reallocated from the DHCP server and the client STA is placed on a new network subnet. The problem with this simple operation is that the client's Layer 3 connections will be lost, which is likely to be disruptive to any network host actions in progress. If a file was in the process of copying from the client to a server, the file copy process would most likely have to be started again from the beginning. The same is true in the reverse scenario, where the file is copying from the server to the client. While this situation is painful for the users, it cannot begin to compare with frustrations of dropped calls due to Layer 3 roaming on Voice over WLAN phones. To solve this problem, *fast, secure roaming* must be implemented, and the APs across which users would roam must somehow be part of the same wired network or provide some other solution to this problem. The access points may be part of the same wired network through tunneling solutions within the infrastructure, or they may simply be connected to the same controller, but they must somehow allow the client STA to maintain its IP address during the roaming process.

Implementing Layer 2 roaming without IP configuration loss is not actually very difficult on open wireless networks. This capability has been available for many years on networks with no wireless security. But previous to 802.11i and 802.11r, WLAN administrators faced difficulties in implementing a standards-based secure wireless network that offered very fast roaming. For enterprise secure wireless networks to support fast roaming, the 802.1X authentication process must be somehow accommodated in a modified, rapid fashion so the user can roam without requiring a complete 802.1X authentication exchange to occur. This is typically accomplished using some form of *key caching*, as discussed in the later sections of this chapter.

Finally, in order for roaming to work in a seamless manner, the coverage cells of the same SSID from WLAN access points must overlap. Though this is more of a design issue than it is security, CWSPs must also understand the importance of proper cell overlap. If there is no point of overlap, the client STA will always lose network connectivity for a brief time as the user moves the STA across the non-covered area. Vendors typically recommend cell overlaps ranging from 15 to 30 percent. Please realize that you can't really measure overlap with precision, but you can determine how many APs can be seen from a given location. Therefore, the goal should be to have at least two visible and usable (meaning sufficient signal strength and SNR) APs at any location where real-time devices may be used.

As a side note, when a user moves his or her laptop around within the coverage area of a single AP, roaming is not required. The user has mobility, but no roaming occurs while the connection to a single AP is maintained.

Wi-Fi Voice-Personal Certification

Wi-Fi Certified Voice-Personal is an optional Wi-Fi Alliance certification that wireless devices may acquire after they are certified for the basic 802.11 standards with which they comply (that is, they have been certified as 802.11n, 802.11ac, or 802.11ax compliant by the Wi-Fi Alliance). The goal of the Voice Personal certification is to provide a certification of compatibility across vendors for Wi-Fi phones and infrastructure devices such as APs. It is referenced here as an example of the network demands required, related to latency or delay features. Per the Wi-Fi Alliance website:

Voice - Personal: Voice over Wi-Fi - extends beyond interoperability testing to test the performance of products and help ensure that they deliver good voice quality over the Wi-Fi link.

As you can see by this definition, the goal is to show that interoperable devices can also perform well with voice communications traveling across the wireless network when the network is in proper order. The Voice-Personal certification requires devices to be tested in a test network consisting of the following:

- A single AP, serving multiple clients which may include PCs, phones, gaming devices, printers, etc. Handoffs between APs are not tested in this certification program.
- Supports at least four simultaneous simulated voice calls from four voice devices associated with the same AP.
- Carries data and video traffic to and from multiple devices, in addition to voice traffic.
- Assigns priority to voice over traffic from other applications.
- Provides security-protected access through WPA2 - Personal.
- Supports power-saving capabilities in the AP and in battery-operated client devices.
- Has a single Internet connection.

Clearly, this test network is not as complicated as an enterprise-class network. But remember, the certification is called Voice-Personal, and not Voice-Enterprise. Eventually, the Wi-Fi Alliance developed a similar certification that tests large scale voice over WLAN deployments called Voice-Enterprise.

In order to acquire the Voice-Personal certification, the devices tested in the network must achieve the following minimum performance requirements:

- Packet loss of less than 1 percent.
- Less than 50 milliseconds of latency.
- Less than 50 milliseconds of maximum jitter.

According to the Wi-Fi Alliance, "If a device does not perform to these levels, the voice call may drop in and out, may end suddenly, or the conversation may suffer from excessive delays, making it unintelligible. Products that do not meet these requirements will not receive Voice-Personal certification."

In addition to the performance requirements within the test network, devices must meet the minimum Wi-Fi Alliance certifications to even be considered for testing:

- 802.11a, 802.11b or 802.11g
- WPA2-Personal (notice WPA-Personal is not allowed)
- WMM
- WMM-Power Save (this is only required for APs and is optional for client STAs)

The Voice-Personal certification only validates the portion of the voice communication that occurs on the wireless link. The end-to-end performance of voice communication as it traverses the wired network is not tested or guaranteed based on Wi-Fi Alliance product certification. This is an important point, as many reported "wireless" problems actually occur elsewhere on the network.

Wi-Fi Voice-Enterprise Certification

Let's discuss the Voice-Enterprise certification briefly, to help you understand the impact of fast roaming requirements when they are added to the business WLAN picture. The Wi-Fi Alliance defines a typical enterprise voice network as:

- APs and STAs are 802.11-compliant members of an enterprise Wi-Fi network.
- The Wi-Fi network is likely to have other traffic while voice calls are active.

- The network consists of multiple APs and a variety of STAs, like voice handsets, PCs, printers, etc.

- The system is designed to support multiple concurrent voice calls.

- Wireless security is high and likely includes a RADIUS infrastructure.

As described for this environment, the target performance of a Voice-Enterprise certified solution is 50ms handovers (roams)- although breaks of up to 100ms may be acceptable. This is based on the reality that 50ms is probably less than four 20ms speech frames.

Additionally, Voice-Enterprise requires that some 802.11r technologies be implemented, as well as 802.11e (QoS) and 802.11k (radio resource measurements) to allow for effective VoIP operations. This performance-specific certification became available in 2012, and it's incumbent upon the WLAN management staff to know whether specific hardware is fully certified or configured according to the certification.

Remember, we're well into the era of Wi-Fi calling, the draw-down of the PSTN, and an overall swell in the use of the WLAN for almost every kind of application, including Voice. Though VoWLAN won't be present on every wireless environment you may support, it is quite common and important to configure, properly given the fundamental importance of voice communications to all organizations.

> Troubleshooting roaming problems for voice communications on Wi-Fi requires specific hardware and software. A protocol analyzer and multiple supported adapters used to simultaneously monitor different channels may be used in such cases, to analyze roaming. Additionally, multiple laptops could be used with later merging of the separate protocol captures.

Preauthentication

Preauthentication is the first way of removing delay from the roaming process. It is used by a wireless station that hears other APs to which it may choose to connect during the scanning process. The full 802.1X/EAP authentication is performed over the Ethernet infrastructure for the purpose of remaining on-channel with its current AP while preparing for connectivity with another AP.

Preauthentication is an IEEE standardized, fast secure roaming (FSR) method. Because of this, interoperability should be effective. Nonetheless, preauthentication has the drawback of requiring a full 802.1X/EAP authentication for each potential AP the client might consider roaming to. This stipulation requires the client to perform predictive authentications, which can add unnecessary traffic to both the wireless and wired mediums, as well as to the backend authentication infrastructure.

CWSPs must remember that preauthentication must occur over the Ethernet medium. EAPoL frames use non-standard, Ethertype values and are treated as standard data frames and forwarded to the distribution system (DS). A special Ethertype value (88-C7) is specified for use by the 802.11 standard for wired-side (Ethernet) communications of the roam.

The strengths of preauthentication are:

- Standardized by the IEEE
- Can be supported on any WLAN architecture
- Performed prior to roaming and allows for preauthentication with many different nearby APs

The weaknesses of preauthentication are:

- Still requires 802.1X/EAP authentication after association
- Is not considered an efficient solution as it preauthenticates to APs it may never touch
- Must happen prior to the roam
- Doesn't scale well
- Only trims 1 to 3ms off of the roam time

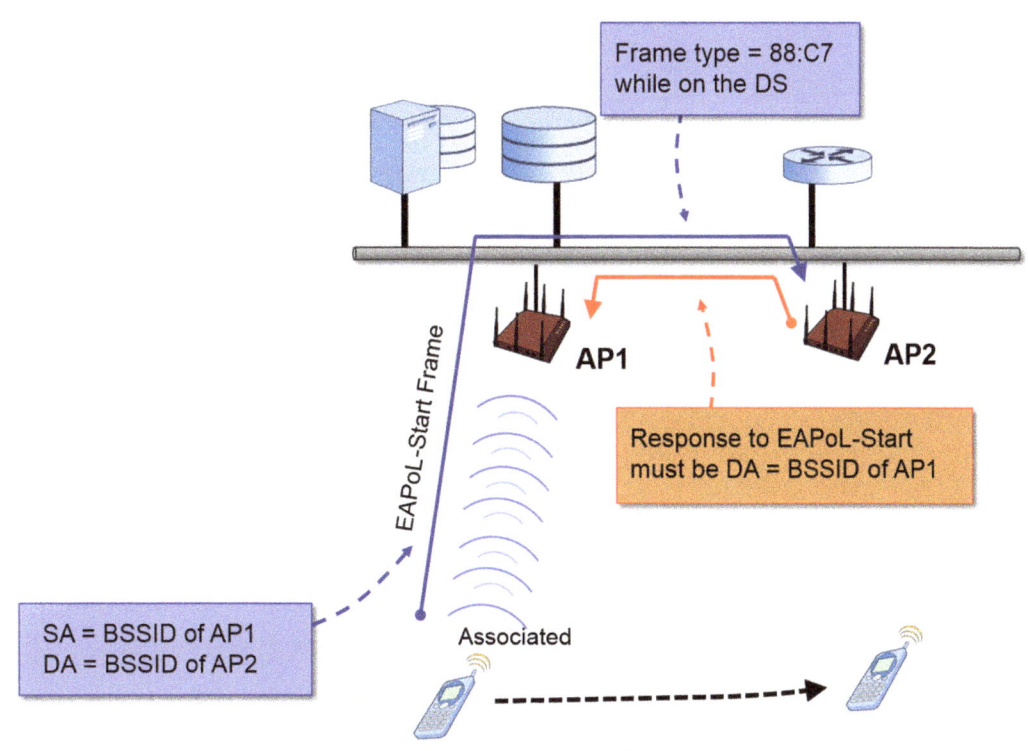

Preauthentication Architecture

PMK Caching

Pairwise master key PMK caching is also known as Fast Roam-Back. In the following image, you can see the steps required for this type of roaming to occur. The following paragraphs explain the process.

The 802.11 standard allows pairwise master key security associations (PMKSAs) to be cached at the AP (or WLAN controller) and on the wireless station for the purpose of fast roam-back. When a PMKSA is built (through a full 802.1X/EAP authentication) with an AP, the station and AP may continue to use that PMKSA at any point in the future when the station might roam back to the AP in which it was previously associated. The purpose of this feature is to avoid the slow 802.1X/EAP reauthentication process. To implement this feature, the client station must include the appropriate pairwise master key identifier (PMKID) in the Reassociation Request frame when it reassociates. Provided the AP still has the PMKSA cached, 802.1X/EAP authentication will be skipped, and the 4-way handshake will immediately ensue.

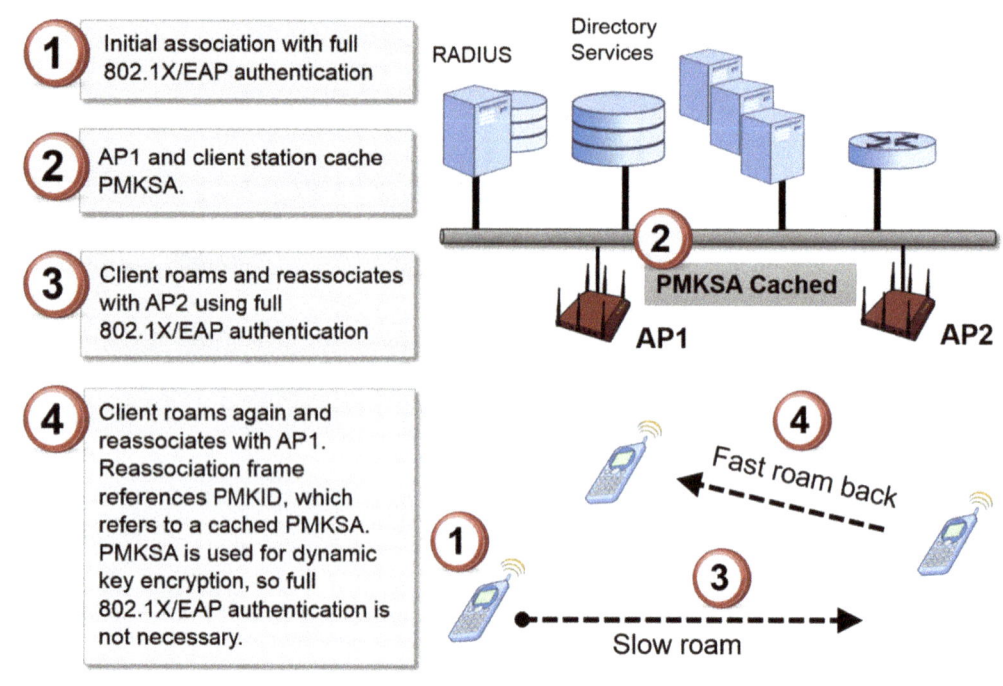

PMK Caching Illustrated

1. Initial association with full 802.1X/EAP authentication
2. AP1 and client station cache PMKSA.
3. Client roams and reassociates with AP2 using full 802.1X/EAP authentication
4. Client roams again and reassociates with AP1. Reassociation frame references PMKID, which refers to a cached PMKSA. PMKSA is used for dynamic key encryption, so full 802.1X/EAP authentication is not necessary.

From IEEE 802.11:

Cached PMKSAs and RSNA key management

"In a non-FT environment, a STA might retain PMKSAs it establishes as a result of previous authentication. The PMKSA cannot be changed while cached. The PMK in the PMKSA is used with the 4-Way Handshake to establish fresh PTKs.

If a STA in an ESS has determined it has a valid PMKSA with an AP to which it is about to (re)associate, it includes the PMKID for the PMKSA in the RSNE in the (Re)Association Request. Upon receipt of a (Re)Association Request with one or more PMKIDs, an AP checks whether its Authenticator has retained a PMK for the PMKIDs, whether the AKM in the cached PMKSA matches the AKM in the (Re)Association Request, and whether the PMK is still valid; and if so, it shall assert possession of that PMK by beginning the 4-Way Handshake after association has completed. If the Authenticator does not have a PMK for the PMKIDs in the (Re)Association Request, its behavior depends on how the STA performed IEEE 802.11 authentication. If the STA performed SAE authentication, then the AP STA shall send a Deauthentication frame. If the STA performed Open System authentication, it begins a full IEEE 802.1X authentication after association has completed."

The robust security network information element (RSN IE) of reassociation frames contains PMKID, which refers to a PMKSA shared between the client and AP. It is important to note that the PMKID Count and PMKID List fields are present only in reassociation frames.

From IEEE 802.11

8.4.2.27.5 PMKID

"The PMKID Count and List fields are used only in the RSNE in the (Re)Association Request frame to an AP and in FT authentication sequence frames. The PMKID Count specifies the number of PMKIDs in the PMKID List field. The PMKID list contains 0 or more PMKIDs that the STA believes to be valid for the destination AP. The PMKID can refer to

a) A cached PMKSA that has been obtained through preauthentication with the target AP

b) A cached PMKSA from an EAP or SAE authentication

c) A PMKSA derived from a PSK for the target AP

d) A PMK-R0 security association derived as part of an FT initial mobility domain association

e) A PMK-R1 security association derived as part of an FT initial mobility domain association or as part of a fast BSS transition.

See 11.6.1.3 for the construction of the PMKID, 12.8 for the population of PMKID for fast BSS transitions, and 11.6.1.7 for the construction of PMKR0Name and PMKR1Name.

NOTE—A STA need not insert a PMKID in the PMKID List field if the STA will not be using that PMKSA."

When multiple PMKIDs are listed in the reassociation frame, the AP (or controller) decides which PMKSA to use. If it does not find an applicable PMKID, it carries on with the full 802.1X/EAP authentication. If it finds a relevant PMKSA, it indicates which PMKID was used, and proceeds with the 4-way handshake.

The strengths of PMK caching are:

- Standardized by IEEE
- Can be supported by any WLAN architecture
- No traffic overhead is introduced and amounts to simple design

The weaknesses of PMK caching are:

- Provides fast roaming only on return to a previously associated AP (Roam back)
- New AP roams still require full 802.1X/EAP authentication

Opportunistic Key Caching (OKC)

Opportunistic Key Caching (OKC) is a key caching method not defined in the 802.11 standard, though it does have some commonality with 802.11r (now incorporated into 802.11-2016). OKC is used both at the supplicant (client device) and authenticator (AP or controller) for fast roaming. The PMK and PMKID are retrieved from the initial AP with which the wireless station associates. An identical algorithm is used on the wireless station and WLAN controller/AP, and a unique PMKID is given to the original PMK when it is passed to each AP. The unique PMKID is based on the BSSID of the AP to which the PMK is sent.

At this point, OKC remains a proprietary and generally undocumented solution. It is important to note that for OKC to function, it must be supported by both the authenticator and the supplicant. The following gets a little heady- read through it a couple of times if need be.

From IEEE 802.11:

"A PMK identifier is defined as:
PMKID = HMAC-SHA1-128(PMK, "PMK Name" || AA || SPA)"

To understand this formula, you should know that AA is the authenticator's MAC address and SPA is the supplicant's MAC address.

Reassociation frames are the only frames that carry PMKID Count and List fields.

In normal OKC operation, reassociation frames sent by the wireless client device, list the PMKIDs to be chosen by the AP. OKC may be implemented such that it gains the PMKID from the reassociation frames, or it may be implemented so that the AP does not require a PMKID from the client. Instead, the AP matches the client's MAC address to its PMKID table to identify an applicable match. If one is found, the PMKID is indicated in the first frame of the 4-way handshake. If none are found, the AP transmits an EAPoL-Start frame and requires a full 802.1X/EAP authentication. CWSPs need to understand the complexities of OKC as listed below.

The strengths of OKC are:

- A good solution until Voice-Enterprise (802.11r) solutions are widely available and implemented
- Scales well
- Only requires a single initial 802.1X/EAP authentication

The weaknesses of OKC are:

- Not standardized
- Not all clients support it
- Not implemented in a compatible way across all vendors

802.11-2016 (802.11r) Fast Transition (FT)

To understand the impact of 802.11r FT on roaming, it's helpful to have a reminder of how 802.11i impacted it. To understand how 802.11i affected roaming, you must first understand the different keys used in 802.11i networks. Remember, the keys are used to secure communications on the wireless link. The following keys are used as you've previously learned:

- **Pairwise Master Key (PMK):** This is the top-level key used in the standard. The PMK is derived from a key generated by EAP, or from a pre-shared key (PSK) in smaller WLAN implementations.
- **Pairwise Transient Key (PTK):** The key derived from the PMK, Authenticator (AP) address (AA), supplicant (client) address (SPA), Authenticator nonce (number used once) (ANonce), and supplicant nonce

(SNonce). A pseudo-random function is used to generate up to five keys. The five keys are the EAPOL-Key confirmation key, the EPOL-Key encryption key, the temporal encryption key, and two temporal message integrity code keys.

- **Group Master Key (GMK):** A supporting key that may be used to generate a group temporal key. The GMK may be regenerated within the AP, periodically, to reduce the exposure of the group temporal key.

- **Group Temporal Key (GTK):** The key used to protect broadcast or multicast MPDUs on a wireless link.

The term *pairwise* refers to two devices associated with each other. A pairwise master key, for example, is a key used between an AP and a client STA to secure the communication. This is where roaming is an issue because the 802.11 security was designed to be between a single AP and an STA pair. Therefore, some method of quickly acquiring a PTK after a roam was needed.

Once an STA is associated and authenticated to the wireless network, with 802.11i, a PMK secure association (PMKSA) exists between the authentication server and the STA. A PTK secure association (PTKSA) exists between the AP and the STA once the 4-way handshake is completed. The problem, when discussing roaming, is that the accomplishment of such a PMKSA and PTKSA takes time. It can take too much time for real-time applications if some additional mechanism is not in play.

The 802.11i solution to the delay caused by establishing a PMKSA is PMK caching. With PMK caching, as you previously read, the authenticator (the AP) and the STA can cache PMKSAs so that regeneration of the PMKSA is not required at the time of roaming. Instead, the first step in the 4-way handshake is that the authenticator specifies the identifier of the PMK in the first message (Message 1) of the handshake. This functionality means that the required process of PMKSA establishment is removed, and only the PTKSA must be established. Always remember this rule: if you can remove steps from a process, you will typically reduce the time required to complete the process. By removing the step of PMKSA establishment at roaming time, we speed up the process or reassociation with the new AP.

The 802.11r amendment was the first standard-based attempt to truly define fast secure roaming in any level of detail. It was ratified in 2008 and is now part of the 802.11

standard as amended (802.11-2016). The 802.11r amendment assumes the 802.11i amendment - as would be expected since 802.11i was ratified in 2004. You must always remember when studying IEEE standards, that an amendment ratified today is based upon the original standard and all prior amendments ratified. If you don't keep this in mind, the standard will become very confusing to you very quickly.

> While 802.11r is standardized, few vendors have existing implementations that use it heavily. Apple's "Fastlane" feature set is a collaboration with Cisco Systems that uses .11r, but whether it can be used or not to improve secure roaming still comes down to a number of variables that will differ across WLAN spaces.

To begin the explanation of 802.11r, a few key definitions from the standard are in order:

- **Fast basic service set (BSS) transition:** A station (STA) movement that is from one BSS in one extended service set (ESS) to another BSS within the same ESS, and that minimizes the amount of time that data connectivity is lost between the STA and the distribution system (DS).

- **Fast basic service set (BSS) transition (FT) 4-Way Handshake:** A pairwise key management protocol used during FT initial mobility domain association. This handshake confirms mutual possession of a pairwise master key, the PMK-R1, by two parties and distributes a group temporal key (GTK).

- **Fast basic service set (BSS) transition (FT) initial mobility domain association:** The first association or first reassociation procedure within a mobility domain, during which a station (STA) indicates its intention to use the FT procedures.

- **Mobility domain:** A set of basic service sets (BSSs), within the same extended service set (ESS), that support fast BSS transitions between themselves and that are identified by the set's mobility domain identifier (MDID).

- **Over-the-air fast basic service set (BSS) transition (FT):** An FT method in which the station (STA) communicates over a direct IEEE 802.11 link to the target AP (AP).

- **Over-the-DS (distribution system) fast basic service set (BSS) transition (FT):** An FT method in which the station (STA) communicates with the target AP (AP) via the current AP.

As painful as it may be, memorizing the preceding list of definitions - at least in your own words - is a key part of preparing for the CWSP exam. Make sure you understand these definitions and what they mean for 802.11 roaming. In addition to these terms, you need to understand that a single PMK is not considered in an 802.11r implementation as a sole entity, such as was introduced in 802.11i. Instead, we must deal with a fast transition key hierarchy. The following definitions will help you understand this hierarchy:

- **PMK-R0:** The first level (or top-level) PMK. The PMK-R0 is derived from the master session key (MSK) when 802.1X/RADIUS is used, or from the pre-shared key (PSK) when personal implementations are used.

- **PMK-R1:** The second level PMK. The PMK-R1 keys are derived from the PMK-R0 key.

Remember this hierarchy. The first level is not PMK-R1, but it is PMK-R0.

The core of what 802.11r is all about is allowing a non-AP STA to preauthenticate with an AP to which it may roam at a later time. During the preauthentication process, in an FT implementation, the PTK is derived from the PMK-R1. It's important to remember that the PTK is not derived directly from the PMK-R0, but that it is derived from the PMK-R1.

Preauthentication is optional. If it is to be used, it must be available and enabled on both the APs and the client devices. Remember that preauthentication is not required of an 802.11-compliant device; however, it will be very useful for wireless networks that must carry voice or other real-time traffic and provide for roaming ability. (This is the sort of optional capability leveraged by Apple and Cisco Networks in the Fastlane initiative.)

As you've seen from the information in chapter 5 (robust security network (RSN) and authentication and key management (AKM)), the key hierarchy follows a process of derivation. If you understand that key derivation process, this part which pertains to IEEE 802.11r fast transition will not be as painful.

Basically, the fast transition (FT) process consists of different levels of the PMK. For example, a WLAN controller may hold PMK-R0, while AP1 has PMK-R1 (#1) and AP3 has PMK-R1 (#2). Both PMK-R1 (#1) and PMK-R1 (#2) are derived from PMK-R0. For standard authentication and key management processes, there is only one PMK that is created for the authenticated session. In 802.11 fast transition, there are many PMK's at different levels in the device authentication hierarchy.

The FT Initial Mobility Domain Association is similar to a non-FT initial association; however, a few new elements are introduced into the frame exchange.

From IEEE 802.11:

Mobility Domain element (MDE)

"The MDE contains the Mobility Domain Identifier (MDID) and the FT Capability and Policy field. The AP uses the MDE to advertise that it is included in the group of APs that constitute a mobility domain, to advertise its support for FT capability, and to advertise its FT policy information."

The MDIE is broadcast by the AP in Beacons and probe response frames. The supplicant includes an MDIE in the (re)association request frame, and the authenticator compares this with its MDIE parameters. If a match is found, the exchange will continue.

Fast BSS Transition element (FTE)

"The FTIE includes information needed to perform the FT authentication sequence during a fast BSS transition in an RSN."

FT Initial mobility domain association

"The FT initial mobility domain association is the first (re)association in the mobility domain, where the SME of the STA enables its future use of the FT procedures.

FT initial mobility domain association is typically the first association within the ESS. In addition to association frames, reassociation frames are supported in the initial mobility domain association to enable both FT and non-FT APs to be present in a single ESS."

FT initial mobility domain association in an RSN

"A STA indicates its support for the FT procedures by including the MDE in the (Re)Association Request frame and indicates its support of security by including the RSNE. The AP responds by including the FTE, MDE, and RSNE in the (Re)Association Response frame. After a successful IEEE 802.1X authentication (if needed) or SAE authentication, the STA and AP perform an FT 4-Way Handshake. At the end of the sequence, the IEEE 802.1X Controlled Port is opened, and the FT key hierarchy has been established."

Over-the-Air FT

The Over-the-Air FT protocol is a reassociation process that expedites reassociations in FT-enabled networks. As you review the following image, compare these contents and processes with a non-FT reassociation process. As you can see, fewer frames are used (8 in a non-FT reassociation; 4 in an over-the-air FT reassociation), and new contents are added to the 802.11 Authentication Request/Response and Association Request/Response.

From IEE 802.11:

Over-the-air FT Protocol authentication in an RSN

"The FTO and AP use the FT authentication sequence to specify the PMK-R1 security association and to provide values of SNonce and ANonce that enable a liveness proof, replay protection, and PTK key separation. This exchange enables a fresh PTK to be computed in advance of reassociation. The PTKSA is used to protect the subsequent reassociation transaction, including the optional RIC-Request."

In an over-the-air FT reassociation, the new PTK is established before the reassociation occurs. Notice that the nonce values are included in the 802.11 Authentication Request and Response frames, which provides the necessary information to create new keys.

Over-the-Air FT

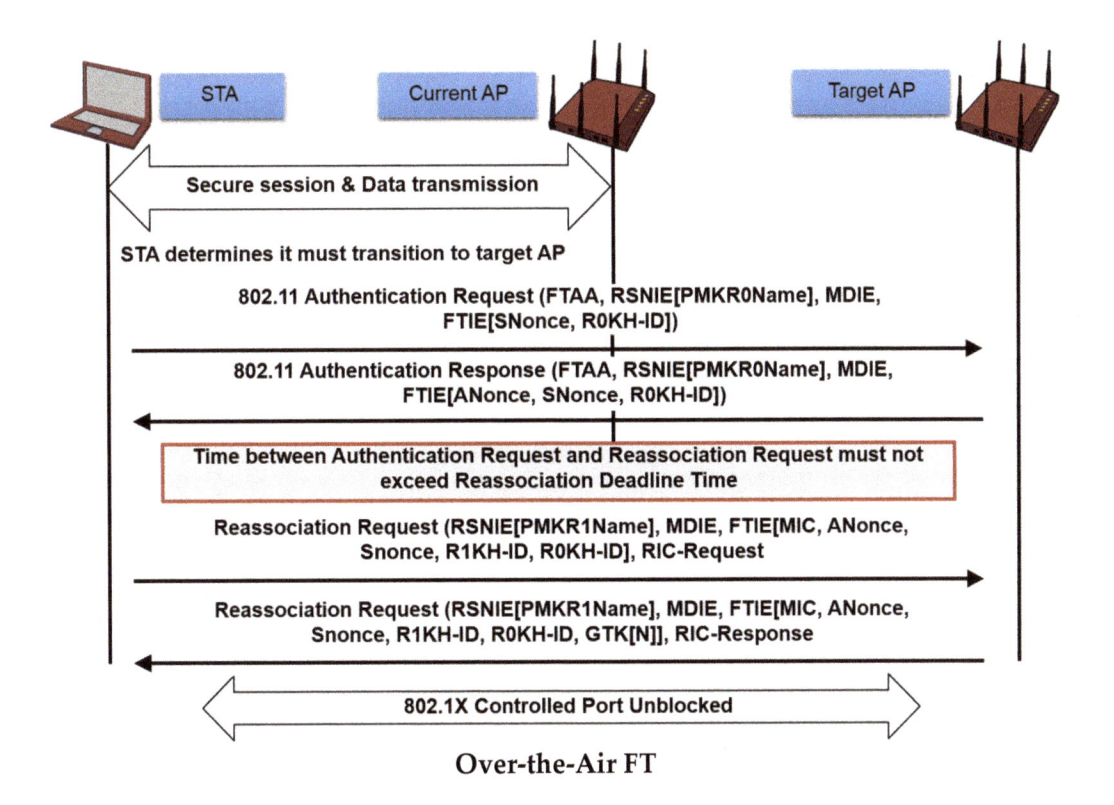

Over-the-DS FT

The Over-the-DS FT protocol is a reassociation process that expedites reassociation in FT-enabled networks. In an Over-the-DS exchange, the open authentication process is established via FT Request and Response Action frames. These frames are transmitted to the current AP, which then relays the frames to the target AP via the current DS. The FT Request and Response frames replace the Authentication Request and Response frames we are all familiar with. After the FT Response is received, new PTKs are created on both the supplicant and target AP (authenticator), and the reassociation may then start via the wireless medium.

The strengths of 802.11 FT are:

- Standard-based fast roaming
- Required by Voice-Enterprise certification

- The most efficient fast roaming method available today
- Eventually, we will see heavy support for it

The weaknesses of 802.11 FT are:

- Has been very slow to market given its 8+ year life
- Wi-Fi Alliance Voice-Enterprise certification only began in 2012
- Introduces many new terms and concepts requiring enhanced education for implementers

Single Channel Architecture

The single-channel architecture represents a proprietary solution to the problem of slow RSN transitions. While the majority of enterprise wireless vendors use multiple non-overlapping channels and a concept known as channel reuse, single-channel architecture (SCA) vendors configure all APs (within a mobility group) on a single channel.

The pioneer and primary vendor with this architecture was Meru Networks (acquired by Fortinet), though a few others offer it as an optional capability. With all APs operating on a single channel, providing pervasive coverage across the service area, clients are not required to roam from one channel to another. Of course, roaming across channels isn't the primary roaming concern. The problem for most networks comes when clients roam from one AP to the next, regardless of channel.

Because they operate on a single channel, SCA APs are able to use proprietary functions to broadcast a single BSSID across all APs. By using a single BSSID for all access points, the client does not realize that there are actually multiple physical APs. Instead, the client sees a single large virtual access point. Of course, the infrastructure must handle client handoffs from one AP to another, but this process is transparent to the client.

A similar method of accomplishing the same result is to create per-client BSSIDs. When new clients join the network, a per-client BSSID is created, and this BSSID is broadcast on each AP. Again, the client does not realize that there are multiple APs but rather sees only a single large and seamless AP. This negates the requirement, from the client's perspective, to roam.

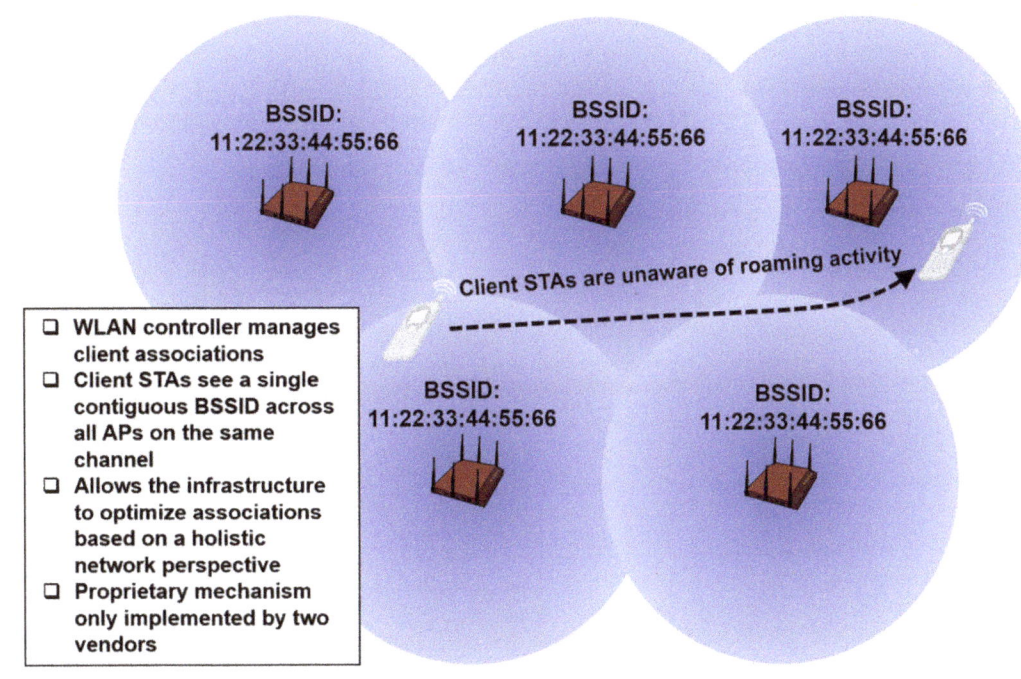

- WLAN controller manages client associations
- Client STAs see a single contiguous BSSID across all APs on the same channel
- Allows the infrastructure to optimize associations based on a holistic network perspective
- Proprietary mechanism only implemented by two vendors

Single Channel Architecture Roaming

Despite the many negative marketing materials from competing vendors regarding SCA solutions, this solution is quite elegant and solves the traditional roaming problem.

The strengths of SCA roaming are:

- One of the best in-use roaming solutions today
- Infrastructure devices have full control over roaming actions
- Transitions are imperceptible to the client STA

The weaknesses of SCA roaming are:

- Proprietary to the given vendor
- Requires SCA, which many feel is an inefficient, obscure architecture in comparison to multiple channel architecture (MCA)

Chapter Summary

In this chapter, you learned about the basic components required to implement fast secure roaming through WPA2. You learned about the applications requiring secure roaming and of the standards-based and proprietary solutions available. You also learned about the 802.11 amendments that relate to roaming, including 802.11i and 802.11r, which are both included now in 802.11-2016.

Review Questions

1. How does SCA help improve fast secure roaming?
    a. The client device sees no SSID
    b. The client device sees a single BSSID
    c. By using Fastlane
    d. By eliminating the need for encryption

2. Which wireless network security type is inherently better when it comes to roaming?
    a. PSK
    b. RSN
    c. WPA2 Enterprise
    d. 802.1X

3. The fundamental risk associated with long roaming times on enterprise secure networks is _____.
    a. Traffic will be buffered
    b. Client devices won't get back on the WLAN
    c. Client devices will have to perform full associations all over
    d. False associations can accumulate

4. Best practices recommend that roam times, in general, should be less than how long?
    a. 500 ms
    b. 300 ms
    c. 200 ms
    d. 150 ms

5. Which of the following isn't a step that is part of the slow roam process?
    a. Open systems authentication
    b. Reassociation
    c. 802.1X/EAP authentication
    d. 4-way handshake

6. VoIP protocols can be described as which of these choices?
    a. Real-time and TCP
    b. Full-duplex and TCP
    c. Real-time and UDP
    d. Half-time and UDP

7. Why do CWSPs most need to understand the nuances of roaming and the effects of roaming on wireless traffic?
    a. Because CWSPs design networks
    b. Because roaming is an unsecured process
    c. Because hackers can jam roaming frames
    d. Because different security features can impact roaming in different ways

8. You are troubleshooting roaming in an SCA environment. What controls client roaming in this WLAN scenario?
    a. The 802.11 standard
    b. The clients themselves
    c. The wireless controller
    d. Enterprise Voice algorithms

9. Which roaming method is currently the most efficient?
    a. 802.11 FT
    b. OKC
    c. Preauthentication
    d. Enterprise roaming

10. Advantages of preauthentication include all of these, except _____?
    a. Is standardized by IEEE
    b. Allows for preauthentication with a single AP
    c. Supported on any WLAN architecture
    d. Performed prior to roaming

11. Target performance of Voice Enterprise includes handovers of how many milliseconds?
    a. 20
    b. 50
    c. 150
    d. 500

12. The capability to classify specific types of network traffic and to give priority to certain traffic types is called what?
    a. Quality of Experience
    b. Fast pass
    c. QoS
    d. Degree of Service

13. Which of the following statements is true about roaming and the 802.11 standard?
    a. The standard is explicit about details of the roaming process
    b. The standard does not address roaming at all
    c. Only pre-standard roaming is supported
    d. The standard includes what should happen during roaming, but not how it should be accomplished

14. What type of roam is in play when a client moves from an access point on Network 1 to an access point on Network 2?
    a. Data Link roam
    b. Layer 3 roam
    c. Layer 2 roam
    d. Out of band roam

15. For proper roaming, cell overlap should be what values, roughly?
    a. 15-30%
    b. 20-40%
    c. 10-25%
    d. 25-35%

Review Answers

1. **B.** With Single Channel Architecture, the client device sees all APs as a single BSSID per channel

2. **A.** PSK is much friendlier to roaming than enterprise security

3. **C.** With too much delay during roaming, full client associations will need to be re-accomplished

4. **D.** Roam times should be under 150ms by established best practices

5. **B.** In a slow roam, a full association has to occur again, so there is no reassociation

6. **C.** VoIP protocols are real-time, and usually UDP for greatest efficiency

7. **D.** Security can impact network operations, so knowing the details of roaming on secure networks will help in troubleshooting

8. **C.** In the SCA, the controller decides when clients will roam to other APs

9. **A.** 802.11 is currently the most efficient roaming method

10. **B.** Preauthentication happens with all nearby APs, not just a single one

11. **B.** Voice Enterprise aims for roams under 50ms, although up to 100ms is allowed

12. **C.** QoS (Quality of Service) is important in wired and wireless networks

13. **D.** Roaming is one of those areas in the 802.11 standard where vendors have a lot of freedom to decide their own mechanisms

14. **B.** When the client leaves one network for another, it has crossed the Layer 3 boundary

15. **A.** Though it can't be easily or precisely measured, the desired overlap is 15-30%

Chapter 9: Network Monitoring

Objectives Covered

2.1	Identify potential vulnerabilities and threats to determine the impact on the WLAN and supporting systems and verify, mitigate, and remediate them
4.1	Understand and implement management within the security lifecycle of identify, assess, protect, and monitor
4.3	Use information from monitoring solutions for load observation and forecasting of future requirements to comply with security policy
4.5	Implement effective auditing procedures to perform audits, analyze results, and generate reports

Wireless security is not a "set and forget" kind of task. After the network has been implemented in accordance with organizational policies and operational goals, the need to monitor kicks in. In addition to network performance monitoring, wireless security also needs to be monitored in various ways, and the management of network devices has to happen using only secure protocols. It's easy to get complacent and occasionally sloppy in your monitoring and management over time, but this is a critical area for long-term compliance with both local policy and regulatory requirements. Let's talk about how to do it right as a CWSP.

Secure Management Protocols

If you administer the individual APs or controllers in your wireless LAN, you should be sure to only use secure methods for management. You may be able to connect to many vendors' components using standard HTTP by default, but this is not a recommended practice. Because HTTP traffic is transmitted as clear text, configuration settings applied via HTTP are easily eavesdropped upon.

The following image demonstrates the vulnerability of using HTTP for systems administration. In this real-world example, we've blocked out identifying information to protect the site owners. You can clearly see the logon is "swettmarden" and the password is "drow1ssap1". This is because the web server does not use HTTPS for the login process and the credentials are passed in the clear.

Of course, this scenario was created completely for this text- but similar situations occur with great frequency.

Understandably, HTTPS should always be used when a web-based interface is employed to manage your APs or controllers. Given the slim chance that the hardware does not support HTTPS, it is best not to use HTTP either, to manage the device. Chances are, you're not dealing with enterprise equipment if only HTTP is available. HTTPS actually uses SSL and requires that a certificate be made available to the server. AP and controller hardware which supports HTTPS already have a certificate installed in the AP. SSL is a Layer 7 encryption technology.

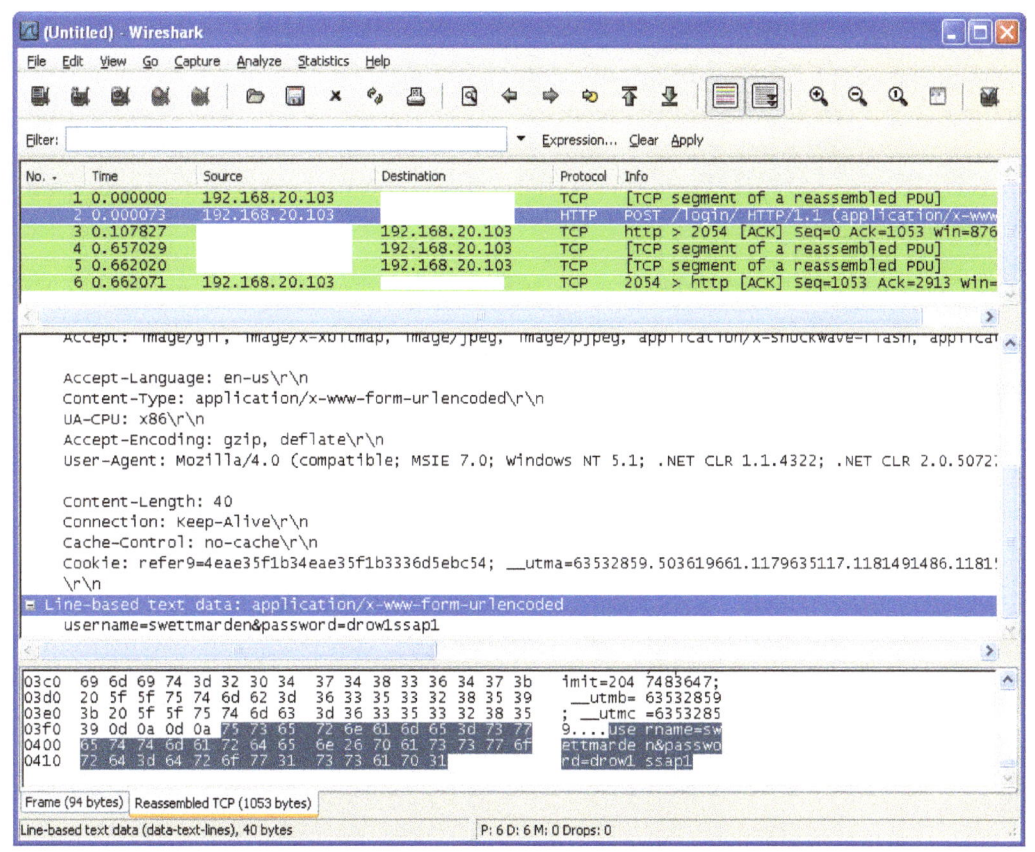

Wireshark showing HTTP Logon Credentials

Another Layer 7 encryption solution is SSH. The first version of SSH has known vulnerabilities and should be avoided, but SSH2 is currently considered secure. It is commonly used to provide command-line interface (CLI) access to managed devices. SSH2 provides the following benefits in a secure networking application:

- Public and private key authentication, or username and password authentication.
- Data signing through the use of public and private key pairs.
- Private key passphrase association.
- Multiple encryption algorithms are supported, such as AES, 3DES, and DES.
- Encryption key rotation.
- Data integrity enforced through hashing algorithms.
- Data compression may be supported.

Occasionally updates are made to SSH in specific operating systems as vulnerabilities are found by researchers, and it's not uncommon to occasionally have to patch or update to get to a more secure version of SSH2.

The *Simple Network Management Protocol* (SNMP) is a standard solution for centrally monitoring and managing network devices. SNMP was plagued by security vulnerabilities early on, and these weaknesses have been addressed in SNMP v3. Version 3 has added authentication and privacy controls to help protect the management information passed on to your network. You should ensure that any device you manage with SNMP uses version 3 or higher of this protocol, if possible. Of course, as is true with any technology, you must be proactive and continually be on the lookout for new vulnerabilities that would impact your network. That which is secure today may be vulnerable tomorrow.

Additionally, if file copy processes are used, secure FTP (SFTP), which combines FTP with SSH, should be used, or the secure copy protocol (SCP), which is common in Linux environments. SFTP is different from FTPS. SFTP is FTP combined with SSH. FTPS is FTP combined with TLS.

Along with secure management protocols, many security-minded network environments will also manage their network gear on one or more private, isolated subnets. This doesn't negate the need to use secure protocols for management but does provide a tremendous increase in security.

In large environments, the CWSP may find themselves doing a fair amount of system administration on the likes of wireless controllers and network management servers which have their own administrative security concerns outside of the 802.11 realm.

WLAN Monitoring

There are a number of ways to monitor enterprise WLAN environments for performance SLAs and security. The monitoring processes will ensure a system maintains the desired performance levels and will provide the required security based on system design and the corporate security policy. Monitoring also is critical to the ongoing management of the Security Lifecycle model (identify, assess, protect, and monitor).

As with the LAN and WAN, wireless network monitoring is an on-going process that is used to gather key system information. That information is used to validate that a system is operating as intended and designed. A well thought out and designed monitoring system will provide valuable metrics and indicators that will allow information technology (IT) professionals the ability to:

- Conduct security audits and locate vulnerabilities
- Maintain regulatory compliance
- Maintain proper performance levels
- Verify network availability
- Possibly assist in troubleshooting client devices

Both manual and automated methods exist that are used with wireless network monitoring. Determining which process and tools you will use really depends on the size of the infrastructure and the number of devices that are used on the network. The staffing structure of a network support organization can also influence which tools get used, as automated testing can augment staff's capabilities if implemented properly. Network vendors often bundle LAN and WLAN monitoring capabilities into a single platform, but it's common to not leverage all of the system options in these sometimes behemoth platforms. Some of the common tools used for network monitoring include:

- Wireless Intrusion Detection System (WIDS)
- Wireless Intrusion Prevention System (WIPS)
- Wireless Network Management Systems (WNMS)
- Protocol analyzer software and hardware
- Spectrum analyzer software and hardware
- Hardware sensors
- Analytics platforms

These tools are each designed to perform specific tasks, as briefly described next.

Wireless Intrusion Detection System (WIDS)

A WIDS is used to gather information about a computer network. Focus on "detection" as a passive activity. The type of information collected will depend on the business model and the requirements of the organization. Hardware sensors distributed around the physical network are used to gather and report information to a physical appliance or a server database. The WIDS will only detect and report any anomalies that are determined from a baseline of the network. WIDS is usually "always-on" and can be set with various levels of automated reporting and alerting. The information gathered will provide the best value when the system is properly tuned for the specific environment.

Wireless Intrusion Prevention System (WIPS)

The WIPS has many of the same characteristics as the WIDS system; however, in addition to detection of threats, a WIPS system may be able to mitigate the threats (an important distinction). Information collected from the sensors is reported to a central server database or network appliance for proper analysis and handling. Alarms will trigger, and alerts will be sent, notifying network personnel of the potential and severity of the intrusion. Depending on how the system is configured, automatic mitigation may be enabled.

Wireless Network Management Systems (WNMS)

The WNMS is a centralized solution that may run on a hardware server, in a virtual environment, or even in the cloud. A WNMS allows a network engineer to centrally manage and control the entire WLAN, including the APs, controllers, RADIUS servers, and location services, depending on the vendor. These centralized management systems are available from many manufacturers for use with their own infrastructure devices. As WLAN systems get more complex, the number of vendor-neutral WNMS solutions that work well with many different manufacturers' equipment is shrinking, but 3rd-party NMS do exist. A WNMS may also incorporate WIPS technology for complete wireless network management, monitoring, and security solution. In these systems, each major feature is typically licensed at additional cost. As "unified" and "software-defined access" networking topologies are marketed, some NMS systems span both the LAN and WLAN.

Protocol Analyzer Software and Hardware

Protocol analysis software is used to capture frames that travel through a network medium. Protocol analyzers are used for both wired and wireless networks. Some analyzers have the capability to function on both types of networks. Protocol analyzers are great troubleshooting tools, and can also be used for discovering potential security issues. Protocol analyzers can be stand-alone, for use with a laptop computer, or other mobile device. They can also be integrated with a wireless AP, a dedicated remote infrastructure device, or part of a WIPS or NMS system. In some cases, like with cloud-controlled WLAN systems, you can do a remote packet capture on far-away access points but save the capture files locally to your PC for analysis. (Packet capture is discussed in depth in the CWAP course.)

Spectrum analyzer software and hardware

Spectrum analyzers are used to monitor the RF environment, and to help identify what type of signals are present on specific frequencies and how strong they are. With respect to wireless networks, a spectrum analyzer will help identify issues and threats that other monitoring tools, such as a protocol analyzer cannot. Spectrum analyzers will vary in size, complexity, and cost, based on the intended use. Some manufacturers build spectrum analyzers to work specifically with standards-based IEEE 802.11 wireless networks, while others may be used for a variety of other RF monitoring purposes. It's common to see some form of spectrum analyzer utility baked into the access points themselves, but the quality and granularity of spectrum analysis can vary widely across devices.

Hardware sensors

Sensors are used to gather needed information by constantly monitoring the open air. These devices can be integrated within a wireless AP or a stand-alone dedicated device. Sensors operate in what is known as monitor mode and are passive devices that will not interfere with other wireless devices which occupy the same radio frequency space. Some access points can be configured as full-time sensors as part of their configuration.

Analytics platforms

Tremendous amounts of WLAN-related data are generated at different points from within the WLAN network. Consider all that goes on at the access point, the controller, and in authentication servers related to individual clients and within the radio environment of busy wireless networks. Analytics platforms attempt to aggregate this data and to use machine-learning to draw conclusions on performance and system health. As with WIDS systems, proper tuning is crucial for the analytics platform to not be just another overwhelming dashboard of limited value.

Rogue AP and Client Detection

Much emphasis has been placed on rogue access points since the earliest days of wireless networking. From the inception of 802.11, rogue devices were recognized as both a security concern and a potential source of RF interference, and they still pose a threat to our networks today. For the purposes of CWSP, a rogue AP can be defined as any AP that is operating in your owned space that has not been authorized by you. The rogue AP may have been placed by an intruder seeking to gain access to your wired network, or it may have been placed by a well-meaning user hoping to make his or her life easier and more mobile while at work. Either way, the rogue AP is a threat to your security and the performance of your WLAN.

There are two primary reasons that motivate an attacker to install a rogue AP in your environment. The first is to gain access to your wired or wireless network by extending it into the air. The second is to attack your valid wireless client stations (STAs).

In the first case, the attacker will usually find an out-of-the-way spot where a live Ethernet port provides connectivity to the wired LAN. She will connect the Ethernet port to the AP using a standard cable and then power the AP with a nearby power outlet. (Some APs may even be powered by battery if the attacker only needs access for a short time.) Once the attacker has the AP in place, she can begin attacking your wired LAN or other WLANs that may be connected to the wired LAN while not needing to be physically close to the AP itself. Of course, the attacker has to be willing to lose her AP in a scenario like this, because she risks not being able to retrieve it after the attack. With physical security being lax in many organizations, however, the retrieval may not be too difficult. Consider hospitals or colleges, where large numbers of people milling around tend to be the norm.

> Many rogue APs are placed by intruders for the sole purpose of gaining high-speed Internet connectivity. They often know that companies have very fast connections to the Internet and, since they will be the only user connected to the AP, they can download a tremendous amount of illegal software, movies, music and more in a short window of time. This malicious use of your network can be protected against by using an Internet proxy server that requires authentication, or by securing your LAN ports in a number of ways.

Protecting against the placement of such APs is important. The first thing to consider is the disabling of all Ethernet ports that are not actually in use. When those ports are needed, they can be administratively enabled through software, usually by using a wired NMS. Additionally, where it makes sense, you should have good physical security in place which deters such behavior (some environments are open to the public purposefully). Even fake surveillance cameras can go a long way here. Install a fake surveillance camera in areas where you think an attacker may attempt to install a rogue AP. The presence of this device – as long as it looks real – will frequently deter the attacker.

The second motivation for rogue AP placement by an attacker is to directly attack your WLAN clients. In this case, the attacker may be using the AP to perform a hijacking attack in an attempt to gain access to the data on the Wi-Fi-connected computers. She may also be attempting to install backdoors on these WLAN clients that will allow her access to the network in the future. In these scenarios, rogue AP detection can be more difficult. The attacker may be a temporary employee who has valid access to the premises and has been granted permission to use her laptop at work. She may be running a software-based rogue AP, a hacking device like the wireless Pineapple, or she may be using a USB-powered pocket AP like the one shown in the following image. While the shown pocket AP is an older 54 Mbps 802.11g AP, this is still sufficient for performing network attacks and is very inexpensive to acquire through websites like eBay. Newer 802.11n and 802.11ac pocket APs are also available now.

D-Link Pocket AP (DWL-G730AP)

Protecting against this type of rogue AP can be challenging. The attacker is not connecting to an Ethernet port and doesn't want to. Therefore, disabling unused ports will not be helpful. CWSPs should remember that the best protection against this type of rogue AP attack is to implement a secure 802.1X/EAP authentication type that uses mutual authentication. This will also help protect your clients from other rogue AP type attacks.

> It is not the intention of this book to suggest that a pocket AP is only useful in rogue AP scenarios. They are used quite frequently in a number of travel scenarios or where legitimate pop-up Wi-Fi might be needed.

Rogue AP detection generally takes place in two ways- through the wired interface and through the wireless interface. Remember that a rogue AP is still an access point, so it will transmit beacon frames at a regular interval. If you use a site survey tool to map the RF coverage in your area and then perform a walk-through on foot with this tool periodically, you can detect the existence of new APs by comparing RF coverage as it varies from the original survey. This is one method of rogue AP detection through the wireless interface.

Another method of detection through the wireless interface would be to keep up-to-date documentation of the number of APs you have installed which can be detected at a given location. Then you can go to that location – and other locations as well – and use a tool like inSSIDer to see if more APs are now present. When you see a new AP, note its MAC address. You can then monitor the signal strength of the beacons from that MAC address while moving throughout the area. You should notice the strength weakening and strengthening as you move around. Using this process and the body-fade technique (using your body to introduce attenuation in one direction), you should eventually be able to find the proximity of the AP, and then the AP itself. It may be an innocent neighbor AP, or it may be a rogue AP installed within your facility.

You can also detect rogue APs through the wired port. Many APs are installed by users who want the flexibility provided by a WLAN, or who want less wireless security or protocol restrictions than the business WLAN enforces. These users will seldom know how to prevent you from detecting the AP through the wired port. Most APs installed by attackers are not configured in such a way to prevent you from detecting them through the wired port either, for that matter. Because these rogue APs are usually cheap SOHO APs or routers that either do not support the disabling of the HTTP management interface on the Ethernet port or are installed by people who wouldn't know how to, they are usually easy to find.

Since you know that an HTTP server is running on most APs but not on most desktop PCs or network servers, you can perform a port scan subnet-by-subnet looking for IP addresses with port 80 open. When you discover an IP address with port 80 open that wasn't there before, it's possible that you've discovered a rogue AP (or some other unauthorized rogue device). A trick you can use is to do the following:

1. When you've finished installing your WLAN, and you know that there are no rogues at this point, do a port scan of every segment and save the output to a text file.

2. Now, every week or so, you can run the same port scan during off-peak hours (if you have them) and save the new scan to a different file.

3. Finally, use any of dozens of file comparison tools to look for differences. Or, even better, write your own script that compares the two files and only tells you of new references to ports 80 (HTTP) and 23 (telnet).

With this methodology, you can build your own rogue detection system very easily. It will not be as powerful as a commercial WIPS solution, but it is better than no detection system at all. If your network supports it (and you have the ability), you could even write your script to disable the Ethernet ports where the new TCP ports 80 or 23 were found, and have it e-mail you a report. You could take action as soon as you receive the e-mail, but at least the script will have disabled the device on the assumption that it is a rogue. This provides you with a form of automatic containment. This model works well in SOHO implementations and smaller SMBs and provides a taste of what commercial WIPS systems can do.

In larger enterprises and larger SMBs, you will need to install more powerful centralized management solutions. For example, Cisco System's Unified Wireless Network solution takes advantage of the fact that all Cisco controllers include a method to automatically detect rogue APs on and off the network. This allows you to spend your time doing more than running scripts and setting up manual solutions.

The old saying reminds us that an ounce of prevention is worth a pound of cure. This is certainly true for rogue APs, given their potential for harm. There are a number of methods you can use to prevent individuals from connecting unauthorized APs to your wired network. These include:

Disabling unused Ethernet ports. Mentioned earlier, this simple solution should not be relied on by itself because people do make mistakes and leave ports open. It can also be hard to effectively manage in large dynamic Ethernet environments.

Using port security on switches. Many switches support port-based filtering by MAC addresses and other parameters. You can specify that the only MAC addresses that can connect to your switch are those in the specified list. Wired MAC addresses are far more difficult for an intruder to find than WLAN MACs, so it's not easy for an attacker to get a rogue AP's MAC into the list of allowed devices.

State clearly that users cannot install APs in your acceptable use policy. This will most certainly not prevent the installation of all rogue APs, but it will deter many from installing them. The effective policy also tackles the human side of operations.

Implement Network Access Control (NAC) technology. This will cause the attacker's computer to go straight to a quarantine area when he or she accesses the network. The NAC device/server would be installed between the switch that provides connectivity to your Ethernet ports and the rest of the network. Any device that connects would require authentication and validation. The operational theory is that this would drive the typical would-be attacker away for fear of being caught by IT staff.

Implement enterprise-capable wireless LAN solutions that automatically detect and report rogue APs and graphically show their locations. Wireless network management systems from a number of market leaders detect and report rogue APs automatically and can display their locations on floorplans in the management solution. Regulatory compliance may require this sort of capability- like in PCI environments.

As you can see, there are multiple methods that you can use to prevent the connection of rogue APs to your wired LAN. Some of these methods are psychological, and others are technological, but a combination of both types usually works best.

WIPS - Features

Enterprise-class wireless intrusion prevention system (WIPS) solutions are used to complement an organization's wireless network encryption and authentication solutions. Commercial WIPS gives you all the rogue detection capabilities referenced earlier plus more in an automated solution. The WIPS can be configured to recognize trusted and known wireless devices installed in a service area and to report changes to the administrator's console. Additionally, a WIPS is capable of providing collected data to a server regarding the overall security and potential recognized threats. Implemented correctly, WIPS solutions can provide a wealth of information as well as protection for your network infrastructure and wireless devices.

WIPS solutions are software-based, hardware-based or cloud-managed, and are capable of monitoring the RF environment through the use of wireless hardware sensors deployed in an "overlay" topology that amounts to one sensor per so many access points. A WIPS typically reports captured information to software programs, to be recorded in a server-hosted database. The WIPS solution will then be able to take the appropriate countermeasures to prevent wireless network intrusions as needed to comply with security policies. These countermeasures are based on identifying the intrusion by comparing the captured information to an intrusion signature database within the WIPS server.

WIPS solutions contain a variety of features, which include:

- Use of hardware sensors for monitoring
- 24x7x365 monitoring
- Mitigation features (containment, blocking, notifications, etc.)
- Notification of threats through a variety of mechanisms
- Detection of threats to the wireless infrastructure, such as denial of service (DoS) attacks and rogue APs
- Built-in reporting and trending systems
- Integrated RF spectrum analysis to monitor and view the RF spectrum
- Compliance validation for corporate security policy and legislative requirements
- Retention of collected data for further forensic investigation
- Location of RF devices

Enforcing Functional Policy

Earlier, you learned about the importance of corporate security policy. CWSPs need to understand that functional policy defines the technical aspects of network security. An enterprise WIPS that has been properly implemented has the capability to provide much of the necessary enforcement of functional policy. The functional policy includes the following components, among others:

- Password policy
- Acceptable use policy
- Authentication and encryption policy
- Wireless LAN access policy
- Wireless LAN monitoring policy
- Endpoint device policy
- Personal device policy

In order to provide enterprise-class WLAN security enforcement, the organization should require the use of an unattended WIDS. Remember, the main difference between WIDS and WIPS solutions is the fact that a WIDS system will detect threats and report those threats whereas a WIPS system has the capability to detect threats and then to mitigate them.

While it is important to write a security policy to document the desired security practices for a network, enforcement of a formal policy is a different challenge. WIPS platforms are designed to complement written policy and to provide monitoring and enforcement of that policy. Monitoring and security auditing are crucial in determining security policy adherence. All companies should perform continuous monitoring - especially those infrequent cases that have a "no WLAN" policy (this can be counterintuitive to some people, but is important to grasp). WIDS platforms stop at the level of intrusion detection and reporting, but WIPS takes it a step further by preventing some threats and assisting in the enforcement of a functional security policy.

Security Monitoring

The initial security audit provides a critical baseline of all active wireless devices (both infrastructure and client) and is used to classify those devices by their roles. The baseline is used to properly identify and categorize devices based on how they fit within the wireless network infrastructure. In all types of network security, a baseline is not a one-time event. To ensure that the security audit baseline remains current, it is necessary to provide on-going monitoring, as most wireless networks have components that frequently change and introduce new technology to the monitored environment. The baseline can be done manually or through the use of automated sensing systems like WIPS. As you now know, WIPS systems are equipped with the ability to isolate and nullify the actions of threatening wireless devices. This activity is referred to as "threat mitigation."

Some common enterprise-class WIPS system manufacturers (at the time this text was written) are:

- AirTight (now Mojo Networks)
- Cisco Systems
- Fluke Networks / AirMagnet Enterprise (Netscout)
- Wildpackets (Savvius) WatchPoint Platform
- Aerohive
- Cisco-Meraki

Reporting and Auditing

Compliance monitoring is essential for many organizations today. Since more businesses than ever are using wireless networks in all facets of their operations, this is an important area to understand as a CWSP. Implementation of legislated regulatory constraints of the WLAN is usually the responsibility of the IT department, or a VAR contracted for this purpose. Compliance with regulatory requirements must be verifiable and auditable by third-party inspectors, and non-compliance can bring fines and reputational damage that leads to loss of business. Technical solutions such as WIPS platforms, help with the automation of many compliance-related tasks.

We touched on these earlier in the text, but it's worth reviewing some examples of legislated security requirements:

- Directive 8100.2 (DoD)
- Health Insurance Portability and Accountability Act (HIPAA)
- Sarbanes-Oxley (SOX)
- Gramm-Leach-Bliley Act (GLBA)
- Payment Card Industry (PCI) Data Security Standard (DSS)

- General Data Protection Regulation (GDPR)

Although all of these are very important, two that get perhaps the most attention are HIPAA and PCI. If a WIPS system is used, it can decrease the administrative overhead in maintaining compliance. The type of business done by an organization will determine if it must comply with one or more regulatory requirements. For example, a healthcare organization may also process credit and debit card payments. This scenario requires compliance with both HIPAA and PCI, by contrast, a small restaurant that takes credit card payments would just need to be PCI compliant.

PCI Compliance as an Example

Payment Card Industry (PCI) compliance is a statement of conformity to the PCI Data Security Standard (DSS). PCI DSS is a set of standards that help to ensure that companies processing payment cards (credit cards, debit cards, etc.) do so in a universally agreed-upon secure manner. The standards encompass payment card processing, account data storage, and information transfer.

The PCI DSS document is a 100+ page document (as of version 3.0) that outlines the process of implementing a secure payment card processing environment. The document covers the following components:

- Building and Maintaining a Secure Network
- Protecting Card Holder Data
- Maintaining Vulnerability Management Programs
- Implementing Strong Access Control Measures
- Regularly Monitoring and Testing Networks
- Maintaining an Information Security Policy

If you've ever been a student of information security, you'll immediately recognize most of these components as standard security best practices. Indeed, the only unique component of PCI is that of protecting cardholder data, and even that could likely be classified under the normal heading of protecting valuable data. In the end, there is really nothing new in the PCI DSS document; however, more and more states and credit card companies are requiring compliance with it in order to process payment cards. The good news is that if an organization already implements security best practices, they'll have very little to change in order to comply with PCI DSS.

The PCI DSS standard lists both recommended and required practices. In the standard, network segmentation is only recommended while the installation of perimeter firewalls between wireless networks and the payment processing segment is required. The standard lists many more requirements for WLAN implementations including (extracted from PCI DSS 3):

> Install perimeter firewalls between any wireless networks and the cardholder data environment, and configure these firewalls to deny or control (if such traffic is necessary for business purposes) any traffic from the wireless environment into the cardholder data environment.
>
> For wireless environments connected to the cardholder data environment or transmitting cardholder data, change wireless vendor defaults, including but not limited to default wireless encryption keys, passwords, and SNMP community strings. Ensure wireless device security settings are enabled for strong encryption technology for authentication and transmission.
>
> Ensure wireless networks transmitting cardholder data or connected to the cardholder data environment, use industry best practices (for example, IEEE 802.11i) to implement strong encryption for authentication and transmission.
>
> Implement additional security features for any required services, protocols, or daemons that are considered to be insecure—for example, use secured technologies such as SSH, S-FTP, SSL, or IPSec VPN to protect insecure services such as NetBIOS, file-sharing, Telnet, FTP, etc.
>
> Ensure that security policies and operational procedures for managing vendor defaults and other security parameters are documented, in use, and known to all affected parties.

> Remember that any system or person involved in payment card processing should be using procedures in alignment with PCI-DSS for full compliance. PCI-DSS is a guideline and not a regulation as it is not enforced by local governments. However, it may be enforced by payment card companies through the disallowance of card acceptance if an organization is found in non-compliance with the guideline.

The following PCI-DSS testing procedures apply to modification of default settings for WLANs:

Interview responsible personnel and examine supporting documentation to verify that:

- Encryption keys were changed from default at installation
- Encryption keys are changed anytime anyone with knowledge of the keys leaves the company or changes positions.

Interview personnel and examine policies and procedures to verify:

- Default SNMP community strings are required to be changed upon installation.
- Default passwords/phrases on access points are required to be changed upon installation.

Examine vendor documentation and login to wireless devices, with system administrator help, to verify:

- Default SNMP community strings are not used.
- Default passwords/passphrases on access points are not used.

Examine vendor documentation and observe wireless configuration settings to verify firmware on wireless devices is updated to support strong encryption for:

- Authentication over wireless networks
- Transmission over wireless networks.

Examine vendor documentation and observe wireless configuration settings to verify other security-related wireless vendor defaults were changed, if applicable.

The following PCI-DSS guidelines apply in relation to the modification of default settings for WLANs:

> If wireless networks are not implemented with sufficient security configurations (including changing default settings), wireless sniffers can eavesdrop on the traffic, easily capture data and passwords, and easily enter and attack the network.

In addition, the key-exchange protocol for older versions of 802.11x encryption (Wired Equivalent Privacy, or WEP) has been broken and can render the encryption useless. Firmware for devices should be updated to support more secure protocols.

The following PCI-DSS testing procedures apply in relation to authentication and encryption on WLANs:

Identify all wireless networks transmitting cardholder data or connected to the cardholder data environment. Examine documented standards and compare to system configuration settings to verify the following for all wireless networks identified:

- Industry best practices (for example, IEEE 802.11i) are used to implement strong encryption for authentication and transmission.
- Weak encryption (for example, WEP, SSL version 2.0 or older) is not used as a security control for authentication or transmission.

The following PCI-DSS guidelines apply in relation to authentication and encryption on WLANS:

Malicious users use free and widely available tools to eavesdrop on wireless communications. Use of strong cryptography can help limit disclosure of sensitive information across wireless networks.

Strong cryptography for authentication and transmission of cardholder data is required to prevent malicious users from gaining access to the wireless network or utilizing wireless networks to access other internal networks or data.

Again, the entire PCI-DSS document is well over 100 pages, but these excerpts give a sense of what to expect when dealing with PCI as a CWSP. In summary, to comply with PCI DSS, a WLAN that is not involved in payment card processing must be segmented from the payment card processing segment using firewalls, vendor defaults must be changed, and strong encryption based on 802.11i must be implemented. It should be no surprise that the standard explicitly states that the use of WEP as a security control is prohibited.

HIPAA Compliance as an Example

The HIPAA regulations require that healthcare organizations (including hospitals, doctor's offices, and any other organization that handles health information) implement policies and procedures to ensure that only authorized individuals may access patient health information in the United States. The theft of personal medical data has been increasing dramatically over the last several years, and poor HIPAA compliance is usually to blame. HIPAA stands for Health Insurance Portability and Accountability Act and was enacted within the U.S. in 2006. Organizations covered by the Act include:

- Health plan providers
- Healthcare clearinghouses
- Any healthcare provider who transmits health information in electronic form

The health information protected by HIPAA includes all individually identifiable health information. This information is identified as information that is unique to an individual and related to the health of that individual. Examples include

- Past, present, or future mental or physical health conditions
- Healthcare that has been provided to the individual
- Healthcare payment information

Information classified as "de-identified" does not require compliance with HIPAA regulations. De-identified information is information that neither identifies nor provides a foundational knowledge base on which a patient may be identified.

HIPAA regulations are somewhat nonspecific, allowing organizations of differing sizes to implement locally appropriate security measures which result in the protection of health information. General requirements include these stipulations:

- Privacy policies and procedures must be documented.
- A privacy official must be designated to oversee the HIPAA regulation implementation and maintenance.
- All workforce members must be trained to understand and comply with privacy policies.
- Mitigation efforts must be taken when privacy policies are breached.
- Effective data safeguards must be implemented.
- Complaint processing procedures must be implemented.

- Patients must not be asked to waive privacy rights, and retaliation against complaints is not allowed.
- Privacy policies and incident documentation must be maintained for six years.

With an understanding of the HIPAA regulations, the only remaining question is this: How do these regulations apply to a WLAN? The answer is simple: they apply to WLANs in the same way they apply to wired LANs. There is no significant difference. Whether wired or wireless, the following five security solutions should be used to effectively comply with HIPAA regulations:

- Authentication
- Authorization
- Confidentiality
- Integrity
- Nonrepudiation

All of these terms were defined earlier in the book. Wired networks do not provide confidentiality by default, and encryption solutions must be used to comply with HIPAA. This requirement is the same for wireless networks. The same is true for the other four requirements, as well. You must comply with all five of these requirements on both wired and wireless networks under HIPAA directives.

Many vendors attempt to differentiate between wired and wireless networks to indicate that their wireless solution is the best when it comes to HIPAA compliance. The truth is any hardware and software combination which allows the implementation of 802.11i security, can be fully HIPAA compliant. Don't fall into the trap of marketing hyperbole when it comes to regulatory compliance.

Auditing and Forensics

In addition to regulatory compliance auditing, WIPS platforms also automate a range of internal network security auditing practices. Data collection is performed by the WIPS sensors and is logged by the WIPS server for easy auditing. When security breaches are detected, forensic analysis may also be done to determine the impact of a network threat. Some WIPS vendors have added a strong forensic analysis component to their WIPS platforms.

The corporate security policy determines how the organization handles auditing and forensics, such as frequency and detail of audits, and how forensics data is logged, stored and archived. By continuously monitoring, analyzing, and logging wireless traffic signatures, WIPS solutions provide automated forensics and security auditing. This can streamline the process and provide accurate results and reports. At the same time, effective WIPS doesn't just happen by purchasing a system. As mentioned previously, proper setup and tuning is crucial to the efficacy of the WIPS investment.

Audit Methods

In order to verify security policy compliance, network administrators, including CWSPs, must be competent with auditing analysis utilities. An understanding of networking concepts is fundamental to this ability. Networking professionals frequently refer to a seven-layer model (the OSI model) when discussing the interactions of the many protocols used to allow reliable, distributed data communications. Documentation for auditing tools often reference these layers and the way they communicate.

As you've learned in CWNA, each layer of the networking model communicates unidirectionally; only with the layers above and below. The communication flow either goes down the protocol stack on its way to be transmitted, or it flows up the protocol stack after having been received. As a quick review - each layer either adds its own interpretive information in the form of a header (if preparing to transmit) or interprets the header information that was added by its counterpart on the remote end (if receiving). When a layer adds its own unique information to data sent from a higher layer, the resulting information field is known as a protocol data unit (PDU). PDUs are distinct for each layer and are only meaningful to the same layer on the remote end of the conversation.

Audit Tools

There are several tools on the market to facilitate the process of audits and reports. Some monitoring tools, like WIDS, NMS, and WIPS, will help automate the compliance reporting process.

Other types of audits include security posture auditing, also known as penetration testing (or simply pen testing). In this more active type of audit, an inside employee and/or outside contractor will test the security posture of an organization in an attempt to expose any weaknesses. Purpose-built tools like Immunity's SILICA are designed specifically to automate the process of discovering and exposing network vulnerabilities. This tool also includes automated reporting of its findings. The SILICA website (www.immunityinc.com/products/silica) lists some of the tasks that can be performed, and information that can be collected including:

- Recover WEP, WPA 1,2, and LEAP keys.
- Passively hijack web application sessions for email, social networking and Intranet sites.
- Map a wireless network and identify its relationships with associated clients and other APs.
- Passively identify vendors, hidden SSIDs, and equipment.
- Scan and break into hosts on the network using integrated CANVAS exploit modules and commands to recover screenshots, password hashes, and other sensitive information.
- Perform man-in-the-middle attacks to find valuable information exchanged between hosts.
- Generate reports for wireless and network data.
- Hijack wireless client connections via AP impersonation.
- Passively inject custom content into client's web sessions.
- Take full control of wireless clients via CANVAS's client-side exploitation framework (clientD).
- Decrypt and easily view all WEP and WPA 1/2 traffic.

Note that this list shows some of what appear to be attacks an intruder would use on a network. It is important to understand that in the context of auditing, these are tools used specifically for auditing purposes. Unfortunately, auditing tools can (and do) end up in the wrong hands and can work against you. That is why understanding what can be collected and used on a wireless network will benefit the network security professional. You'll be able to better secure the network if you understand what tools potential intruders are likely to use.

Other tools are available beyond SILICA, including compilations of security auditing tools, like the Kali Linux project. This free software includes several of the best penetration testing tools available for Linux distributions.

Auditing tools included in Kali Linux are:

- AirCrack-NG

- ASLEAP
- Bully
- Cowpatty
- EAP-MD5-Pass
- Fern-WiFi-Cracker
- FreeRADIUS-WPE
- GenKeys
- GenPMK
- GisKismet
- Kismet
- MDK3
- nmap
- WiFiARP
- WiFiDNS
- WiFi-Honey
- WiFiPING
- WiFiTap
- WiFiTE
- WireShark
- zenmap

Enterprise WIPS Topology

We talk about WIPS a lot in this chapter, for a good reason, given its capability to defend the WLAN. Let's get a little deeper into how WIPS is deployed. Enterprise-class WIPS systems usually consist of a centralized server which runs the main application, a remote console, and a number of remote sensors located at various locations throughout the organization's facilities. The sensors send a constant low-bit rate stream of data to the server application over tunneled LAN or WAN connections. The central server accumulates, logs, and reports on the data from the various sensors. The remote console can connect to the server and review the state and alarm conditions.

Enterprise WIPS may be configured to work with some popular WNMS solutions, or integrated with a given WNMS. Enterprise WIPS may also be configured to recognize and work with popular WLAN controller systems.

WIPS sensors are configured as passive—unless they are actively mitigating a threat—devices that quietly listen to all in-band radio traffic in a service area. These readings are forwarded upstream to the WIPS server. Some manufacturers enable autonomous and lightweight APs to be converted to full-time or part-time WIPS sensors, but additional licensing may be required to take advantage of this option.

It is important that the WIPS sensors use the properly monitored radio frequency bands, and that the WIPS system has the same capabilities as the installed network infrastructure wireless APs. If the sensors are not configured correctly or do not parallel the network capabilities of the specific WLAN being monitored, some events may go unnoticed and result in potential unchecked security issues.

Integrated vs. Overlay
The two high-level WIPS deployment techniques are known as *integrated* and *overlay*.

In an integrated WIPS solution, WIPS functionality is integrated into AP hardware with additional licensing required. The hardware performs dual roles as both a wireless AP and a WIPS sensor. There are different implementations of integrated solutions. Some use dedicated WIPS radios for full-time scanning, while others use part-time scanning with the same radios used for client access.

Integrated WIPS solutions often use part-time scanning to make the most of the existing AP's radios and negate the need for the complexity of an overlay approach. In this setup, the radio alternates between client access and off-channel scanning. While there is often an element of cost-effectiveness with integrated WIPS solutions, the scanning capabilities can sometimes be severely limited compared to dedicated WIPS hardware

Many AP manufacturers currently offer the option to configure an AP radio as a WIPS sensor. In most cases, this would be a dual-radio AP in which one radio is configured to provide client access, and the other provides WIPS functions. Only a few vendors to date have developed tri-radio APs where dual-radio client access is provided with the third radio configured as a WIPS sensor. This is the most robust integrated solution, but the additional radio also adds cost. Some manufacturers build "band unlocked", or software-defined APs. This means you would have the flexibility to specify which radio frequency band the AP radio would operate in. If the infrastructure consists of dual-radio APs and you were only using the 2.4 GHz band for the wireless infrastructure, for whatever reason, the second radio could also be configured for 2.4 GHz and used as a WIPS sensor.

In most cases, integrated solutions use part-time scanners. The advantages of this model are cost savings and simplicity. An AP is already cabled for Ethernet connectivity to the network, so there is no need for additional cabling, power, or mounting. (In some environments, the cost of cabling for an AP or sensor can exceed the cost of the device itself.) This also means that the WIPS solution is integrated with the AP solution, which usually indicates a shorter learning curve for the WIPS infrastructure administrator. The drawback of this solution is that WIPS scanning is only part-time. The same radio must perform both client access and WIPS scanning; thus, there is often a tradeoff between frequency/length of scans and availability for associated clients. It's worth noting that in cases where VoWiFi is supported, most part-time scanners cease scanning altogether to accommodate the latency-sensitive voice client traffic. Further, when a wireless threat is detected during an off-channel scan, the radio has limited time resources to dedicate to threat mitigation. Client access could be compromised if rogue mitigation is prioritized.

Pair these drawbacks with the number of available Wi-Fi channels for scanning, and it is easy to see why dedicated WIPS radios provide much greater security and are preferred in the highest-security environments.

Third-party WIPS vendors are often focused on wireless security as their core offering, and often provide high quality, sophisticated dedicated overlay products. The obvious drawback of this solution model is that an overlay means more hardware and wiring to that hardware. Dedicated WIPS appliances and dedicated hardware sensors can add significant cost to a wireless deployment. However, for those customers who are particularly security conscious, an overlay WIPS solution typically provides the greatest protection. Overlay sensors are often dual-radio and dual-band, which is a significant advantage for maximum scanning and threat detection.

When deploying dedicated WIPS sensor hardware, it is important to consider both the security requirements of the network and the features available with the solution, to know how many sensors to deploy and where to mount them. Since WIPS sensors listen passively to network traffic, and collision domains are not an issue, sensor radios are often configured to receive at full sensitivity. This allows for an AP-to-sensor ratio that is generally between 1:3 and 1:5, meaning you need fewer sensors than you have access points. Of course, each deployment is unique, so some situations may call for more sensors and some for less. Customers should always consult with qualified integrators or the WIPS vendor documentation to determine best practices for sensor deployment locations and quantities.

WIPS also has a strategic role in location services. When location services are desired, higher quantities of sensors provide greater location accuracy for client devices or property tracking tags. This application may drive the placement sensors at the edge of the desired access area, where you generally might not locate APs.

By deploying more sensors, you can ensure that channels are being scanned more frequently or for longer intervals, which will improve the likelihood of detecting an attack. The trade-offs are increased cost and administrative overhead, but some environments warrant both, depending on the value of the data in play on the WLAN.

Defining WIPS Policies

WIPS allows an organization to define the allowed usage policies for their WLAN within the monitoring capabilities of the WIPS. For example, your organization may have a security policy that only allows WPA2-Enterprise using 802.1X/PEAP with CCMP/AES. If this is the case, the WIPS will alarm and/or report upon seeing WEP, WPA-Personal, etc.

The WIPS can use various methods including manual configuration, to determine the identity and intention of the wireless devices which reside in a service area. If the parameters set by the manufacturer do not suit your environment, in many cases, they can be manually manipulated. For example, the manufacturer may set a deauthentication frame threshold of 10 frames in 1 minute. Any more than that, they report a deauthentication frame attack. Perhaps you want to reduce the chance of false positives, so you might set the threshold to 20 frames in 1 minute. WIPS platforms come pre-configured with an extensive catalog of attacks and attack signatures, and usually get periodic updates as new threats are discovered.

The administrator can customize the rules which govern acceptable usage of the organization's WLAN. When conditions are met within the defined rules, automated actions can be taken by the WIPS if configured to do so. For example, if a client station's MAC address starts with anything other than 00:40:96, then the WIPS should alarm and take steps to contain that station as a rogue or intruder. The customer's network needs and security policy will dictate the type of response that accompanies a policy violation or network attack. Policies should specify these actions.

Similarly, you may desire a performance report to analyze network utilization every 2 weeks. A WIPS can automate this report and have it emailed to you as a PDF.

Responses to alarms can also be customized and automated. Alarm filter configurations can be highly granular if so desired by the administrator, although default settings for many of the alarms are usually sufficient to start using the solution.

Enterprise WIPS allows the centralization of management alarms, which extends enterprise security policy to all of the organization's locations, including branch and remote offices. It also allows the organization to view trends in security policy violations and performance issues over an extended period of time.

WIPS, when present in an organization, should be running constantly, to track security policy violations. Organizations having a "No Wi-Fi" policy should use a WIPS to ensure policy cooperation. Despite the popularity of WLAN, you very well might support environments where Wi-Fi is prohibited. The best way to implement and assure an organization's "No Wi-Fi" policy, is by using an enterprise-class WIPS. Sensors can be distributed around a business' premises so that all WLAN conversations in the 2.4 GHz and 5 GHz bands are monitored in real-time. Of course, tuning may be needed, so neighboring Wi-Fi environments don't create false alarms.

Classifying Devices
After the WIPS has been configured, it will perform its initial discovery of the radio service area. At first, all of the discoverable devices will typically be considered unknown and threatening, as the WIPS hasn't yet been tuned. The administrator will be required to perform a manual identification of all known devices. Although each manufacturer uses their own classification terms to describe the devices discovered within a wireless service area, the following terms can serve as a general hierarchy for these classifications:

- Classified:
  - Friendly
    - Internal/Trusted – Authorized and supported by this organization

- ✓ Neighbor/Known/Interfering – Neighboring system that has a right to be in the same air space, but does not fall under the jurisdiction of this organization
- Rogue – Demonstrates aggressive activity. Could be wirelessly attacking the network, or connected to the network's wired backbone
  - Unclassified – Until a device can be effectively categorized, it should remain unclassified.

Classification of devices is both an initial and an ongoing WIPS configuration task that must be performed to ensure proper operation and application of WIPS policies. This is a "quality in, quality out" paradigm that requires time and effort to get right.

> Wi-Fi client devices are uniquely identified based on their MAC addresses. If a user inserts a USB Wi-Fi adapter and uses it instead of the built-in wired adapter of a laptop, for example, he may not be able to access the network without administrative intervention. This is because the WIPS system may not recognize the client laptop, but it may recognize the USB adapter in that laptop as a potentially "bad" device.

Establishing a Baseline

In order to properly configure the WIPS, it is important to effectively characterize the existing radio environment belonging to an organization. In some cases, the organization will be isolated from other WLAN users, but in most cases, there will be legitimate outside WLAN activity coexisting with the devices supported by the organization. This normal activity should be monitored, and trends established before WIPS policies are set. If you rush the WIPS into service before a meaningful baseline has been established, you'll more than likely have to start over at some point when the WIPS proves unreliable or too sensitive. Worse yet, the WIPS may cause problems for innocent neighboring networks.

Event Logging and Categorization

Enterprise-class WIPS can monitor all radio activity on a given channel in the service areas being scrutinized, on a 24x7x365 basis.

The WIPS will track, categorize, and log the wireless activities of APs, client stations, and any stations operating in ad hoc (IBSS) mode. WLAN activities which lead to vulnerabilities, and in some cases, degraded performance will be monitored and reported for further action by the administrator.

Activity Reports

Data from the remote WIPS sensors can be accumulated, sorted, and compared to the acceptable usage thresholds which were defined during system configuration. The WIPS can also gauge and predict trends in WLAN usage. All WIPS systems have a dashboard which summarizes what is happening with both security, and performance, across all sensors. The dashboard also typically allows the administrator to define areas where sensors are deployed (e.g., City, Building, and Floor). This provides a snapshot of events, per location, and an administrator can drill down to view specific activity details with a geographical frame of reference.

We've mentioned that WIPS can often be used to track performance as well as security metrics. While most customers deploy WIPS for their security benefits, the ability to gain oversight of network performance from WIPS is also significant. (It's worth mentioning that vendors including 7Signal, Cape Networks, and Wyebot all sell performance monitoring overlays, also using dedicated sensors. Performance can also be monitored through analytics platforms like Nyansa's Voyance.) In the performance monitoring role, the WIPS or WIPS-style overlay can provide extremely granular insights into hundreds of key performance indicators.

Measuring Threats

A well-configured WIPS makes the identification of unknown wireless devices clear. Though unknown devices within the range of the WIPS sensors are not necessarily hostile, they should be monitored for behavior, which indicates their intention. Even unknown devices that don't display outward signs of aggression should be viewed suspiciously, since they may be eavesdropping on network traffic in an attempt to steal information or to determine vulnerabilities in the WLAN before engaging in an active attack. We've established early in this text that you cannot detect eavesdropping with a WIPS solution (or by any other means), but you may receive an alert that an unknown wireless device exists in a specific area. Of course, a more skilled attacker would ensure that no wireless signals are transmitted from his or her device so that it could not be detected by a WIPS.

Despite the threatening connotations of the word "rogue", not all rogue devices are hostile. Recall that the generally accepted definition of a rogue network device is one that is considered unauthorized for where it is located. Remember also, that employees with no malicious intent are often behind rogue devices, installing them for a number of reasons which likely all violate the security policy. In this scenario, the rogue is not hostile or malicious, but it does still constitute a threat.

Given that each network vulnerability poses a different threat, it is helpful for WIPS to categorize threats in accordance with their severity. A rogue client performing deauthentication DoS attacks is a major threat. By contrast, the presence of a new client that is not associated to any APs and is not transmitting frames other than probe requests, is not a major threat, assuming that strong authentication and encryption is in use on the WLAN (for example 802.1X/EAP). While the administrator may want to be notified of this new unclassified client, it is not a severe problem and will likely be considered a minor threat in comparison to an active attack.

These examples illustrate different methods of mitigating threats – one, a rogue AP, and the other, an accidental association. *Threat mitigation* is a general term that includes all the different types of WIPS responses. *Rogue containment* and *port suppression* are two specific terms that may be used to identify specific WIPS response actions.

The use of threat mitigation tactics should be performed cautiously, especially using automated mechanisms. Some WIPS systems intentionally isolate suspicious APs by continuously deauthenticating all clients that associate with the intruding device. This technique renders the suspicious AP useless, as no client device will be able to maintain an association with it. Because it is part of the WIPS, it may be very tempting to liberally use this sort of mitigation technique. However, if the suspicious device should turn out to be a legitimate and harmless neighboring AP, there could be serious, civil repercussions.

Because WLANs use unlicensed frequency bands, it is legal for anyone to use the channels allocated for 802.11 WLANs. It is not legal for one WLAN user to disrupt the legal networking activities of another WLAN user. Ironically, WIPS makes it possible to do just that through intruder mitigation services. If an organization decides to implement these mitigation services to curtail the activities of an unknown wireless device operating within the same radio service area that they occupy, then they must be certain that the target device is not a legitimate neighbor device. WIPS intruder mitigation services are, in themselves, aggressive attacks against WLAN systems and their usage could result in prosecution or civil litigation. There have been a number of fines levied by the FCC against companies using rogue mitigation tools provided by their WIPS vendors, against new "rogues" like Mi-Fi devices. The smart CWSP should follow these events closely as the laws and technology try to figure out how to handle one another in our rapidly evolving WLAN world.

If an unknown WLAN device exhibits aggressive behavior against an organization's infrastructure, the response should be anticipated and defined within the enterprise WLAN security policy. If the organization allows for the enactment of self-defense mechanisms, then the policy should also define the circumstances which would predicate the use of mitigation tactics as well as how to establish a chain of evidence that supports the decision to contain or nullify the attacker.

Intrusion mitigation consists of a targeted attack against either the unauthorized device (client or AP) or an organization's own client stations in an effort to prevent successful, unauthorized associations. In this case, the nearest sensor will issue the targeted commands which are used to isolate the intruding device. Generally, this would be the only time that the sensors operate in anything other than a strictly passive (listening) mode.

Consider the three common phases of rogue management as a summary of this section:

Rogue detection – Rogue devices detected by radio scanning, their attempts to associate, or with RF spectrum activity. The different WIPS solutions will do this in varied ways.

Classification – rogues can be classified as wired or unwired by many systems. For example, in Cisco solutions, the Rogue Location Discovery Protocol (RLDP) can be used to determine if the device is connected to the wired network or not.

Mitigation/Containment – switch ports can be shut down, the location of the rogue can be identified, and the rogue can be contained – usually through the use of deauthentication frames.

Compliance Reporting

Recall that the use of a WIPS greatly simplifies the requirement to provide legislated security compliance. Various compliance reports may be pre-formatted and included as part of the WIPS' report manager sub-system. This allows the administrator or security officer to generate a near real-time compliance report which can then be supplied to visiting auditors or inspectors.

Forensics

WIPS can also retain continuous logs of all known activity on a 24x7x365 basis. Since this information is automated, it can be used as part of the evidentiary chain during criminal or internal corporate security investigations. Some WIPS even include a forensics analysis component which can be valuable in the event that the organization decides to pursue litigation or prosecution against an apprehended intruder.

Device Location & Tracking

Some WIPS platforms can provide device location identification by using sophisticated radio techniques, including triangulation, RF Fingerprinting, and Time Difference of Arrival (TDoA). Understanding the basic differences of each will serve you well as a CWSP.

Triangulation and RF Fingerprinting are the most common techniques for device tracking. With triangulation (or trilateration), multiple sensors (at least three) that have a view of the intruding device, measures signal strength metrics and sends this data to the WIPS server. The WIPS server compares these signal strength measurements and performs a calculation based on known path loss formulas that can identify the location of the intruder. The location of the device can then be graphically plotted on a floor plan.

A modern technique known as RF Fingerprinting may also be used by some WIPS as a method to provide a more accurate device location solution. RF Fingerprinting requires that a detailed analysis survey be performed in advance. After a WIPS solution has been installed, a manual walkabout is performed with a client device, to calibrate the WIPS system. Later the radio signature of a moving target is tracked throughout the premises, and the resulting signal strengths are logged to a database. When used in conjunction with the triangulation or TDoA information, this RF Fingerprint detail can improve the accuracy of the WIPS, allowing it to locate the offending intruder within a relatively precise area.

In TDoA systems, a WIPS platform uses the known speed of radio wave travel to locate a device. As the WIPS server processes frames from the sensors, it uses time stamping to mark the first instance of a specific frame. Then, as subsequent instances of this same frame are recorded by other sensors, the WIPS server can compare the time delay of the same frame as received by different sensors, to determine the distance of the transmitting device from the sensors. Again, the goal is for WLAN security staff to rapidly locate the intruder device and remove it from the premises, and discreet time measurement of frames is beneficial to that goal

> The latest trend in location systems is Angle-of-Arrival (AoA) and Angle-of-Departure (AoD). AoA fully functions on the locationing device, such as an AP. AoD requires a located device that supports it. AoA and AoD take advantage of the multiple antennas in a system to locate devices. These locationing methods are more accurate (to the 10-centimeter range) than traditional methods. Bluetooth Low Energy (BLE) devices running the latest versions of Bluetooth may support this method and it is a locationing method you will see utilized much more in the coming years.
> -Tom

Integrated Spectrum Analysis

Another advanced feature which may be found within a limited number of enterprise-class WIPS is an integrated spectrum analysis engine. With this feature, a number of the remote sensors contain an integrated spectrum analyzer chipset which can upload its data to the central WIPS server.

This capability allows the remote administrator to view the exact state of a remote radio environment through the management console. Remember that spectrum analysis is associated with Layer 1, and using a spectrum analyzer allows the accurate diagnosis of spectrum problems, such as Layer 1 denial-of-service attacks, from the remote console.

New 802.11 Challenges

As the 802.11 specification continues to expand, new features that create new problems, are introduced. 802.11n, ac, and ax can be problematic for older WIPS systems.

With the new PHY frame formats of 802.11n/ac/ax and the 40 MHz, 80 MHz and possibly 160 MHz wide channels, legacy IEEE 802.11a/b/g WIPS systems will not be able to recognize and/or interpret some newer standard's transmissions. This means that some attacks could be conducted by 802.11n/ac devices that are not identified by those dated WIPS systems.

IEEE 802.11w and WPA3 introduce new frame protection features that provide management frame authentication. When these features are enabled, only securely associated stations will be able to terminate the session with a disassociation or deauthentication frame. Before 802.11w, a deauthentication or disassociation frame was a notification and could not be refused. This functional operation made it possible for any WIPS to terminate an active association with a deauthentication or disassociation frame. However, when these management frames require authentication (MIC validation), some rogue containment measures will no longer work. 802.11w is a double-edged sword, so the decision to use it or not should be carefully considered where it is optional.

Monitoring in the Cloud

Cloud-managed WLAN systems continue to gain popularity. This technology is becoming widely used and popular for many different markets including enterprise wireless network deployments. In addition to the cloud, some manufacturers create on-premises solutions so they can have the features of a cloud solution in their own data centers. Cloud-managed infrastructures are sometimes referred to as controller-less solutions because they operate without the need for a hardware controller. Some companies manufacture infrastructure devices that are only cloud-managed and some previously built controller= only solutions now also have some cloud-managed models. These companies primarily provide cloud-managed solutions:

- Aerohive
- Mojo Networks
- Cisco-Meraki
- Mist Systems
- Open Mesh

Cloud-managed monitoring and WIPS solutions have many (if not all) of the same features like hardware and software-based WIPS solutions. The main difference is that there may not be a physical presence within the local network. Instead, the information collected is accessible from anyplace with an active Internet connection. Many of these solutions offer the following features, and more:

- Alarm management
- Automatic device classification
- Event logging and categorization
- Location tracking features
- Auditing and forensics
- Rogue AP detection
- BYOD policy enforcement
- Traffic analytics
- Security monitoring
- Regulatory compliance
- Reporting

With some of these solutions, the cloud-managed AP can act as a part-time or full-time WIPS sensor based on administrative configurations. The same software that is used to manage the wireless infrastructure is also used for the WIPS functionality, and usually, a common interface provides tight integration between the two functions. The main benefit to these solutions is that there is no hardware, server or appliance that needs to be purchased or installed. This can be a significant cost savings to many organizations. It also allows for smaller networks such as small office home office (SOHO) and small and medium-sized businesses (SMBs) to be able to easily incorporate a WIPS solution into their infrastructures. Their "reachable from anywhere" paradigm is also a significant selling point,

WNMS Security Features

A Wireless Network Management Systems (WNMS) can often be used to provide much of the functionality found in some WIPS solutions. For instance, a WNMS can be used to identify trusted, known, and rogue devices. However, a WNMS usually differs from a WIPS in that they do not have the ability to use dedicated, remote, hardware sensors. Instead, they use APs as sensors. You may have to buy additional licenses for the WNMS to leverage WIPS functionality, and APs in use can function as dedicated full-time sensors or split their operations between WIPS and providing client access.

A WNMS can identify and display authentication types in use on a per-association basis. A WNMS can be configured with graphical floor plans to encompass multiple floors of multiple buildings. The WNMS can then display coverage maps and identify user locations, rogue device locations, information on interferers, and a range of other indications that help show the status of system security and health.

WLAN Controllers

Controller-based WLANs may allow some of their APs to be recommissioned as dedicated, remote hardware sensors. Some WLAN controller systems also allow their controller-based APs to multitask these duties, switching briefly between sensor and AP operations. However, in this scenario, it may be possible for the sensor to miss some crucial information when it is busy functioning as an AP.

While controller-based WIPS functions are usually not as robust as dedicated third-party WIPS products, they still offer substantial security benefits and configuration options. The cost savings of integrated WIDS/WIPS functionality may be significant when compared to dedicated WIPS options. Beyond smaller WLAN environments, WLAN controllers are typically managed by a WNMS. This includes the WIPS capabilities of both, which are integrated through licensing and the UI of the WNMS.

Distributed Protocol Analysis as a Monitoring Solution

Wireless LAN Protocol Analyzers may have the ability to use distributed sensors to accumulate and report packet/protocol capture detail from remote locations to an administrative desktop application. This allows the administrator to view live, Layer 2 data from different points in the WLAN, in a remote console.

Dedicated WIPS platforms are also improving traffic analysis functionality in newer implementations by providing remote frame captures and decodes.

Protocol analysis includes frame exchanges and frame decoding.

Protocol analyzers can be important wireless security analysis tools for both network security administrators and intruders alike. Protocol analyzers can be used to capture and save wireless traffic in formats that can be imported by attack applications such as password crackers. The use of protocol analyzers is a major component of the Certified Wireless Analysis Professional (CWAP) certification and a routine part of the job for WLAN administrators.

Not all wireless protocol analyzers can decode every OSI layer. Some protocol analyzers only display 802.11 MAC Layer networking information while others can capture, filter, decode, and display all network traffic, including user data from layers 2-7. Most protocol analyzers allow the insertion of a preshared key, so that captured, encrypted traffic can be unencrypted and displayed in real-time or saved and decoded later. This feature is a must-have on PSK networks but is not practical when 802.1X/EAP is in use.

Some protocol analyzers can capture and reconstruct TCP sessions into their application layer information (layers 4 - 7) while others can generate and transmit customized 802.11 frames (Layer 2). Distributed protocol analyzers can be placed throughout an enterprise and be configured to supply constant data captures of wireless frames from each location to a centralized console and database. This capability can be invaluable for forensics work.

Popular Protocol Analyzers used for 802.11 (at the time this was written):

- Wildpackets - Omnipeek (Savvius)
- AirMagnet – WiFi Analyzer (Fluke Networks/Netscout)
- Tamosoft - Commview for WiFi
- Network Instruments – Observer Analyzer
- Wireshark

When using a WLAN protocol analyzer, special interface drivers are typically used. These drivers may not offer the full supplicant feature set of the normal use-case drivers. After performing protocol analysis, it is important to revert drivers back to the standard use-case drivers for network access, or the computer running the analyzer may not be able to access the WLAN

Working with IEEE 802.11 Frames

Knowledge of the different 802.11 frame types, their formats, and usage are crucial to being able to interpret protocol analyzer captures and decodes. Competency in protocol analysis is expected from any network administrator that intends to provide a secure, high-performance, wireless network. For more detailed information on this, see the CWAP Official Study Guide.

Unlike Ethernet, the 802.11 protocol uses many different frame types, but all of them are based on a general frame format. 802.11 uses specialized frames for:

- Data

- Control
- Management

IEEE 802.11 Frames vs. 802.3 Frames

Frames used by Project 802, which is the IEEE project inclusive of all 802 standards, all have similar structures. They are all defined as fields made up of bits, which together form octets, and ultimately, frames. This similarity in frame structure makes for easier conversion from 802.3 networks to 802.11 networks and vice versa. For example, a frame originating from a wired client and destined for a wireless client will first be transmitted on the wire as an 802.3 frame. Then the access point will strip off the 802.3 headers and reframe the data unit as an 802.11 frame for transmission to the wireless client. The access point bridges 802.11 and 802.3 networks.

The first difference between 802.3 and 802.11 frames is the frame size. 802.3 frames support a maximum MSDU payload size of 1500 bytes or octets (There are "jumbo" frames allowed in 802.3 Ethernet networks that are larger than the standard-defined 1500 bytes, but these are beyond the CWSP course). 802.11 frames support a maximum MSDU payload size of 2304 bytes (or larger in 11n and 11ac). It seems that extra processing will have to occur in order for 802.3 and 802.11 networks to coexist because of the different frame sizes. It's a reasonable assumption that a certain amount of frame fragmentation will have to occur to convert 802.11 frames to 802.3 frames. However, since TCP/IP is the most commonly used protocol, and since IP packets are usually no larger than 1500 bytes, the vast majority of data units passed to the 802.11 MAC Layer will be 1500 bytes or smaller anyway, allowing for easy conversion to the 802.3 format.

Most network engineers are familiar with the IP Maximum Transfer Unit (MTU) of 1500 bytes that exists because of Ethernet networks. Wireless LAN standards were designed to easily bridge with Ethernet. For this reason, the IP MTU on a WLAN is held to 1500 bytes even though it could be as high as 2304, minus 8 for the LLC SNAP header. A 1500-byte IP PDU becomes a 1508-byte LLC PDU with the LLC SNAP header and this, in turn, becomes the wireless LAN MSDU. The LLC 8-byte header is often ignored and results in the varied and confusing byte or octet size values you read and hear about.

The second difference between the 802.3 and 802.11 frames is the MAC address fields. Both frame types use the same standard for MAC address structuring based on Clause 5.2 of the IEEE 802-1990 standard. But, 802.3 frames have only two MAC address fields, whereas 802.11 frames have one, two, three, or four. These four MAC address fields can contain four of the following five MAC address types, and the contents will be dependent on the frame subtype:

- Basic Service Set Identifier (BSSID)
- Destination Address (DA)
- Source Address (SA)
- Receiver Address (RA)
- Transmitter Address (TA)

The 802.11 standard documents the frame types supported by the 802.11 MAC. According to the standard, there are three frame types supported in 802.11 networks: management frames, control frames, and data frames. The *Type* subfield in the *Frame Control* (FC) field of a general 802.11 frame may be 00 (management), 01 (control) or 10 (data). The *Subtype* subfield determines the subtype of frame, within the frame types specified, that is being transmitted. For example, a *Type* subfield value of 00 with a *Subtype* value of 0000 is an association request frame; however, a *Type* value of 10 with a *Subtype* value of 0000 is a standard data frame. Understanding all the details about the frame structures and formats is not required of a CWSP; however, it would be of great benefit to you to review the 802.11 standard, which defines each frame and each frame field with diagrams. Should you advance to CWAP, this information will be invaluable.

Management frames are used to manage access to wireless networks and to move associations from one access point to another within an Extended Service Set. Control frames are used to assist with the delivery of data frames and must be able to be interpreted by all stations participating in a Basic Service Set. This means that they must be transmitted using a modulation technique and at a data rate compatible with all clients participating in the Basic Service Set. Finally, data frames are the actual carriers of application-level data. These frames can be either standard data frames or Quality of Service (QoS) data frames for devices supporting the 802.11e amendment. To round out the discussion, you should know that the type value of 11, with any subtype, is a reserved frame type. This simply means that it is not used today, but is reserved for any future needs.

Frame Exchanges

While you will not be required to understand every detail of the frames and frame exchanges that occur on a WLAN in order to become a CWSP, you will need to understand the basics of frame exchange sequences and the frame-level flow of creating a WLAN, accessing a WLAN and disconnecting from a WLAN. Though much of this is a review at this point in the text, it is important to reiterate. You should be able to define and explain the following basic MAC Layer functions:

Scanning – Before a station can participate in a Basic Service Set, it must be able to find the access points that provide access to that service set. Scanning is the process used to discover Basic Service Sets or to discover access points within a known Basic Service Set.

Synchronization – Some 802.11 features require all stations to have the same time. Stations can update their clocks based on the timestamp value in Beacon frames.

Frame Transmission – Stations must abide by the frame transmission rules of the Basic Service Set to which they are associated. These rules are defined for standard operations (DCF) and QoS solutions within the 802.11 standard.

Authentication – Authentication is performed before a station can be associated with a Basic Service Set.

Association – Once authentication is complete, the station can become associated with the Basic Service Set. This includes the discovery of capability information in both directions – from the station to the access point and from the access point to the station.

Reassociation – When a user roams throughout a service area, they may reach a point where one access point within an Extended Service Set will provide a stronger signal than the currently associated access point. When this occurs, the station will reassociate with the new access point (provided the station's drivers are behaving).

Data Protection – Data encryption may be employed to assist in preventing crackers from accessing the data that is transmitted on the wireless medium.

Power Management – Since the transmitters/receivers (transceivers) in wireless client devices consume a noteworthy amount of power, power management features that assist in extending battery life by causing the transceiver to sleep for specified intervals, are provided.

Fragmentation – In certain scenarios, it is beneficial to fragment frames before they are transmitted onto the wireless medium. This type of scenario most often occurs due to intermittent interference.

RTS/CTS – Request to Send/Clear to Send is a feature of 802.11 that will help prevent hidden node problems and allow for more centralized control of access to the wireless medium.

Spectrum Analysis

Spectrum analyzers capture raw RF signals (Layer 1) and display visual representations of ambient signals in terms of frequency and signal strength. Spectrum analyzers are the most useful tool for performing RF-specific security audits (e.g., locating RF DoS attacks) and finding RF interference types and sources. These analyzers give the wireless network professional a unique monitoring visibility that can be leveraged to accomplish many performance and security-related tasks.

Some spectrum analyzers have the ability to identify suspicious activities and devices, and to hone in on the trespassing devices based on signal strength comparisons and known device RF signatures. Hardware and software combinations can also automatically classify many common types of RF sources such as Bluetooth devices, wireless video cameras, microwave ovens, and cordless phones based on signal signatures.

Spectrum analyzers may have distributed sensors and be configured to send their reports to a centralized console. This information can be used for forensic purposes if an attack should occur. The spectrum analysis information can be searched and analyzed to provide clues and evidence as to the time, location, and possible identity of the perpetrator. Careful handling of this information is important if the analysis is intended to be used as legal evidence in court. For more information on this aspect, the corporate legal department should be consulted and have input into the security policy formation.

Spectrum analysis will facilitate the discovery of intentional and unintentional DoS attacks by RF interferers. As an added benefit, being able to effectively use a spectrum analyzer will make you better at general wireless network support, too. In fact, given the importance of RF to Wi-Fi, it's fairly impossible to be good at wireless without being able to use spectrum analyzer tools.

Physical Layer Defenses

The only defenses against a Physical Layer Denial-of-Service (DoS) attack are great distances between the WLAN and potential attackers or to RF-harden the building or room where wireless communications are in use. Likewise, the way to keep RF signals from propagating outward is to RF-harden the coverage areas. Some methods used to RF-harden an area against RF DoS attacks include:

- TEMPEST protection –
    - Government/Military strength RF leakage protection.
- Anti-RF paint / wallpaper
- Faraday cage / Faraday Shielding
- Faraday shielding, in which the metal mesh spacing is less than one-half the wavelength being used, can effectively restrict that radio wave.

Due to the expenses involved, installation of anti-RF paint or Faraday shielding is not typically a practical solution to wireless security vulnerabilities for most organizations but is not uncommon in certain military and government facilities. Also, shielding related to X-ray rooms, and other hospital similarities can be challenging for WLAN networking as RF gets blocked quite efficiently in these locations.

Examples of spectrum analyzers include the AirMagnet Spectrum XT device (and software), the MetaGeek Wi-Spy dBx (and software, which includes the MetaGeek software or CommView for WiFi, which can include spectrum analysis from a Wi-Spy dBx device), and Ekahau's Sidekick. Oscium's WiSpry series uses the lightning connector on Apple mobile devices, with a compatible app, to provide a unique spectrum analysis option. AirMagnet Spectrum XT is a USB form factor protocol analyzer for use with laptops (or desktops, if you have a need for stationary analysis). The features of the AirMagnet Spectrum XT include:

- USB form factor for use with practically any modern computer
- Combining spectrum analysis with traffic analysis (traffic analysis required a compatible Wi-Fi adapter)
- Automatically identify WLAN and non-WLAN interference sources
- Real-time RF spectrum and WLAN graphs

- Integration capabilities with other AirMagnet solutions such as Survey PRO and Wi-Fi Analyzer PRO
- Recording and playback of spectrum analysis sessions
- Support for both the 2.4 GHz and 5 GHz bands

Laptop-Based Intrusion Analysis

It is possible to combine protocol and spectrum analysis with expert security knowledge to manually perform WLAN monitoring with a single laptop. In many smaller networks, administrators will choose to perform manual analysis. This decision is often based on smaller budgets. At the same time, larger networks may also do spot checks with "boots on the ground" in addition to running enterprise-class IDS or IPS solutions. Manual analysis is performed with spectrum and protocol analyzers.

Examples of protocol analyzers include Wireshark, AirMagnet Wi-Fi Analyzer Pro, OmniPeek, and CommView for Wi-Fi. When selecting a protocol analysis laptop-based solution, the engineer must consider the following:

- Protocol analysis software does not work with every wireless adapter or chipset, and an adapter that works with the software (usually specified by the software vendor) must be selected
- Protocol analysis software has a wide range of features which varies between vendors, and the engineer must carefully analyze available options to choose the right solution
- All of the mentioned protocol analyzers can import PCAP file formats so captures can be performed using external tools/software and then analyzed using the specific wireless protocol analyzer software of choice

The right adapter must be chosen to perform protocol analysis with a given software solution. The majority of internal adapters do not work with most protocol analyzers for Wi-Fi frame captures, and USB adapters are often used for this reason.

The following types of intrusions can be located with a laptop analyzer:

- Rogue APs
- Unauthorized clients
- Denial of Service attacks
- MAC Layer wireless attacks

To locate a physical DoS attack source, you can use the laptop analysis software to find the location where the signal is strongest. Then, you can simply look around until you find the RF generating source. Tools like AirMagnet Spectrum XT often include device locator tools with graphical elements like that represented in the following image:

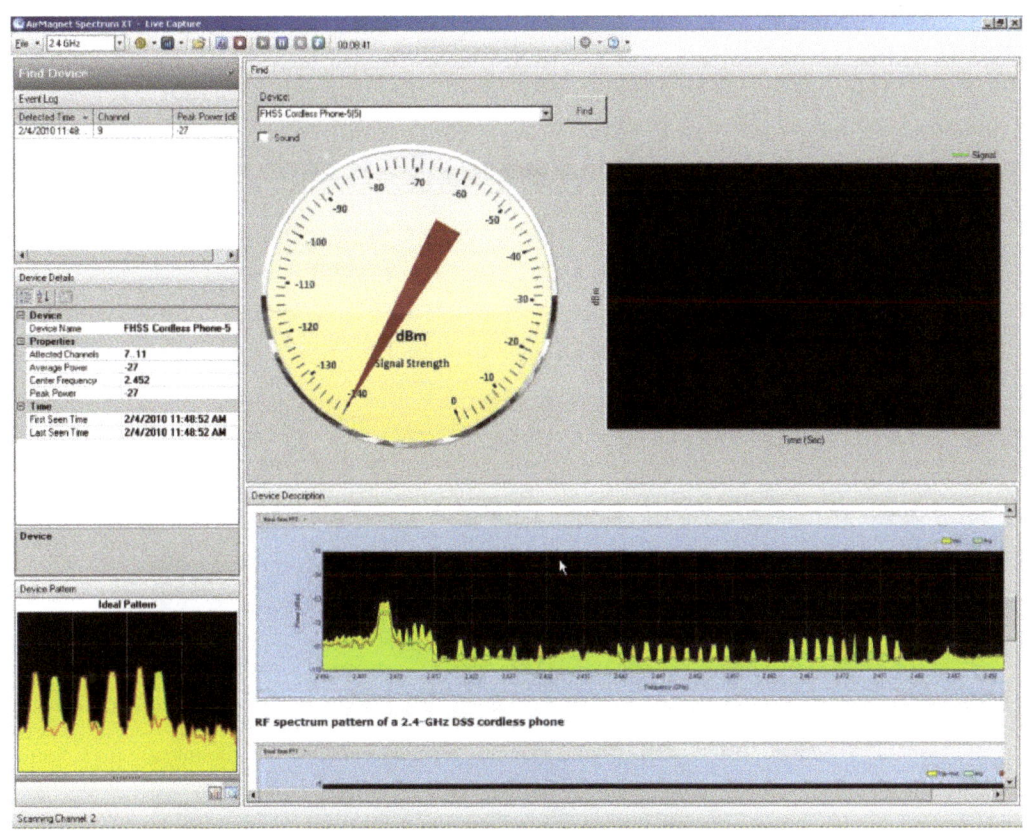

AirMagnet Spectrum XT Device Locator

In most cases, laptop-based spectrum analysis products are used for auditing purposes. Audits can expose security risks and accidental interference sources. Devices like AirMagnet Spectrum XT can locate and identify non-WLAN interference sources such as:

- Baby monitors
- Cordless phones
- Microwave ovens

- Bluetooth devices
- Wireless cameras
- Game controllers
- Digital video devices

Intentional security attacks can also be detected during an audit. You can locate RF jammers using Spectrum XT and you can locate laptops being used to flood WLANs with "junk" frames (frames that do not contain meaningful data), thanks to the ability to perform parallel protocol analysis as well as the spectrum analysis. When AirMagnet Spectrum XT locates an interference source, it can provide detailed information about the source including:

- Peak and average output power
- Center channel frequency
- Channels impacted by the interference
- First and last times the source was detected
- Number of times the source was detected

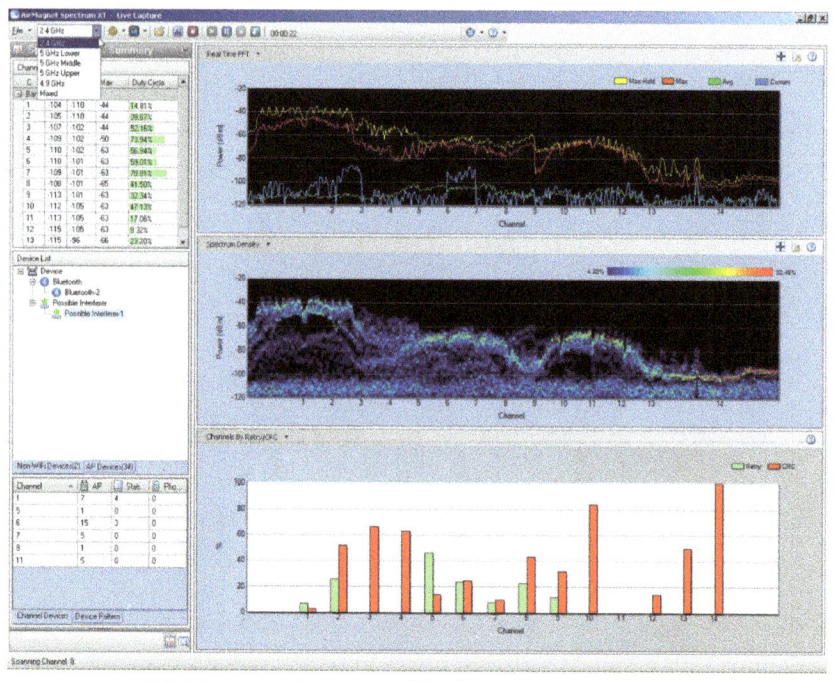

AirMagnet Spectrum XT Spectrum Analysis

When used with a directional antenna, spectrum analyzers can often pinpoint the location of the interference source down to a few meters. The following figure shows the AirMagnet Spectrum XT application running a standard spectrum analysis view with the ability to select different frequencies. The tool can be used for 802.11n and 11ac analysis since it supports both 2.4 GHz and 5 GHz bands.

The following image shows AirMagnet Spectrum XT displaying non-Wi-Fi devices. In this image, a Bluetooth device and a potential interference device is shown.

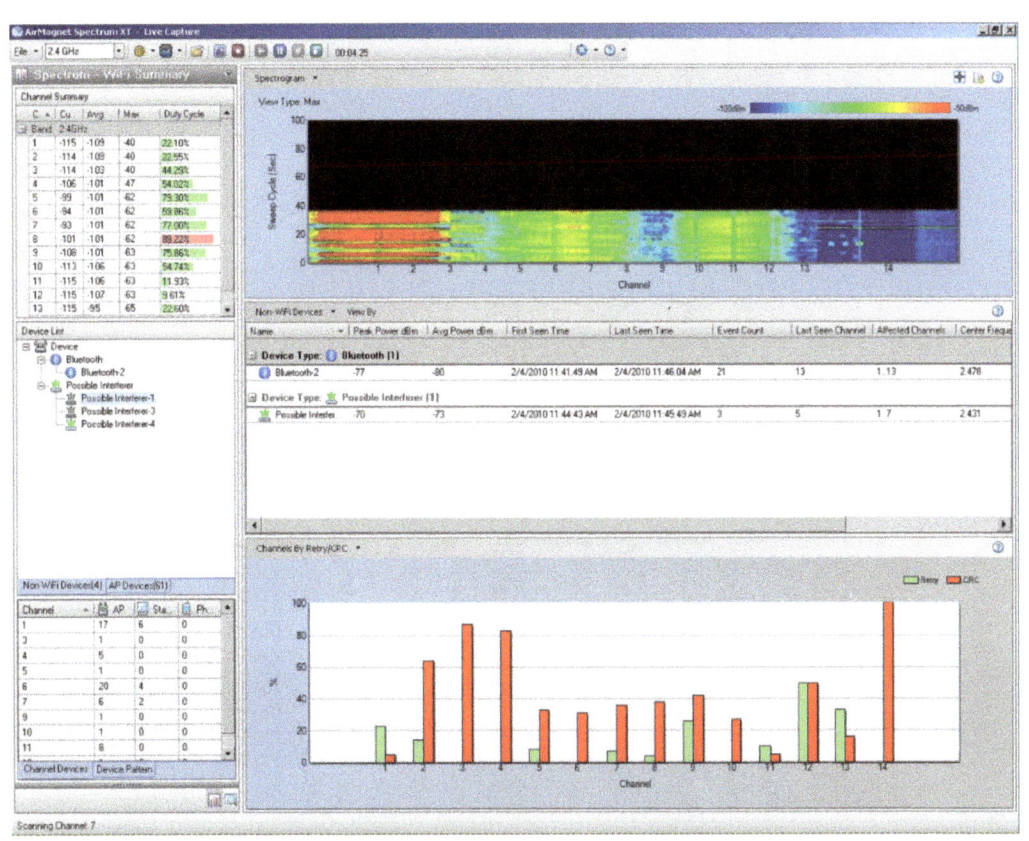

AirMagnet Spectrum XT Showing non-Wi-Fi Devices

Specialty Analysis Devices

The Netscout AirCheck G2, and to a certain extent the LinkRunner G2 represent a class of hand-held testing platforms that aggregate many functions provided by other analysis and monitoring tools. For example, the AirCheck G2 can do a wireless packet capture, wireless environment characterization and device identification, rudimentary spectrum analysis, and even baselining if used properly. This sort of non-PC, non-appliance, and highly portable tool is a tactical Swiss Army Knife of capability that can certainly be used in security-oriented WLAN work.

Chapter Summary

In this chapter, you learned about intrusion monitoring solutions. Wireless intrusion detection systems (WIDS) detect potential intrusions and may log or report the intrusion. Wireless intrusion prevention systems (WIPS) detect potential intrusions and react to the attempted intrusion to prevent further damage. Common features of WIPS solutions include rogue AP and unauthorized client detection as well as detection of DoS and MAC Layer wireless network attacks. You also learned about using a laptop-based spectrum protocol analyzer, or mobile stand-alone specialty analysis tools to monitor smaller environments or to locate and detect specific kinds of attacks in environments of any size.

Review Questions

1. As you talk with a client about wireless security, she asks you the difference between WIPS and WIDS. How do you answer?
    a. Both can detect threats, but WIDS can also do threat mitigation
    b. Both can detect threats, but WIPS can also do threat mitigation
    c. WIDS can only detect, WIPS can only mitigate
    d. WIPS is only used in PCI environments

2. You are auditing the WLAN security practices of a stadium that uses wireless Point of Sale terminals. Which regulation are you most concerned with regarding credit card operations?
    a. FERPA
    b. HIPPA
    c. Sarbanes-Oxley
    d. PCI

3. You are troubleshooting a WIPS system used to secure an 802.11ac WLAN that doesn't seem to be working as expected. Which of these is not a potential reason for the trouble?
    a. The WIPS system wasn't staged right
    b. The WIPS system uses 802.11a/g sensors
    c. The WIPS system has one sensor for every 3 access points
    d. The WIPS sensor isn't part of the IT change control process

4. A fellow CWSP is seen doing wireless protocol analysis with a laptop that has three wireless USB adapters in use. What might she be doing, specifically?
    a. Troubleshooting VoIP roaming
    b. Calibrating a PCI system
    c. Setting up mobile WIDS
    d. Tuning the WIPS for voice traffic

5. What type of WIPS might stop scanning in the presence of voice traffic?
    a. Dedicated WIPS
    b. Full-time WIPS
    c. Integrated WIPS
    d. Distributed WIPS

6. You have been asked to provide a customer site with the ability to detect passive eavesdroppers. Which tool do you recommend?
    a. WIDS
    b. Protocol Analyzer
    c. IoT Sensors
    d. None, as passive eavesdropping can't be detected

7. You've been hired to conduct clandestine "security posture auditing" at a corporate site. What's another description for what you've been hired to do?
    a. Penetration testing
    b. Installing NAC agents on each host
    c. Network scanning
    d. Social engineering

8. Which of the following can enterprise WIPS be configured to work with?
    a. WNMS
    b. WLAN controllers
    c. Sensors
    d. All of these

9. A WLAN that employs which type of security should be able to meet HIPAA requirements?
    a. 802.11ac
    b. 802.11i
    c. 802.11dss
    d. 802.11avc

10. When it comes to WLAN security, Sarbanes-Oxley, PCI, and GLBA are all examples of what?
    a. Security best practices
    b. Optional enhancements
    c. Legislated security requirements
    d. Security overlays

11. One important outcome of an initial security audit is the _____.
    a. Baseline
    b. Security Policy
    c. Inventory
    d. Executive summary

12. How can you protect against rogue access points being installed from the wired network perspective?
    a. Keep doors to common spaces locked
    b. Power down network switches until needed
    c. Disable individual network ports not in use
    d. Do not make patch cables easily available

13. Which of the following will best help the CWSP identify security threats at Layer 1?
    a. Spectrum analyzer
    b. Protocol analyzer
    c. Integrated analyzer
    d. WNMS

14. WLAN monitoring tools include all of the following except which of these?
    a. WIPS
    b. WNMS
    c. Client utilities
    d. Protocol Analyzer

15. You've been asked to advise on the purchase of a WIPS system. Your customer has a modern, robust WLAN, but budget is a significant concern for a potential WIPS project. Knowing this, what do you recommend?
    a. Dedicated WIPS
    b. Stand-alone WIPS
    c. No WIPS because of expense
    d. Integrated WIPS

Review Answers

1. **B**. Knowing this fundamental difference between WIPS and WIDS is important
2. **D**. PCI is all about retail environments that process credit cards
3. **C**. A ratio of 1:3 to 1:5 WIPS sensors to APs is generally accepted design
4. **A**. Troubleshooting roaming uses multiple adapters to listen to different channels as a client moves between cells
5. **C**. Integrated WIPS often defers to voice traffic above all other functionality
6. **D**. No system will detect passive eavesdropping
7. **A**. Many wireless security professionals work in the pen-testing realm
8. **D**. Some WIPS systems can integrate with a variety of other systems
9. **B**. 802.11i-compliant networks should have all of the ingredients to be HIPPA compliant
10. **C**. Many aspects of wireless security are based on legislated requirements
11. **A**. A security baseline is an important outcome of the initial security audit
12. **B**. Disabled network ports won't support rogue APs
13. **A**. Spectrum analyzers work in the physical layer domain
14. **C**. WIPS, WNMS, and protocol analyzers are all WLAN monitoring tools, whereas client utilities are single-device views and configuration tools
15. **D**. In this case, integrated WIPS will provide benefit while saving money, versus stand-alone/dedicated WIPS

Chapter 10: WPA3 and OWE

Objectives Covered

3.1 Select the appropriate security solution for a given implementation and ensure it is installed and configured according to policy requirements

4.4 Implement appropriate maintenance procedures including license management, software/code upgrades, and configuration management

The existence of wireless networks is made possible through the use of electromagnetic waves being transmitted and received via various types of antennas. Data is digitized and placed on carriers within these waves and sent across the airwaves with the intent of passing information from one device to another. This is a brilliant process of getting information sent quickly at much more affordable costs with one major downfall. It is, in fact, possible for anyone on the same frequency and channel to listen/receive all data transmitted across these wireless transmissions if they too have the antenna and other equipment to do so. This is why wireless network security is constantly scrutinized, evaluated and revised in order to stay ahead of those individuals looking to compromise data privacy. Encryption processes, key lengths, algorithm choices, and identity verifications all continue to be enhanced, as cracks within the defensive mechanisms become evident, revealing critical vulnerabilities in security.

WPA3 is not a wireless security standard. It is, however, a wireless security certification like its predecessors, WPA, and WPA2. WPA3 certifies the implementation of standards-based protocols and algorithms such that they will be compatible with other WPA3-certified compliance implementations. Opportunistic Wireless Encryption (OWE) is a standard that is not part of WPA3 certification; however, it is being implemented by many WLAN vendors and must be considered in this chapter on recent enhancements to 802.11 wireless security.

This chapter will expand on the descriptions provided above and, in the process, detail the inner-workings of WPA3-certified systems as well as OWE systems. Both solutions have an important role to play in modern wireless security.

WPA3 is discussed first in this chapter because it is likely to be more widely deployed and supported than OWE. It will be included in all new 802.11-based systems released until another security solution supplants it. WPA3 is the latest in the line of security certifications, as of 2019, for general 802.11 wireless access offered by the Wi-Fi Alliance. It implements stronger encryption, better authentication, and improved compliance with some government requirements. The news of its availability was announced on June 25, 2018 and vendors began implementing it in solutions shortly thereafter. It was enhanced in April of 2019 to address the Dragonblood vulnerabilities exposed by Mathy Vanhoef and Eyal Ronen which relates to WPA3-Personal (password-based) authentication.

> The *Dragonblood* vulnerability allows attackers to use a side-channel attack to gain small amounts of information related to the authentication process. The attacker can perform the attack several times, and the cumulative small amounts of information can reveal the password used. All WPA3 systems, even those released before the publication of the vulnerability, can be patched through software alone to prevent the attack.

The Need for WPA3

The predecessors to WPA3 include Shared Key Authentication, WEP, WPA, and WPA2. Of these early security solutions, Shared Key Authentication and WEP are by far the weakest. Both WPA and WPA2 can be cracked under the right conditions, but Shared Key Authentication and WEP can be cracked in any conditions and at this point should not be deployed on your wireless network. Surprisingly, more than a decade and a half after 802.11i-2004 was ratified, we are still seeing some WEP secured networks in the wild (if you really want to call it "secured"). WPA, as you learned in preceding chapters, was a stop-gap solution until hardware could be replaced to support WPA2. WPA2 is the most widely used 802.11 security solution in 2019 and, for this reason, in the following section, we will focus on vulnerabilities of WPA2 that led us to the need for WPA3.

The actual Man-in-the-middle attack that exploited the weaknesses of WPA2, better known as KRACK or Key Reinstallation Attack takes advantage of resetting over-the-air nonces in order to reuse encryption keys. Unfortunately, handshakes take place in many security components such as Wireless Network Management, Fast Basic Service Set Transition, Tunneled Direct Link Setup and several others. By resetting replay counters and manipulating nonces, weaknesses in WPA2 allows an attacker to reinstall Pairwise Transient Keys, Group Temporal Keys, and several other keys involved in encryption setup processes, with the end result being a client device that is assumed to be transmitting securely encrypted information, may, in fact, have all of its transmitted and received data intercepted and decrypted through this man-in-the-middle attack.

FEATURES	WPA2	WPA3
STANDS FOR	Wi-Fi Protected Access 2	Wi-Fi Protected Access 3
WHAT IS IT?	Security protocol developed by the WI-FI Alliance for use in securing wireless networks.	Next generation of WPA2 and has better security features.
RELEASE YEAR	2004	2018
ENCRYPTION	WPA2 uses the Advanced Encryption Standard (AES) with CCMP standard.	AES-GCM encryption & Elliptical Curve Cryptography of CNSA Suit B.
SESSION KEY SIZE	128-bit	192-bit
HANDSHAKE PROTOCOL	Pre-Shared Key (PSK) exchange protocol.	Uses the Simultaneous Authentication of Equals (SAE), also known as Dragonfly Key Exchange, with Forward Secrecy feature.
SECURITY MODES	WPA2 Personal: Pre-shared Keys (PSK) WPA2 Enterprise: IEEE 802.1X (Radius)	WPA3 Personal: 128-bit SAE (Optional 192-bit) WPA3 Enterprise: 192-bit SAE
AUTHENTICATION	Uses 802.11x Open Authentication & Extensible Authentication Protocol (EAP)	Opportunistic Wireless Encryption (OWE). OWE also protects open "unsecured" networks. e.g. Wi-Fi at libraries or cafes.
DATA INTEGRITY	CBC-MAC having 64-bit Message Integrity Code (MIC)	Secure Hash Algorithm-2 for each input.
WIRELESS CONNECTION PROTOCOL	Wi-Fi Protected Setup (WPS) – Vulnerable	Wi-Fi Easy Connect using Device Provisioning Protocol (DPP) – Secure.
PROTECTED MANAGEMENT FRAMES FOR IMPROVED RESILIENCY	Mandates support of PMF since early 2018. Older routers with unpatched firmware may not support PMF.	WPA3 mandates use of Protected Management Frames (PMF).
VULNERABLE TO KRACK ATTACKS	Yes.	No, due to SAE key exchange.
VULNERABLE TO OFFLINE DICTIONARY ATTACKS	Yes.	Blocks authentication after a certain number of failed log-in attempts.

WPA2 vs. WPA3

Some enterprise users still entertain the belief that only WPA and WPA2 personal is vulnerable to these key reinstallation attacks, but this just isn't the case. The attack isn't performed on the authentication method being used to get onto the network. The attack is performed on the frame exchanges during handshakes that generate encryption keys. This includes Enterprise networks utilizing various deployments of AES. No matter which version of authentication is being used, there are always a series of frame exchanges that takes place in order to generate encryption keys. This is where the problem resides. It is important to note that the Wi-Fi Alliance released a security update in October of 2017 that helps with the vulnerability exploited in KRACK. The Wi-Fi Alliance also has provided a detection tool for this vulnerability, and it is available to members for download here: https://www.wi-fi.org/security-update-october-2017

Let's take a look at the 4-way handshake and what happens in the exchange:

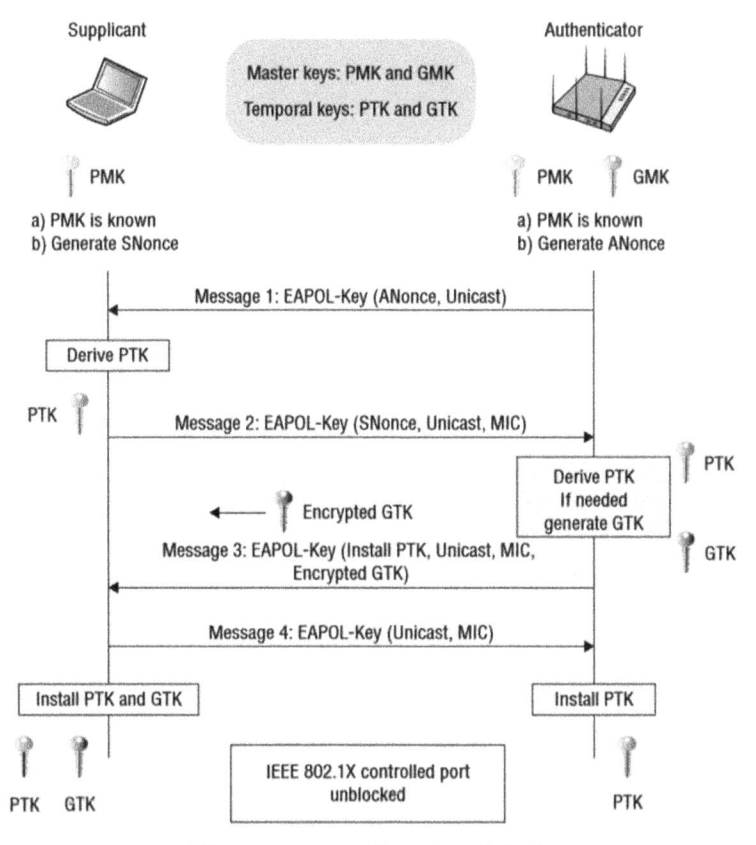

Figure 10.1: 4-Way Handshake

371

In the illustration shown in Figure 10.1, we see 4 distinct messages being transmitted between the authenticator and the supplicant.

- **Message one (1)** is sent from the authenticator, and it contains the ANONCE (Authenticator Nonce) sent in unicast to the supplicant. The supplicant will use this as seeding material to derive a Pairwise Transient Key (PTK) as well as the SNONCE (Supplicant Nonce) that it will transmit back to the authenticator in the next frame.

- **Message two (2)** is unicasted from the Supplicant to the Authenticator, and this message contains the SNONCE generated from the message previously sent, the Robust Secure Network Element (RSNE), and a Message Integrity Check (MIC). The MIC is set to one (1), and the authenticator confirms the MIC upon arrival. In this frame, the RSNE is visible and can be clearly seen if the packet exchanges are being monitored or captured.

- **In Message three (3),** the authenticator derives its PTK, verifies the MIC from the supplicant, and also unicasts the encrypted Group Temporal Key (GTK) over to the supplicant.

- **In Message four (4),** the supplicant notifies the authenticator that it has installed the keys, and it is ready to begin sending and receiving secured data. It is important to note here that the controlled port on the authenticator is opened, allowing access to secured information.

Now let's take a closer look and see how and where the threat exists:

In Wi-Fi, packets get lost and/or dropped in transition all the time. When this happens, packets can be resent several times by the device transmitting them. In the handshake scenario demonstrated in Figure 10.1, we can see the key installation as being part of message three (3). The previously mentioned packet retransmission tactic is used by the attacker when it captures and continuously resends message three (3) as if the authenticator is not getting the response it is looking for, sent in message four (4), from the supplicant. Every time the "FAKE" authenticator (Man-in-the-middle) sends message three (3), the supplicant installs the same encryption key and thus resets the Nonce. This is a great place to note that Nonce is short for "**N**umber used **once**." The Nonce is used for initializing the encryption algorithm just before the encryption of data, so the Nonce should only be used one time. This is similar to the use of random

Initialization Vectors (IV) where the IV is used along with a key to help ensure the integrity of the encryption process. In the KRACK scenario, by resending message three (3) of the 4-way handshake, the supplicant will always reinstall the same key again, resulting in the resetting of the Nonce. With this, in the process of creating encryption keys, the same keystream is always getting retransmitted and this allows the attacker to recognize repeated values and predictable patterns in the keystream until enough information is gathered to totally recreate the encryption keys, thus leading to the same ability to encrypt and decrypt data passing between the supplicant and the authenticator. All the security that AES is supposed to be providing is now lost. This being said, we can see how important it is for the Nonce to be a single random number and never duplicated.

The threat doesn't just stop at data getting hijacked; the data could also be manipulated or even held for ransom. Once the hijacker has gained access, any type of invasion could be carried out including the installation of malware. The only good side of this is that the attack cannot be performed remotely as the attacker has to be within range of the client being targeted using an active antenna to gain connectivity. This doesn't give a lot of comfort though, given the fact that so many antennas will allow for attackers to be in adjacent offices or close proximity hiding places and still pull off this attack. Thus, the reason for seeking out a more robust form of security.

The Solutions in WPA3

WPA3 provides added layers of protection and a more complex way of deriving encryption keys as a more robust form of defense against the previously mentioned weaknesses. In this section, we will go over the components in WPA3 that resolve past security weaknesses discovered in WPA2 and later in the chapter we will see how these solutions are used in the new WPA3 authentication processes.

In January of 2018, the Wi-Fi Alliance announced its latest security certification, WPA3, as a replacement to WPA2. The new certification calls for new and more secure capabilities for personal and enterprise deployments as well as IoT wireless networks. There is also a new addition with this protocol known as Easy Connect for simplicity in onboarding wireless devices. One major adjustment is that WPA3 only allows the use of AES, leaving TKIP and other legacy protocols to no longer be an option.

In personal mode, 128-bit encryption is standard, while the enterprise mode utilizes 192-bit encryption. Along with the new standard, a new feature in cryptography, known as forward secrecy (FS), is utilized. This assures us that session keys will not become compromised even if the server keys have been compromised. FS also protects past sessions against current or future attacks, meaning that encrypted data in past sessions cannot be compromised in the event passwords and keys are compromised in the future. Another feature concerning passwords is that WPA3 offers a more robust level of security when users disregard complex password recommendations and create really simple passwords. As an added protection, WPA3 mandates the use of Protected Management Frames.

WPA3 Core Components Explained

In the past, WPA and WPA2 personal have used a Pre-Shared Key that was pretty much used as the PMK which was seeding material in the 4-way handshake. In WPA3, the PMK is secretly calculated by both parties without sharing key data used in the process. Pre-Shared Key is replaced with Simultaneous Authentication of Equals, which is widely used in the deployment of wireless mesh networks. This makes guessing at even the simplest of passwords much more difficult to crack with dictionary and/or brute-force attacks. SAE comes from the IETF Dragonfly key exchange that is highly resistant to dictionary and brute force attacks. The hacker now has to perform interactions rather than computations in order to gain any ground achieving a hack. In other words, if the hacker cannot simply guess and physically enter the correct password or passphrase then they will not be able to obtain any part or parts of the password or passphrase by any other means.

As noted previously, the technology was introduced into the wireless world for use in mesh networks as defined through the IEEE 802.11s protocol and has been around for a good while now. This protocol does a couple of things worth noting here, and that is node discovery and MAC-based routing. The client or node has to know the password the access point is using, and the access point has to recognize the MAC Address of the client or node before it is allowed to join the network. In the Dragonfly Key exchange, this node discovery works for the client (supplicant) just as a mesh node would discover the network. It ensures that each node joining the wireless network will be categorized based on how it will interact with other nodes or devices on the network. Devices identify themselves as a client or a mesh node being used to extend the network while passing WLAN traffic through to other clients and/or nodes on the WLAN. This technology prevents hackers from detecting a mesh deployment, powering up their own rogue access point in mesh mode and connecting to the network. Now let's apply these same methods to a Wi-Fi client joining a wireless network using these same protocols and standards in the new WPA3 certification.

By utilizing this technology in WPA3, SAE requires the use of the Diffie-Hellman Key Exchange Method (which seems to defy all common sense) while also adding an authentication element where the shared key is created through a NIST elliptical curve cryptography (explained in more detail below). Here, the client (supplicant) connects to the access point (authenticator) by performing a successful SAE exchange, then each (supplicant and authenticator) will independently derive a very strong cryptographic key which the encryption keys will be derived from.

The NIST elliptical curve cryptography (ECC) is a sophisticated public-key system, much like RSA. Each user has both public and private keys that are utilized in the calculation of hidden keys only known to the two parties involved. Let's try a fun game to explain how the key calculation process works, but we'll make it simple:

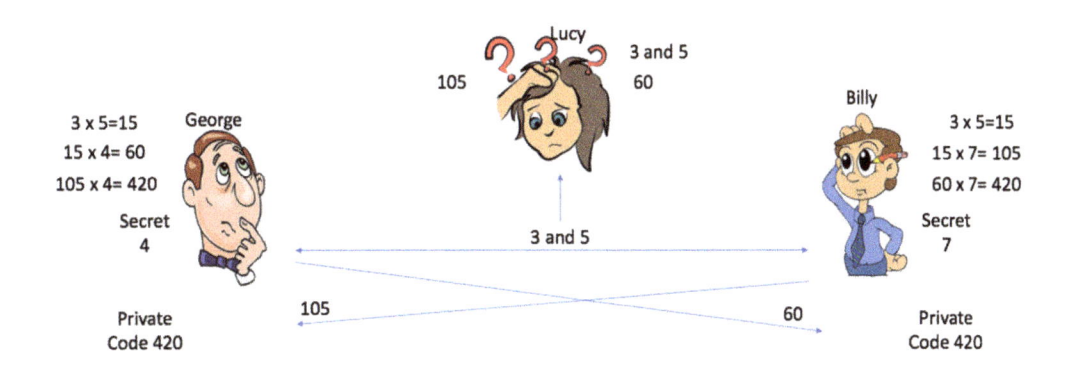

Figure 10.2: Multiplication Keep Away

Three friends get together to play a game of multiplication keep away and Lucy is the player in the middle. George and Billy think up 2 random numbers that they agree to start the game with and then speak them to each other, allowing Lucy to hear the numbers. George and Billy also think up a secret number that they do not share with anyone. To make this even more fun, Lucy has thirty (30) seconds to figure out the private code after she hears the results of the first calculation from both Billy and George.

George, Billy, and Lucy all quickly multiply three (3) times five (5) and will all come up with fifteen (15) easily. All of a sudden, George yells out "sixty" (60) and Billy yells out "one hundred and five" (105). Lucy quickly goes to work trying to find the numbers they both multiplied the number fifteen (15) by before they perform their next multiplication. While Lucy is dividing both numbers by fifteen (15) to get the secret numbers, George and Billy are both multiplying each other's shared numbers by their own secret number and quickly writing down the private code of four hundred and twenty (420). Now, in a matter of seconds, Lucy figures out that George's secret number is four (4) and Billy's secret number is seven (7) and so now she multiplies 60 x 7 to get 420, and she multiplies 105 x 4 to get the same private code as George and Billy.

In the above exercise, we see and understand a very elementary process of private and public key without the use of any form of hashing. In the next illustration, we will change our game up a bit.

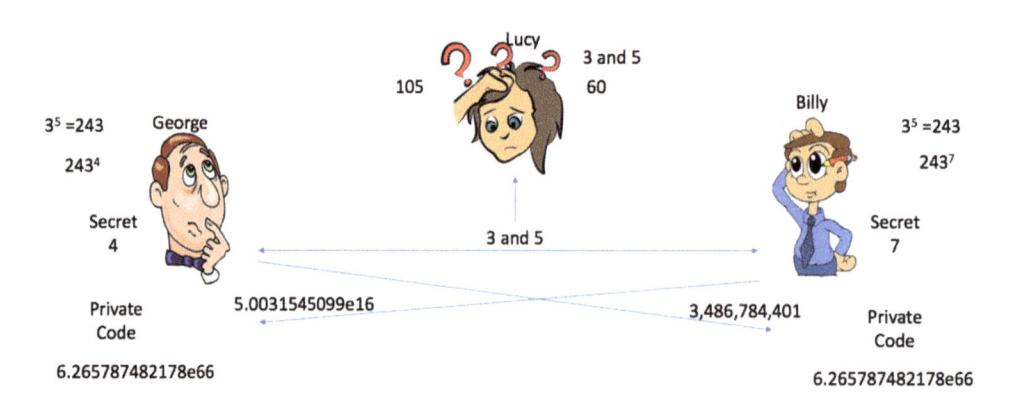

George — $3^5 = 243$, 243^4, Secret 4, Private Code 6.265787482178e66, 5.0031545099e16

Lucy — 105, 3 and 5, 60, 3 and 5

Billy — $3^5 = 243$, 243^7, Secret 7, Private Code 6.265787482178e66, 3,486,784,401

Figure 10.3: Game Changer

In Figure 10.3 the game changes from multiplication to powers of ten (10). This time, George and Billy come up with their agreed-upon public numbers of three (3) and five (5) and start the game. Using a calculator, all participants can come up with two hundred and forty-three (243) rather quickly but it is the next stage that becomes very difficult for Lucy because the figures Billy and George are sending to each other are very large leaving Lucy to start running guesses at how many times each player multiplied two hundred and forty-three (243) times itself to get the number each one shared out loud. By the time Lucy can guess the two (2) secret numbers so she can perform the next step to obtain the private code, the game is over.

In both of the previous illustrations, we used two different mathematical options in order to cause Lucy difficulty when looking for our secret numbers and private codes. In multiplication alone, she was able to obtain the values quickly, but when using powers of ten (10) the job became much more difficult and time-consuming because of forcing Lucy into a guessing game. Sure, with unlimited guesses, Lucy could eventually guess the secret numbers used by George and Billy to finally come up with the private code, but how much time is it going to take? If we added yet another mathematical factor or even two (2) more and tossed in a limit on how many guesses Lucy gets, what is her chance of ever winning the game? Welcome to WPA3.

WPA3-Personal

WPA3-Personal (Mandatory in all 802.11ax devices), also known as WPA3-SAE comes in two (2) different modes, one being WPA3-SAE Transition Mode. This mode is configured on the access point for the purpose of supporting WPA2-PSK and WPA3-SAE on the same SSID. This allows for the same passphrase to be used on both WPA2-PSK and WPA3-SAE connections. The main difference for the WPA3 user is that the passphrase will grant a hacker access to the network, but not grant them the ability to decrypt traffic on any of the WPA3 sessions. Protected Management Frames are optional in this mode and not mandatory. The network administrator can turn this mode on or off if needed, until all WPA2 client devices are updated.

The second mode in WPA3-Personal operates strictly in WPA3-SAE mode with no connectivity for WPA2 certified devices. In WPA3-SAE mode, Protected Management Frames (802.11w) is required, which helps to prevent spoofed management frames, and connections are more secure through a unique cryptographic exchange process.

> When PMF is required, it means that both the AP and the client have PMF set as mandatory, and if one or the other doesn't have PMF set as mandatory, then it is not a certified client, and it will not connect.

With the use of a Diffie-Hellman key exchange and the NIST elliptical curve cryptography (ECC), an attacker can know the password and still not be able to decrypt traffic because it isn't used as a credential in the authentication protocol. The password is only used to index a secret point on an elliptic curve and that point on the curve becomes the generator for use in the cryptographic exchange known as the Dragonfly Key Exchange (See Figure 10.4). The result is a 32-byte PMK, unknowable to the attacker. If an attacker were to be passively observing this exchange, knowing the password, he/she wouldn't be able to discover or calculate the session's PMK, leaving the encryption keys unknown and unavailable to the attacker.

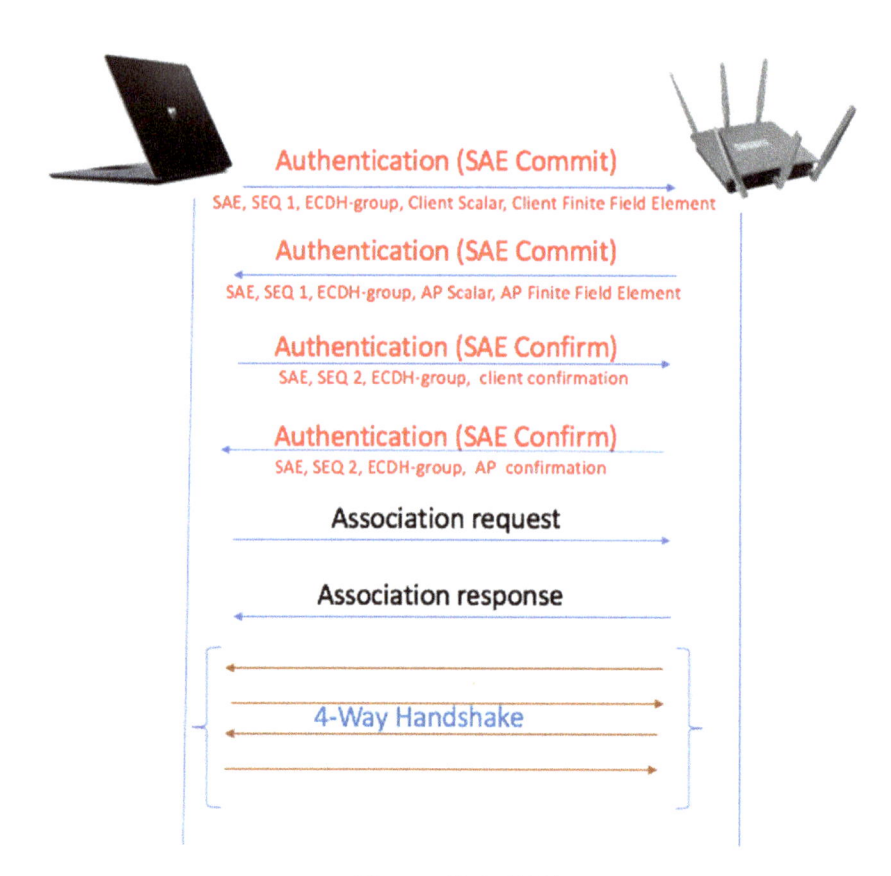

Figure 10.4: SAE

WPA3 refuses authentication after a certain number of attempted log-ins. This added security helps mitigate Brute-Force-Attacks. This attack is mitigated in WPA3-Personal by using tokens to limit the number of connection attempts. As noted in Figure 10.4,

WPA3-Personal authentication has four (4) frames instead of two (2) found in WPA2-Personal. The last two (2) WPA3-Personal authentication frames contain a confirmation token. When the access point gets too many SAE requests, it uses the tokens to limit how many simultaneous connections can be attempted, providing protection against Brute-Force-Attacks.

Let's take a short look at the use of the NIST elliptical curve cryptography (ECC) as it pertains to the password or passphrase.

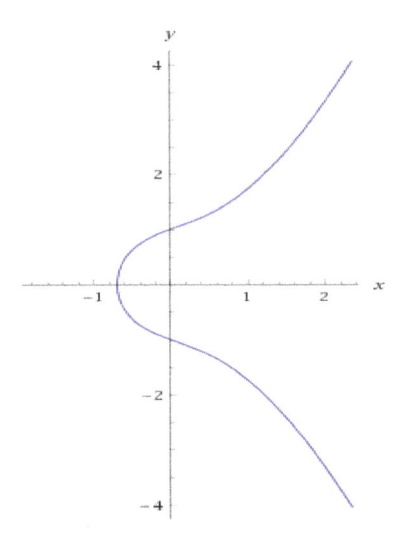

Figure 10.5: Elliptical Curves

If we take a look at Figure 10.5, we see an XY axis in which the elliptical curve function takes place. The actual math used in the indexing is well known as Clock Arithmetic and it is a major part of how the PMK is derived in such a secure manner. Since the actual password is not used in deriving the PMK but instead used to index a point on this elliptical curve, then every different character in a password calculates out a result that graphs to a different point on the axis. This point becomes the generator in hashing out the PMK. This being said, using a dictionary attack to guess the password would be impossible to use against this process. Every time one character gets changed, the secret point on the elliptical curve would change and create a different generator to be used in the cryptographic hash, taking the end result of the algorithm way out in left field. On top of this, both the client and the access point each secretly generates a private key that is applied to the cryptographic function, giving another unknown variable for an intruder to search for. The intruder would be forced to physically guess at the password and see if the guess was correct or not. If the intruder was lucky enough to guess the password, the cryptographic process still keeps the attacker from reverse engineering encryption keys and decrypting data. This unique process provides each client/access point session with a zero-knowledge proof PMK only known to that particular session.

Let's explain the Dragonfly Key Exchange a little further:

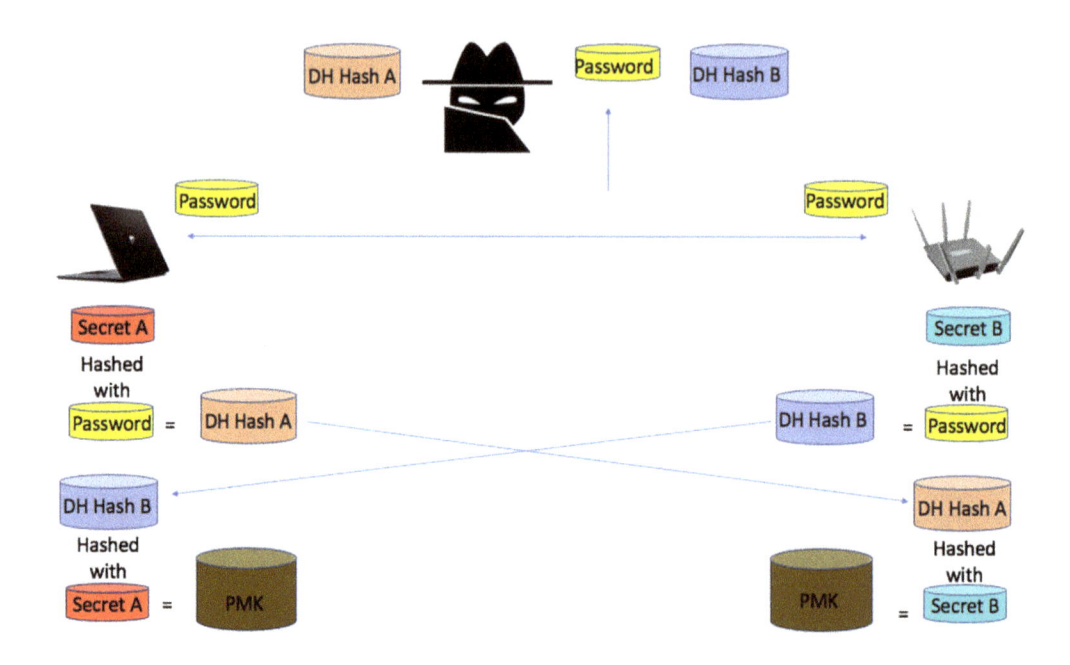

Figure 10.6: Dragonfly

In Figure 10.6, we see a client device and an access point as they enter into the process of establishing a secure PMK. Please note that the password is already known by the hacker. The purpose of the Dragonfly Key Exchange is to create a unique Pairwise Master Key for each individual session negotiated on the network, and not directly use the password or passphrase to derive the PMK. Both the client and the access point have derived private keys that are not shared with anyone and are shown as Secret A and Secret B. Through the cryptographic exchange, each side uses their public and private information in the Diffie-Hellman hashing process and derive DH Hash A and DH Hash B and then exchanges these two hashes with each other over the air. As shown, the hacker intercepts this information when exchanged. Now that the client has the access point's DH Hash B and the access point has the client's DH Hash A, they both hash their respective private keys with the exchanged hash results. They end up with matching results that serve as the PMK, while the hacker still does not have enough information to get started on finding the PMK.

At this point, the client and access point begin the 4-way handshake, get the keys installed (deriving the Pairwise Transient key and the Group Temporal Key), and they're on their way to sending and receiving encrypted data.

If you would like to look further into the development of the Diffie-Hellman key exchange and the mathematical functions involved, copy this link into your browser to view the video: https://www.youtube.com/watch?v=YEBfamv-_do

WPA3-Enterprise

In the new WPA3-Enterprise certification, Protected Management Frames are added to WPA2-Enterprise. WPA3-Enterprise states that all WPA3 connections SHALL negotiate PMF. That statement, concerning WPA-2 Enterprise clients, can get confusing. Simply put, if a WPA2-Enterprise client is attempting to connect to a WPA3-Enterprise network and they are capable of successfully negotiating the use of PMF, the client now becomes a WPA3-Enterprise certified client. When we look back in time at the WPA2-Enterprise 802.1X process, we see that there are many flavors of EAP to be used in the authentication process. There are certificates, tokens, passwords, and PACs in many forms with many options. We also know many of the weaknesses and vulnerabilities found in most of these options. In the new WPA3-Enterprise certification, the provision made to allow WPA2-Enterprise clients on a WPA3-Enterprise network through the negotiation of PMF, also allows the client to utilize all the WPA2-Enterprise EAP options in their 802.1X authentication process. With this, the only difference between WPA2-Enterprise and WPA3-Enterprise is mandating the use of PMF.

WPA3-Enterprise 192

One of the biggest reasons for the development of WPA3-Enterprise 192 is to ensure the best level of security throughout every stage of authentication, association and data encryption. The level of security is found in the name itself, "WPA3-Enterprise 192".

There are 192 bits of security when operating strictly in this mode, and not as a WPA2-Enterprise client utilizing PMF. There are some factors that get us to the 192-bit security beyond the type of AES encryption being used. We have AES-128-GCM, AES-192-GCM and AES-256-GCM available to use for encryption and data authentication, but the goal is to provide the highest level of cryptographic strength so AES-256-GCM is chosen for this certification. Along with encryption and data authentication, there are other things (such as hashing, key establishment, and digital signatures), which could all provide a vulnerability, so the following are also required:

- AES-256-GCM for encryption and data authentication
- SHA-384 for hashing
- ECDH-P384 for establishing keys
- ECDSA-P384 for digital signatures

The overall combination of the above-mentioned cipher suites equals 192 bits of encryption strength.

- AES is an Advanced Encryption Standard with a symmetric-key algorithm, and
- Galois/Counter Mode (GCM) is the mode of operation for AES and is known for its high-speed throughput rates.
- GCM is defined for 128-bit block ciphers.
- The AES block cipher is 128 bits and is available in three (3) different key lengths: AES 128, AES 192, and AES 256.
- Secure Hashing Algorithm (SHA)-384 is a Federal Information Processing Standard (FIPS) published by the National Institute of Standards and Technology (NIST). It has an output size of 384 bits and 192 bits of security against collision attacks.
- P-384 is a 384-bit elliptic curve in the NSA's Suite B Cryptography, specifically used in the Elliptical Curve Diffie-Hellman (ECDH) and Elliptical Curve Digital Signature (ECDSA) algorithms.

Although costly on processing resources, the Rivest–Shamir–Adleman (RSA) cryptosystem can still be used as long as the key size is 3K bits or more. Alternatives such as ECC 256 or ECC 384 may be used instead of RSA to preserve processing resources, as long as the cryptographic strength is sufficient. Take a look at the permitted cipher suites listed below:

- TLS_ECHDE_ECDSA_WITH_AES_256_GCM_SHA384
  - (ECDHE and ECDSA utilizing the 384-bit prime modulus curve or P-384)
- TLS_ECDHE_RCA_WITH_AES_256_GCM_SHA384
  - (ECDHE utilizing P-384 and RSA utilizing the 3072-bit modulus or larger)
- TLS_DHE_RSA_WITH_AES_256_GCM_SHA384
  - (RSA and DHE both utilizing the 3072-bit modulus or larger)

WPA3-Enterprise 192 requires the above-mentioned cipher suites to set a mandatory high standard of security in the new certification, and you may commonly hear it referred to as Suite B or the Commercial National Security Algorithm Suite (CNSA) which replaced suite B in 2018.By mandating the use of these TLS suites, WPA3-Enterprise 192 rises high above the standards set in WPA3-Enterprise and WPA2-Enterprise.

In WPA2 and WPA3-Enterprise, all flavors of EAP are still used while WPA3-Enterprise 192 only allows the use of elliptical curve certificates. This is controlled by the authenticator and the RADIUS server. In the past, the authenticator did not present the actual client security information to the RADIUS server. During open authentication, the client and access point negotiate several requirements before allowing the client device to associate to the access point, then 802.1X begins and minimal information about the client device itself is forwarded to the RADIUS server. WPA3-Enterprise 192 requires the authenticator to present a client's negotiated security attributes to the RADIUS server in order to ensure that the client meets the required security standard before continuing with the authentication. The authenticator receives a request from a supplicant, then turns around and notifies the radius server of the authentication and key management (AKM) in use. The RADIUS server examines the EAP message, looking for a required TLS cipher suite. If the RADIUS server sees a required TLS cipher suite is used, it will allow the EAP message and continue the authentication process. If the RADIUS server examines the EAP message and sees that a required TLS cipher suite isn't being used, it will reject the EAP message and deny the client access. This added verification ensures the use of properly sized, valid elliptical curve certificates and also prevents the downgrading of security in the session.

Let's take a look at Figure 10.7 and see where the WPA3 -Enterprise 192 changes provide increased security in the 802.1X authentication process.

If you think back to the 802.1X authentication process in WPA2-Enterprise, you will see several similarities in the WPA3-Enterprise 192 process. In Figure 10.7, we show added levels of security enhancements that provide the 192-bit security we covered earlier in this section. It is important to know that the user doesn't experience any changes in their connection process as the security enhancements are all in the EAP Exchange and not in user actions.

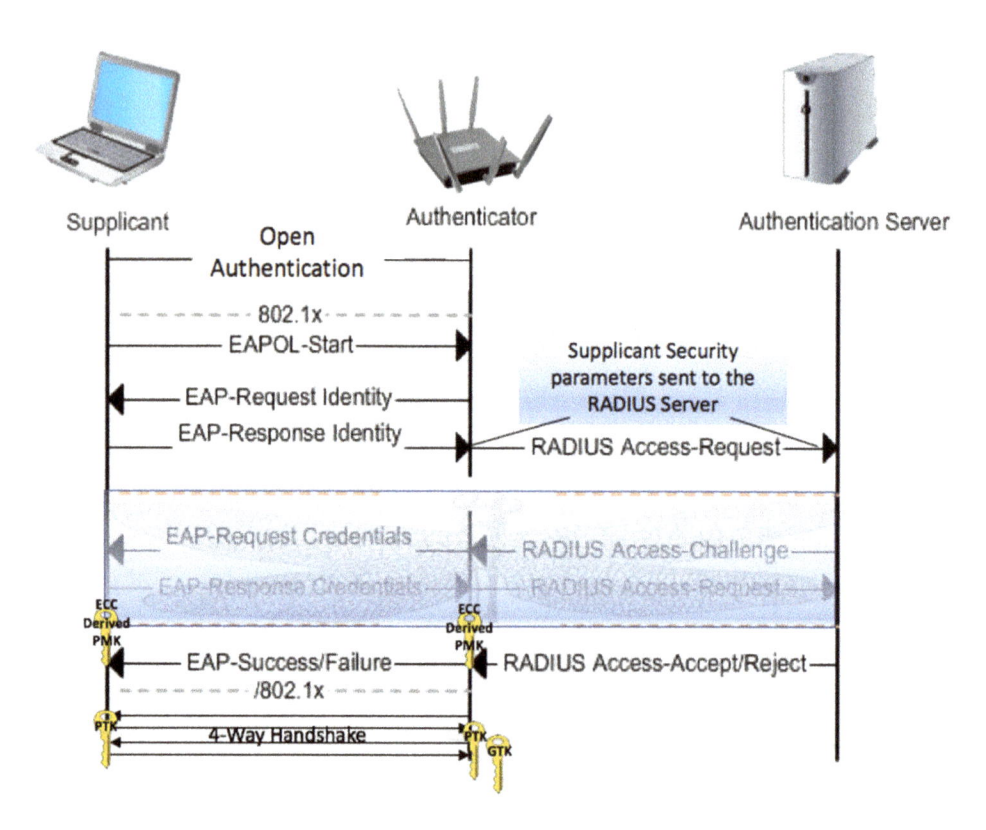

Figure 10.7: WPA3-Enterprise

We see that the open authentication (or OWE) process is still required to provide 802.1X with an authenticator. When the EAP-Request Identity frames take place, WPA3-Enterprise 192 requires the authenticator to provide the supplicant's security parameters to the RADIUS server. Once the RADIUS server sees that the supplicant meets the required security parameters, it will accept the access request, and the mandated AES-GCM encryption and Elliptical Curve Cryptography begin hashing out the Pairwise Master Key. At the end of the 802.1X process, the encrypted PMK is used in the 4-way handshake to derive the Pairwise Transient Key (PTK) and the Group Temporal Key (GTK).

Wi-Fi Easy Connect

Wi-Fi Easy Connect was designed to both make connecting devices to wireless networks less complex, and at the same time, utilize strong security standards. Provisioning is made as simple as scanning a quick response (QR) code and running the device provisioning protocol (DPP). Wi-Fi Easy Connect is different from Wi-Fi Protected Setup (WPS), which requires the pressing of a physical button to initiate the process. Let's first look at what Wi-Fi Easy Connect brings to the table:

- Standardized onboarding method
- Easy provisioning through QR Codes
- Easy setup for IoT devices
- Supports WPA2 and WPA3 provisioning
- User-chosen device for network management access
- Secure authentication through the use of Public key cryptography
- Works with devices with little to no user interface (screen)
- Access Points can be replaced without re-enrolling all devices to the new access point

Since the inception of IoT devices, questions concerning security have followed the technology. Wi-Fi Easy Connect provides beyond that, the home use IoT products, at times, left users struggling with the do-it-yourself setup. Wi-Fi Easy Connect greatly simplifies the user experience by allowing users to utilize one simple user interface as a console to set up all of their home use IoT devices. The user can choose a device such as a tablet or a smartphone as their primary point of configuration, known as the configurator, and all other devices added become enrollees on the network. The user establishes a secure connection to enrollees by using their primary device to scan a QR Code or by entering a passcode or string associated with each enrollee device being provisioned on the network. By doing so, the DPP automatically runs and sets up the proper credentials for each enrollee device. The user's smartphone or tablet is used as a management interface for the IoT network. This is very simple to use and there is no struggling to remember lengthy passwords to add additional devices to the network. See Figure 10.8 for a visual in provisioning.

How the protocol works

Wi-Fi Easy Connect utilizes four (4) distinct steps in the protocol's onboarding process:

- Bootstrapping
- Authentication
- Provisioning
- Connectivity

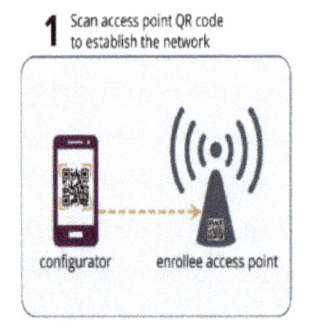

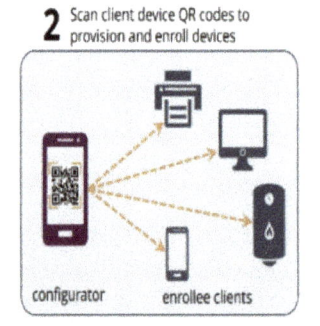

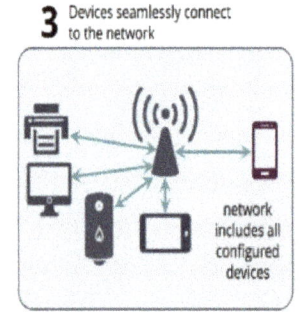

Figure 10.8: Wi-Fi Easy Connect

Wi-Fi Easy Connect security depends on the public-private key technology and the public keys are used for the identification and authentication of every device connecting to the network. Trust is established between the mobile app (used for configuring enrollees and the enrollee itself) by setting up public keys. This is the bootstrapping process and is performed using QR Codes or manually entered strings already assigned to the device onboarding, prior to performing the Wi-Fi Easy Connect protocol.

Authentication takes place as the mobile app on the configurator and the product being configured authenticate each other by proving ownership of the public keys. Mutual authentication is optional as the mobile app doesn't have to provide its public key to device onboarding. However, the product or device onboarding is strongly authenticated because the configurator (mobile app) is guaranteed to receive the device's public key through the scanning of the QR code or the entry of the string.

Provisioning will only take place if the authentication process went through successfully. This stage is always initiated by the connecting device. Here the mobile app will provide credentials to the onboarding device, and these credentials will be used to establish connectivity. The mobile app also provides the access point with credentials during the initial setup between the mobile app and the access point.

The final step is when the device proves to the access point that it has been authorized to join the wireless network. This step is also initiated by the connecting product and it proves its authorization by using the credentials given to it by the mobile app in the provisioning stage. Once the credentials are proven to the access point, the new device can successfully communicate on the wireless network and begin functioning as intended.

OWE (Enhanced Open)

In our past open networks, all connectivity was unsecured unless the user installed a third-party solution that provided their client device with security such as Virtual Private Network (VPN), for example. Take a look at the image below to see the simplicity of open authentication, as we have been utilizing until now.

In Figure 10.9, we see the four frames exchanged during open authentication. A client device (supplicant) wishes to connect to a BSSID so it transmits an authentication request frame over to the access point (authenticator) in hopes of being allowed to join this BSSID and use the network. The authenticator looks at some information elements from this request and basically verifies that the supplicant meets the minimum requirements, then replies back to the supplicant with an authentication response frame. This frame exchange simply verifies the two are speaking the same language and the supplicant could indeed operate on this network as the authenticator's requirements dictate. These requirements could simply be that you must be able to function, at minimum, within the 802.11n protocol or newer because restrictions on this network deny access to clients only capable of utilizing 802.11abg protocols. Once the client is authenticated, the supplicant will then tell the authenticator that it would like to join the BSSID by sending over an association request frame. The authenticator responds back with an association response frame to tell the client device, "Welcome to my network". At most free Wi-Fi locations, the user can now freely roam about the internet. At best, there will be a shared password given to everyone who comes in, but there is still nothing protecting your precious packets as they transport your sensitive information into the wild and dangerous world of cyberspace. The problem OWE was created to solve is, passive eavesdropping. It is meant to provide privacy and not be mistaken as a replacement or option for network security.

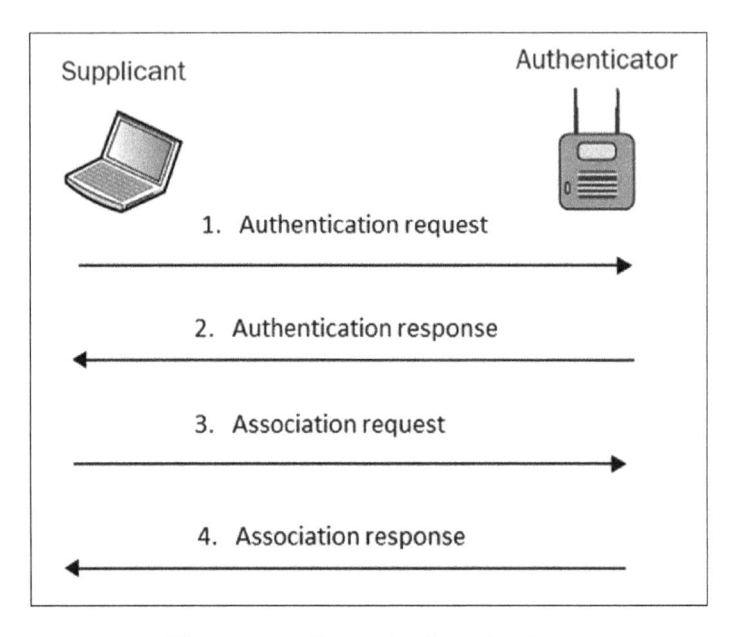

Figure 10.9 Open Authentication

Most companies who issue mobile devices have mandated the use of a VPN when utilizing company-owned devices anywhere other than when actually present at the corporate office. This mandate is critical when using free Wi-Fi hotspots. The reasoning behind this is because of the huge risks involved in using open networks. The main concern is sensitive data being transmitted over the air with no protection or encryption applied, making it easy for an amateur hacker to capture the packets and possess the sensitive material.

Another concern that has made itself apparent is that most employees, when issued company-owned mobile devices, still do not utilize the provided VPN while traveling, even when it is mandated to do so. The issue is not that they are insubordinate and simply just trying to sabotage company data. The truth is, many employees just do not train themselves to connect to open Wi-Fi networks through their VPN, and every connection they make at a coffee shop or hotel leaves them at risk of displaying sensitive information in front of the eyes of malicious Wi-Fi packet collectors just dying for the chance to capture something they can use as a trophy in their hacker's club or even worse, holding it as ransom or selling it to corporate competitors. Now, it is obvious that many of us think this will never happen to us and these dark knights of the Wi-Fi underworld are not as plentiful as so many companies and training instructors lead on, but there are many more wolves in sheep's clothing out there than any of us are aware of.

Try to picture a hacker as a die-hard fisherman who goes out daily looking for the best fishing holes on the water in order to hook the biggest and most fish each time they go out. This fisherman will use wireless equipment to continuously scan beneath the water's visible surface in order to see everything going on beyond what the eye can see. This wireless equipment allows the fisherman to understand what kind of fish are in each location as well as how many are available to them. Once the fisherman locates a few really nice schools of fish, it's like catching fish in a barrel. When we turn this analogy towards a would-be hacker in search of vulnerable Wi-Fi users, mischievous minds allow the temptation to get the best of them and they go out and get some wireless equipment, either free or very cheap to purchase, and they go fishing. At first, they are just curious as to what is out there until they begin recognizing some of the information they are seeing and how this information could possibly be used for personal gain and/or causing damage. Once this happens, curious George crosses over to the dark side, never to return, and many more wireless users are now at risk of having their personal Wi-Fi traffic intercepted and used with malicious intent.

The previously mentioned scenario was mainly described using the corporate world and mobile users as an example, but they are not the only ones at risk. Millions of college students, siblings, and moms and dads connect to the internet everywhere, and most of them will access social media sites, perform online purchases, fill out employment documents, or participate in internet banking activities. Credit card, social security, and bank account numbers, as well as passwords, addresses, and birth dates, get sent across the open airwaves through all of this activity while hackers lie and wait for the perfect catch. No one mandates the use of a VPN for this crowd, and all too often this is the group that is hit the hardest because of the lack of threat awareness and their own vulnerabilities.

There is a better option for connecting to open networks which help protect information for even the most naïve hot spot Surfers, and that option is Opportunistic Wireless Encryption (OWE).

The Solutions in OWE

The problem with open authentication is that without any other form of security, all network traffic can be seen at some level. With the new enhanced open authentication, OWE, there is an overlay of security taking place as the four (4) open authentication frames take place. With the use of short-term public keys, the client and the access point undergo a Diffie-Hellman exchange in order to derive a unique Pairwise Master Key, undetectable to a hacker. After association, this unique PMK is used in a 4-way handshake to generate encryption keys to protect traffic for this session between the client and the access point. Let's look at the hot spot scenario now:

- A Wi-Fi user walks in and sits down with their laptop.
- The hacker anticipates the new opportunity to collect sensitive information.
- The Wi-Fi user's laptop is OWE capable.
- The Wi-Fi password is displayed on a sign.
- The Wi-Fi user enters the password and begins surfing the web.
- The hacker begins passively scanning for their packets.
- The Wi-Fi user logs into their company email and begins sending and receiving messages.
- The hacker sees hundreds of encrypted packets flowing to and from the user.
- The Wi-Fi user attaches financial documents to an email and sends them.
- The hacker can't figure out how to decrypt the traffic.
- Wi-Fi user finishes and leaves.
- Hacker fails.

With the enhanced overlay of protection, even when the user doesn't take extra precautions before sending sensitive data, the user's wireless traffic gets encrypted behind the scenes, providing session privacy, and the user may not even be aware of the data protection taking place. However, the client nor the access point have authenticated identity. This means that the data is the only thing protected and end-to-end security is not provided by OWE.

Note - This is a great place to interject a word of caution concerning man-in-the-middle attacks. If the hacker above had set up their own AP software on the same SSID as the hot spot and caused the above Wi-Fi user to connect to the "man-in-the-middle" upfront, then the hacker's machine now has the secretly generated PMK on it and the encrypted data is not so safe anymore. The good news is that Protected Management Frames are required with OWE so once the user has successfully associated to the network (not through a man-in-the-middle), he/she is not going to be vulnerable to de-auth frames used by MITM attacks.

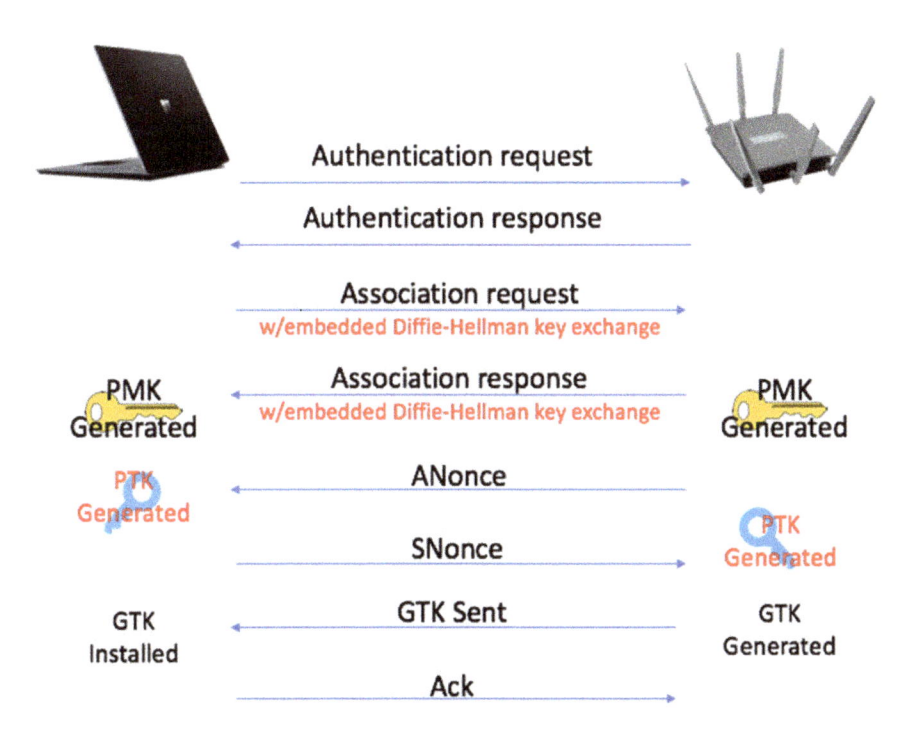

Figure 10.10 OWE Enhanced Open Authentication

Figure 10.10 shows the added enhancements that OWE adds to open authentication. As noted, the authentication request and response frames are still present and followed up by the association request and association response frames, which are both

391

enhanced with the Diffie-Hellman key exchange process. This generates a Pairwise Master Key on both the client and the access point, then sets up the 4-way handshake we are used to seeing in pre-shared key networks. Only this key was generated on both sides and not pre-shared.

Even with "No Password Required" hotspots, all wireless traffic gets encrypted with OWE capable networks and clients.

OWE Connection Processes

Let's first look at how our devices become OWE capable. The good news is that we do not have to buy all new equipment or devices. OWE can be implemented through minor software changes and can even run on legacy equipment. Because OWE is optional and not a mandate in WPA3, open system authentication will function both ways if you want. In the open authentication process, there is an added information element signaling OWE client devices to associate to a hidden BSS. If a client device isn't OWE capable, then it simply ignores the OWE information element in the open system authentication process and the four (4) frames of open authentication is all that is performed. If a client device is OWE capable, it will recognize the OWE information element and the client will be sent to a hidden BSS that performs the OWE process. Here, the client and the access point will initiate the Diffie-Hellman key exchange during open authentication, then utilize the generated PMK to start the 4-way handshake. Once the 4-way handshake is complete, encryption keys are generated on both the supplicant and the authenticator and all traffic during this session gets encrypted. There is also the option to have your access point(s) function in OWE only mode, not allowing non-OWE clients to associate.

> Because of the way OWE derives its PMK, OWE is actually considered safer and provides more privacy assurances than that of WPA2-PSK.

Let's compare the two against this claim.

Using the two (2) frame exchanges found in Figure 10.11, the one on the left side of the page is WPA2-PSK and the one on the right is OWE with the Diffie-Hellman key exchange. As you can clearly see, both look very similar with the exception of the embedded Diffie-Hellman. Given the same environment, we are going to say here, that the password in both settings is known by everyone, including the hacker in the room.

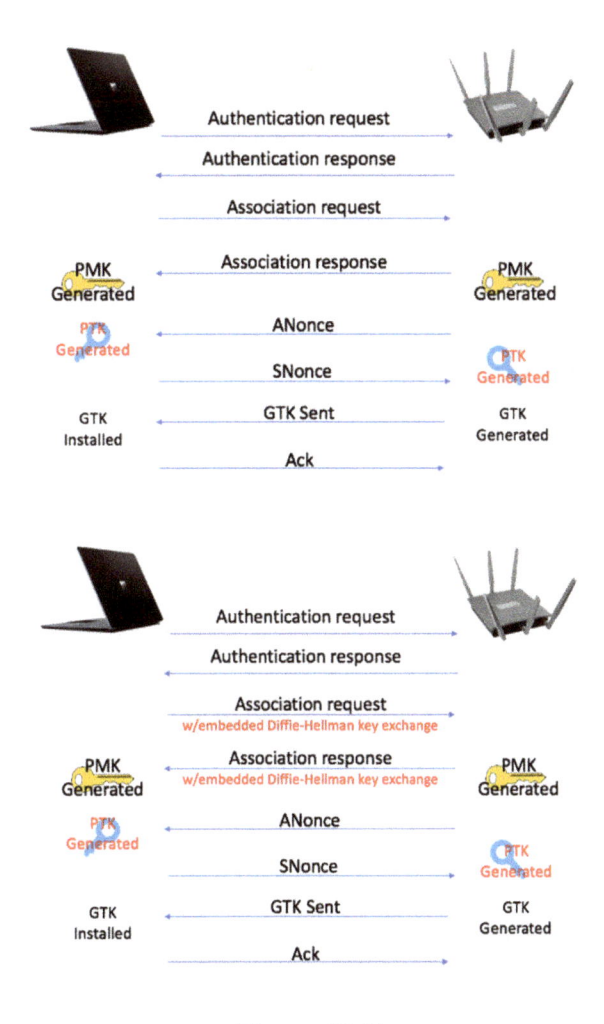

Figure 10.11

In the WPA2-PSK environment, the password itself is used to generate the PMK and the formula for getting that key is known, so the open authentication frames performed in the WPA2-PSK environment actually helps the hacker find enough information about the association so the PMK can be reverse-engineered. With the PMK known, the hacker follows the following frame exchanges and gathers enough information to generate the PTK. Now the hacker has the ability to capture all traffic on

393

the network and decrypt it. Also, if the hacker captured encrypted frames and later on discovered the encryption key, the encrypted data could be decrypted at a later date.

In the OWE environment on the right, the password is known to everyone as well, but now the open authentication frames exchange an information element, letting the access point know the client is OWE capable and the frame exchanges change along with the PMK generation process. During the association request frames, the password is not used to derive the PMK, and the Diffie-Hellman key exchange produces two (2) very private keys that are unknown to the hacker. These two keys are used in the calculations to derive the PMK rendering the attempt to find the PTK useless. In addition to this, if a successful man-in-the-middle attack was performed on a single client device on the network (the MITM was the first association for the client), then only that client would be affected, and all others would have totally different encryption keys. Lastly, if the hacker did eventually discover the keys to another client, all that client's past data is still kept private through forward secrecy, thus making OWE more secure than WPA2-PSK.

Chapter Summary

In this chapter, we covered the need for better security standards due to vulnerabilities in past security measures. We provided insight into the new measures taken to secure wireless traffic in both WPA3-SAE and WPA3-Enterprise and their various modes of operation. The Diffie-Hellman Key Exchange With Elliptical Curve Cryptography, used in the new security measures, provides much better security due to the fact that the passwords in WPA3-Personal and OWE are not used as part of the pairwise master key generation, and each client device associating to one of these type networks generate their own, unknowable PMK per session. The process now used to derive encryption keys found in WPA3 and OWE networks, provides forward secrecy which protects the integrity of encrypted data in past sessions in the event encryption keys are later discovered. Lastly, we covered Wi-Fi Easy Connect and how it has simplified and standardized the onboarding process of devices, provides secure authentication through the use of Public key cryptography and supports WPA2 and WPA3 provisioning.

Review Questions

1. During the key re-installation attack executed against the WPA2, 4-way handshake, which key is forced to be reinstalled over and over?

    a. The pairwise Master Key generated on the client device during the authentication process.

    b. The Pairwise Transient Key generated on the access point just before the 4-way handshake.

    c. The Group Temporal Key Generated and sent to the supplicant on the third frame of the 4-way handshake.

    d. The Pairwise Transient Key generated and installed on the supplicant during the 4-way handshake.

2. To ensure the integrity of the encryption process, which of the following is used to initiate the encryption algorithm prior to the encryption of data?

    a. The 4-way handshake

    b. The Nonce

    c. The PTK

    d. The PMK

3. What cryptographic feature is introduced in WPA3 that protects encrypted data in past sessions from being compromised in the event that passwords and keys become compromised in future sessions?

    a. Enterprise 192-bit encryption

    b. Prime Modulus and Generator

    c. Forward Secrecy

    d. Enhanced 4-way handshake

4. Opportunistic Wireless Encryption serves as a better option for open authentication because it _____.

   a. Adds a 4-way handshake to the 4 frames of open authentication so encryption keys can be derived to protect all wireless traffic.

   b. Encrypts all wireless traffic through an overlying Key exchange process followed by a 4-way handshake, as long as the user knows the password.

   c. Generates unknown strong passwords for OWE capable clients through a Diffie-Hellman key exchange so the PTK derived in the 4-way handshake can't be cracked.

   d. Encrypts all wireless traffic by adding a Diffie-Hellman key exchange that creates an unknown PMK used in the 4-way handshake to generate the encryption keys.

5. During the configuration of Wi-Fi Easy Connect, there are two (2) distinct members that perform the onboarding procedures set forth in the protocol. These two (2) members are the _____ and _____.

   a. Master and slave

   b. Access Point and client

   c. Mobile device and Mobile app

   d. Configurator and enrollee

6. An out-of-band process for providing trust between the mobile app and the connecting device, utilizes public keys, also known as QR Codes and/or strings, in order to identify and authenticate connecting devices before performing the Wi-Fi Easy Connect protocol. This mechanism is known as _____.

    a. Provisioning

    b. Authentication

    c. Bootstrapping

    d. Connectivity

7. The authentication process in Wi-Fi Easy Connect is capable of performing mutual authentication although it isn't mandatory to do so. However, the enrollee is guaranteed to be strongly authenticated because_____.

    a. The mobile app always provides the enrollee with its public key

    b. The configurator and the enrollee perform a Diffie-Hellman key exchange

    c. The mobile app always receives credentials from the access point

    d. The configurator always receives the enrollee's public key

8. OWE was designed to provide public and open network users with _____ by solving the problem of _____.

    a. End-to-end security, passive eavesdropping

    b. Data privacy, man-in-the-middle attacks

    c. Traffic encryption, weak passphrases

    d. Data privacy, passive eavesdropping

9. Opportunistic Wireless Encryption is a great alternative to open authentication due to the fact that it prevents users from associating to simulated access points used in man-in-the-middle attacks.

    a. True

    b. False

10. Given a known password, Enhanced Open or OWE is _____.

    a. More secure than WPA2 using 802.1X TLS because the WPA2 password is the PMK and the PTK can be discovered leaving all session traffic accessible to a hacker.

    b. More secure than WPA3-PSK because the WPA3-PSK PMK and PTK can be discovered and all session traffic can be decrypted.

    c. More secure than WPA2-PSK because the WPA2-PSK PMK and PTK can be discovered leaving all network traffic accessible to a hacker.

    d. More secure than WPA3- Enterprise when PMF isn't on because the key exchange is no longer performed inside protected frames.

11. The way legacy clients and OWE capable clients connect to the same SSID is _____.

   a. An information element in the AP beacon tells the OWE capable client to associate to a hidden BSS.

   b. An information element in the AP beacon tells legacy clients the AP is OWE and the client downloads a software patch from the AP before associating to it.

   c. An information element in the client's beacon tells the AP that it is OWE capable and the AP performs OWE instead of open authentication.

   d. The AP sends out a broadcast beacon that contains an information element which allows OWE capable clients to associate to the same BSS as legacy clients.

12. Although the password is still used as part of the PMK generation process in WPA3-SAE, it cannot be used by a hacker as a way to derive the PMK because_____.

   a. The password gets hashed by a Diffie-Hellman encryption before calculating the PMK against private keys on the client and AP.

   b. The Password only indexes a secret point in an elliptical curve and the secret point is used as a generator in calculating the PMK.

   c. The password is partial seeding material in the Diffie-Hellman Cryptographic exchange making the PMK unknowable to the hacker.

   d. The password is encrypted along with the private keys on the AP and client, before the Diffie-Hellman key exchange is performed.

13. The WPA3-Personal certification is mandatory for which 802.11 protocol?

    a. 802.11bg

    b. 802.11ac

    c. 802.11n

    d. 802.11ac wave2

    e. 802.1ax

    f. 802.11ax

    g. 802.1X

    h. 802.11a

14. The distinct difference(s) between WPA2-Enterprise and WPA3-Enterprise is_____.

    a. Only EAP_TLS AKMs are allowed in WPA3-Enterprise

    b. WPA3-Enterprise requires client devices to use PPM

    c. EAP Frames are protected by PMF in WPA3-Enterprise

    d. WPA3-Enterprise requires client devices to use PMF

    e. A and D

15. WPA3-Enterprise 192 is also referred to as Suite B or the _____.

    a. Commercial National Security Algorithm Suite

    b. Commercial NIST Security Association Suite

    c. Cryptographic NIST Security Algorithm Suite

    d. Cryptographic National Security Algorithm Suite

Review Answers

1. **D**. The Pairwise Transient Key generated and installed on the supplicant during the 4-way handshake.
2. **B**. The Nonce is used for this purpose.
3. **C**. With forward secrecy, a future breach does not expose past transmissions.
4. **D**. Encrypts all wireless traffic by adding a Diffie-Hellman key exchange that creates an unknown PMK used in the 4-way handshake to generate the encryption keys.
5. **D**. Configurator and enrollee
6. **C**. Bootstrapping
7. **D**. The configurator always receives the enrollee's public key
8. **D**. Data privacy, passive eavesdropping
9. **B**. False
10. **C**. More secure than WPA2-PSK because the WPA2-PSK PMK and PTK can be discovered leaving all network traffic accessible to a hacker.
11. **A**. An information element in the AP beacon tells the OWE capable client to associate to a hidden BSS.
12. **B**. The Password only indexes a secret point in an elliptical curve and the secret point is used as a generator in calculating the PMK.
13. **F**. 802.11ax
14. **D**. WPA3-Enterprise requires client devices to use PMF
15. **A**. Commercial National Security Algorithm Suite

Chapter 11: Penetration Testing

Objectives Covered

4.2   Use effective change management procedures including documentation, approval, and notifications

4.5   Implement effective auditing procedures to perform audits, analyze results, and generate reports

This chapter will discuss the principles of Penetration Testing. It will introduce you to the concept of hacking and penetration testing, specifically in the wireless world. You will also learn about vulnerability databases and how to assess the risk and impact of a vulnerability.

Enterprise networks are vulnerable to attack. They contain information that can be very valuable to an attacker – whether it be on a personal level, or with a view to selling for compensation. Stolen information is big business. According to the Insurance Information Institute, it is estimated that, in 2017 alone, the value of the information stolen as a result of cybercrime was approximately $16.8 billion US.

There are multiple ways that an attacker can try to access a system. This chapter is designed to teach the engineer structured ways in which weaknesses in the system can be evaluated.

Penetration testing is most commonly used as an auditing function to appraise the security of a network.

What About Your Hat?

Penetration testing is exactly what it sounds like. The pentester will attempt to ascertain how far they can penetrate into a system. If you want to become a good pentester, you need to start with the basics: the first rule of penetration testing that you must always remember and observe is "Only with Permission". You only use your skills on your own network, or a customer's network, with written permission from an authorized representative.

Penetration testing is a little bit like martial arts. By engaging in this course of study, it is as though you have decided to learn a new martial art. In martial arts, you turn up at the dojo, and they make you practice punches, kicks, and blocks. These are the basic building blocks that, over time, can be combined to create complex kata. It is important to understand that you must become familiar with these movements – the potential attacks, and defenses. It is the same in the penetration testing world: you must learn the basics, and then combine them to build a repertoire of attacks and defenses.

The second thing to remember is that the term "hacker" scares people. (In fact, in many circles, for example, government, or city/county/state management, the word hacker is considered to be a dirty word.) The problem is that, when people think of hackers, they immediately think of people in the shadows, dressed in hoodies to hide their identities, who are up to no good. Professionals use the more refined term "penetration tester," which sounds much safer and is more generally accepted.

You should also be aware that there are various types of hackers. They are named after the hats worn by cowboys in the old wild west movies: White, Black, Gray. If you watch the old movies, the bad guy invariably wore a black hat. The good guy – the sheriff – usually wore a white hat. It is by the color of their hat that you could define their role in the film. You may recall that the Lone Ranger always wore a white hat… because he was a "good guy"!

In this industry, you will hear the term "White Hat Hacker". This is someone who has learned hacking skills but only uses them in a penetration testing role. Most White Hat Hackers are quite adept, and skillful at what they do… and they are the good guys.

The term Black Hat Hackers generally refers to hackers who carry out illegal activities. These are, in general, the bad guys. However, many - because of their vast knowledge and from a desire to earn money from a legal source - reform to become White Hat Hackers and earn a good living using, and teaching, their skills in industry and corporations.

Grey Hat Hackers, are a little bit of both… perhaps they work in the cybersecurity industry, but partake of dubious activities outside of their day job? Superhero by day, but villain by night!

Hacking Process Phases

The hacking process has been formalized and documented for the purpose of teaching the pentesting methodology.

The white hats, or pentesters, will take the skills of the hacker and "attack," within specific rules of engagement, a network to see how it withstands those attacks. Issues discovered, such as continuing to use WEP, or the PSK key is "12345678," will be highlighted and reported back to the customer with recommendations for fixes.

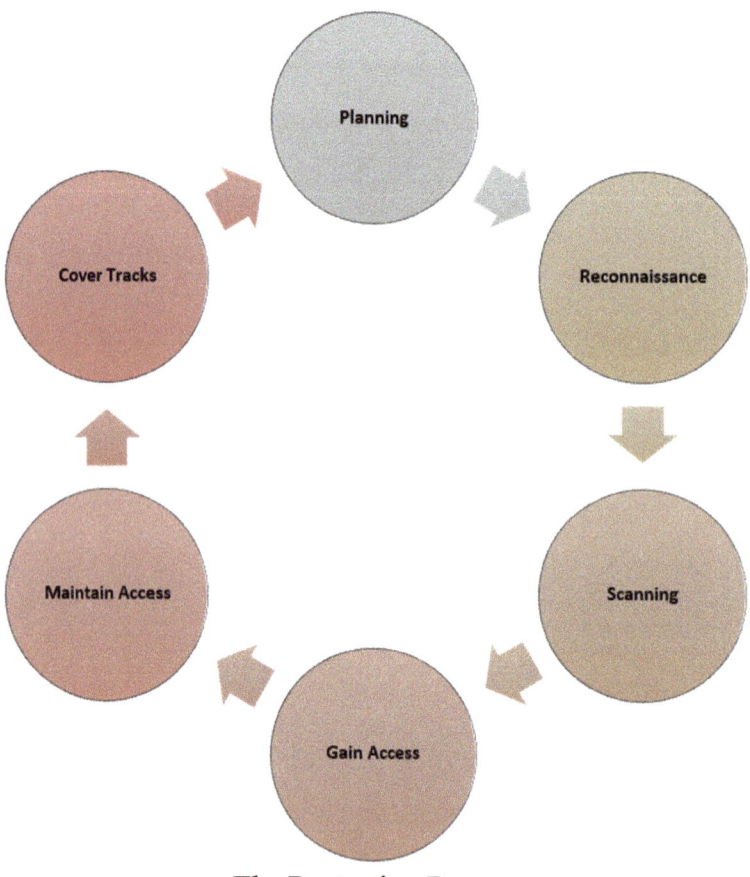

The Pentesting Process

The hacking process can be described using the following steps:

- **Planning:** Failing to plan is planning to fail. Make sure you have planned out your actions carefully. Specifically determine those things for which you will test, and how you will test them.
- **Reconnaissance:** This is the planning phase. You will research and study your target.
- **Scanning:** Also called enumerating. In this phase, you will scan the network to see what vulnerabilities can be found.
- **Gain Access:** Here, you simply gain access to the target network's resources.
- **Maintain Access:** Now that you have access, you can install backdoors or other malware to make it easier to reconnect.

- **Cover Tracks:** Here you hide your attack, maybe by deleting or manipulating log entries. This way, your client will not be able to detect your intrusion.

It is important to have a strategy or plan when considering doing penetration testing.

There are certifications in pentesting, just like CWNP has certifications for wireless skills. Certified Ethical Hacker (CEH) is one such certification that teaches the basics of hacking. You can then move on to EC-Council Certified Security Analyst (ECSA) or Licenced Penetration Tester (LPT) to formalize a methodology of pentesting.

These certifications tend to be more for generic system pentesting, and not so much focused on wireless pentesting.

In a generic pentesting job role, you normally carry out the following phases:

- **Planning:** Here, you plan the engagement and define the rules.
- **Scope Out:** In this phase, you will identify what systems the customer is using.
    - **Reconnaissance:** monitoring, and gathering information on the customer.
    - **Scanning:** using tools to discover the customer's network and environment.
    - **Enumerate:** expand, and gather more details on the customer network.
- **Attack:** Here, you attack the network and the servers.
    - **Gain Systems Access:** The goal is to get access to the customer's systems.
- **Report:** This is where penetration testing differs from hacking, here you create a report notifying the customer of the vulnerabilities in their system.

Wireless Penetration Testing

In a wireless pentesting job role, you are normally limited to only interacting with the wireless systems. You would generally carry out the following tasks:

- **Planning:** prepare, and define the work.
- **Discovery:** monitor what is visible.
- **Attack:** attack the infrastructure.
- **Report:** report on your activities.

We will now dig deeper into each of the four steps.

Planning

Planning is one of the most important parts of the pentest process. This is where you and the customer clearly define what is and what is not allowed. Areas to test are clearly defined, and the scope and rules of engagement are clearly spelled out.

The planning phase is used to define the Statement of Work, that you will work against, so you can invoice for the project. You should NEVER partake in a penetration testing project until the scope and rules of engagement are clearly defined and documented, and a signed statement of work has been created.

Your motto as a pentester should be "only with permission."

It is recommended that you modify this to be "only with permission, in writing, clearly defined by an authorized member of the staff of the customer."

It is very important that you understand that the only thing stopping you from being prosecuted is the SoW, that defines what you are allowed to do. If you go beyond that or do the work on a "handshake", you leave yourself open to potential prosecution by the company. For example, the CSO may want a pentest performed, but the CEO may not have been properly consulted, and is unhappy because something has stopped working, and they decide to blame you!

In some states and countries, it may be a federal offense to participate in any form of hacking without prior, clearly defined permission. You must take this part of the process very seriously.

Discovery

In this phase, you are going to discover the client's vulnerabilities.Primarily you will search for networks and clients. You will be looking for answers to questions like:

- What networks are visible?
- Are the networks hidden?
- What type of security is being used?
- What distance away from the company perimeter can the networks be detected?
- What clients can be seen?

On your first pentest, this is the point when you will realize how vulnerable wireless networks can be. It might be a real eye-opener to you, or you may already have an idea of how easily wireless networks can be intruded upon.

The customer's wireless signal can sometimes be easily seen across a parking lot, or across an open space. This means attackers do not even need to be close to the premises in order to mount an attack!

Attack

The attack is the focal point of the pentest. Working within the SoW, you will attack the wireless networks, and see how far you can get.

Some pentest SoWs want you to simply test getting onto their network; others may involve trying to gain access to resources inside the company. This will be defined in the planning phase and must be adhered to, no matter how tempting other options may seem.

You may also be instructed to just go at the network with full force or to try and be stealthy. The purpose of your pentest may not just be to see if you can penetrate the network, but may also be to test the company defenses like WIDS or WIPS. A stealth type of attack may be used to test the functionality of the WIDS and/or WIPS.

You will use the relevant tools in your arsenal to try to attack, and break, the wireless protocols in use. It may simply be a dictionary attack, or you may be required to implement a full-blown man-in-the-middle redirection.

Again, all of these different requirements will be specified in the planning phase.

Reporting

The reporting phase is the final part of the pentest process. This can be a vital part of the process. Many an awesome pentest has been ruined by an inadequate report.

It is important that you give the customer lots of useful and detailed information in this phase. You have done the work, you have done the testing, you now need to tell them what a great job you did, and here are the results.

You may want to consider multiple techniques to soften bad news to the customer. No-one likes it when you call their "baby" (network) ugly. Here is one example of a bad summary: "The pentesting team discovered that woefully inadequate security was present on the company network." Even if it is accurate and succinct, any customer will find this quite blunt and offensive.

Here is another example, with a positive suggestion for a remedy included: "The network has clearly been designed with security in mind; however, recent developments in security have made available more advanced forms of security methods. We recommend that the company network be upgraded to stay up-to-date with current defenses and to ensure compliance with regulatory governance."

As you can see, the second method is preferable. It is positive, constructive, and sounds much more professional - and is far more likely to keep your customer as an actual customer.

The format of the report should be professional. If the report is more than a few pages, you may want to consider a table of contents. Always include an Executive Summary, it shows respect for the executives' time by summarizing the findings quickly and professionally, and usually wins their favor. The executive is unlikely to read anything more than a 2-3-page report. Most likely, they will task someone else with that job. However, if you give them an Executive Summary, they are quite likely to read that. They will also have a level of respect for you as a professional.

An Executive Summary would state the problem to be fixed, what you did and why you did it, and what you found, along with a brief synopsis of your recommendations for fixes.

Following the Executive Summary, you should include the technical details, and then your findings. This is the main focus of the report, where you explain your discoveries, and recommendations for action to resolve issues.

The report contents should list detected networks, any hidden networks, and clients discovered. Any rogue APs should be noted and identified. Coffee shops, neighboring networks, and so on, should be clearly identified. There should never be a case of "we don't know who that is!"

Each vulnerability, with details, and the risk and potential impact should be clearly laid out for the customer. Ideally, a countermeasure or solution to the issues should be provided.

Tools and commands used should be listed for reference.

It is advisable to list appendices and references to add value to your discoveries and advice.

You must remember that this customer hired you because they couldn't do it for themselves. All they probably knew was that something wasn't quite right, but they might not have been able to figure out what that was. When you write your report, you need to be patient and thorough in explaining the issues (what they are, how you found them, and what caused them) and suggesting resolutions for them.

Reporting Tools in Kali Linux
Kali Linux has tools that can assist in note keeping and report writing.

- KeepNote is a useful note-taking utility available in Kali Linux.

- Dradis is a very comprehensive reporting tool, that comes in a free version, and a more powerful paid version.

These save time by helping you build a play-by-play account of your activities and can speed up the summarizing process. The report you will need to write should be very detailed, so any applications that can help you gather needed information (data, documentation, etc.) will definitely be of use to you.

A wealth of information will help you write a more comprehensive report which will, in turn, enable you to more easily, and comprehensively, address all the vulnerabilities and mitigations related to the situation. Of course, you could always just take screenshots and use your favorite word processing tool as well.

Common Vulnerabilities and Exposures

A vulnerability is a flaw or weakness in a system, that may leave it open to some form of attack. There are many vulnerabilities in the technologies we use today. Vendors strive to reduce or remove vulnerabilities from their products. Occasionally, a significant amount of time may pass before a known vulnerability is fixed or "patched." Some vendors offer "bug bounties" where anyone discovering a vulnerability can report it and receive payment for its discovery.

Some bodies have decided to make a database of discovered vulnerabilities. The Common Vulnerabilities and Exposure (CVE) system is an online database of publicly known cybersecurity vulnerabilities. It is maintained by the Mitre Corporation with funding by the US Government.

Mitre is a not for profit organization, supporting several government agencies. Since 1999, Mitre has maintained a database of Common Vulnerabilities and Exposures (CVEs) and is now the primary CVE Numbering Authority (CNA). There are almost a hundred CNAs, spread over more than fifteen countries. CNAs can include vendors, vulnerability researchers, national and industry CERTs, and bug bounty hunters.

CVEs are assigned by a CNA, usually Mitre, or can be assigned by another CNA, e.g., a vendor (Cisco, Microsoft, Oracle, etc.). It is possible for a vulnerability to not appear as a CVE immediately; it can take time for the vulnerability to be identified, classified and entered into the database system. The majority of weaknesses or vulnerabilities with wireless systems can eventually be found in the CVE database.

A CVE will take the form CVE-yyyy-nnn. Where yyyy represents a year, and nnn represents a number. It is common that a vulnerability may have more than one CVE allocated to it.

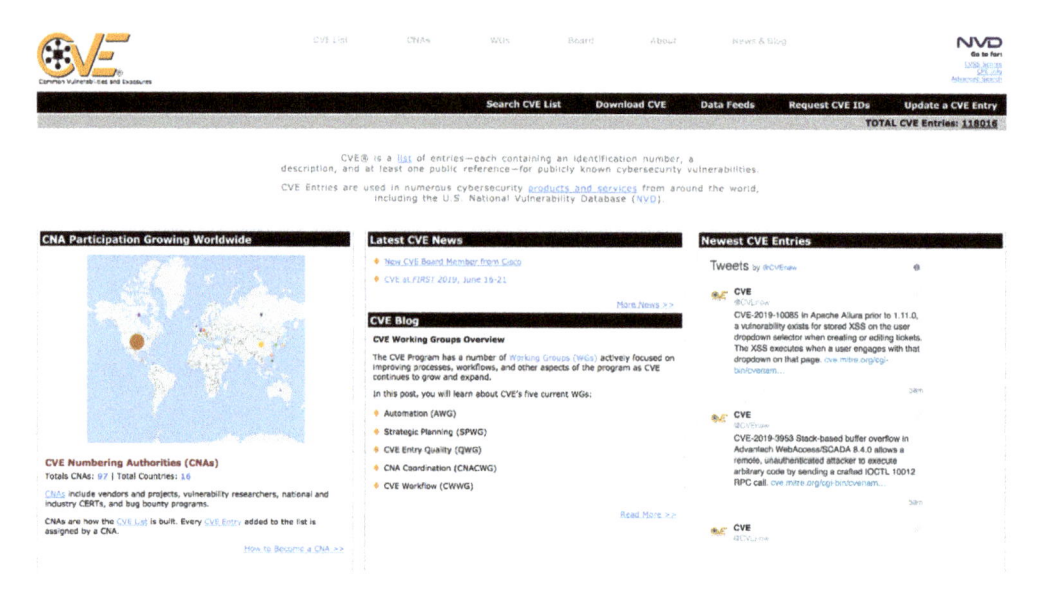

National Vulnerability Database

NIST (National Institute of Standards and Technologies) maintains another database – the National Vulnerability Database (NVD).

From the NVD Website:

The NVD is the U.S. government repository of standards-based vulnerability management data represented using the Security Content Automation Protocol (SCAP). This data enables automation of vulnerability management, security measurement, and compliance. The NVD includes databases of security checklist references, security-related software flaws, misconfigurations, product names, and impact metrics.

Originally created in 2000 (called Internet - Categorization of Attacks Toolkit or ICAT), the NVD has undergone multiple iterations and improvements and will continue to do so to deliver its services. The NVD is a product of the NIST Computer Security Division, Information Technology Laboratory and is sponsored by the Department of Homeland Security's National Cyber Security Division.

The NVD performs analysis on CVEs that have been published to the CVE Dictionary. NVD staff are tasked with the analysis of CVEs by aggregating data points from the description, references supplied and any supplemental data that can be found publicly at the time. This analysis results in association impact metrics (Common Vulnerability Scoring System - CVSS), vulnerability types (Common Weakness Enumeration - CWE), and applicability statements (Common Platform Enumeration - CPE), as well as other pertinent metadata. The NVD does not actively perform vulnerability testing, relying on vendors, third party security researchers and vulnerability coordinators to provide information that is then used to assign these attributes. As additional information becomes available CVSS scores, CWEs, and applicability statements are subject to change. The NVD endeavors to re-analyze CVEs that have been amended as time and resources allow to ensure that the information offered is up to date.

The NVD builds upon the information included in CVE entries to enhance the information provided. NVD provides advanced search features so you can search by OS, vendor name, product name, etc.

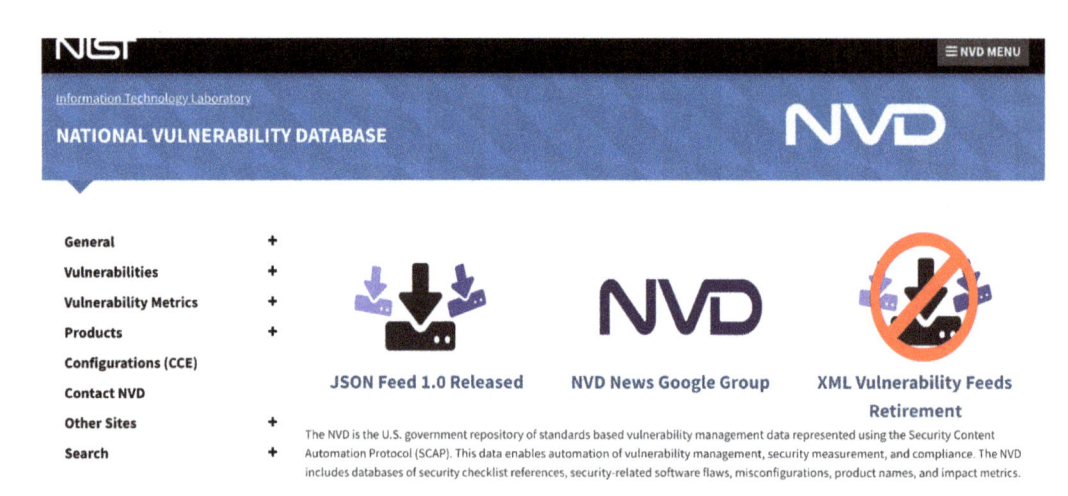

The NVD Database

> A good way to remember the relationship between the CVE and the NVD is to remember that the CVE feeds new vulnerabilities into the NVD. Note the sentence that reads, "The NVD performs analysis on CVEs that have been published to the CVE Dictionary."

CISA and US-CERT

The Cybersecurity and Infrastructure Security Agency (CISA) is another agency in the United States. It provides extensive cybersecurity and infrastructure security knowledge and practices and shares knowledge to enable better risk management.

On November 16, 2018, the Cybersecurity and Infrastructure Security Agency Act of 2018 was signed into law. This legislation created the CISA, which includes the National Cybersecurity and Communications Integration Center (NCCIC). Prior to the establishment of CISA, NCCIC realigned its organizational structure in 2017, integrating like functions previously performed independently by the U.S. Computer Emergency Readiness Team (US-CERT) and the Industrial Control Systems Cyber Emergency Response Team (ICS-CERT).

The CISA may assign a number (VU#xxxxxx) to a vulnerability.

This agency sends out regular bulletins, by email, alerting enrollees to newly-discovered vulnerabilities. The information, which comes on a regular basis is important for pentesters to be aware of.

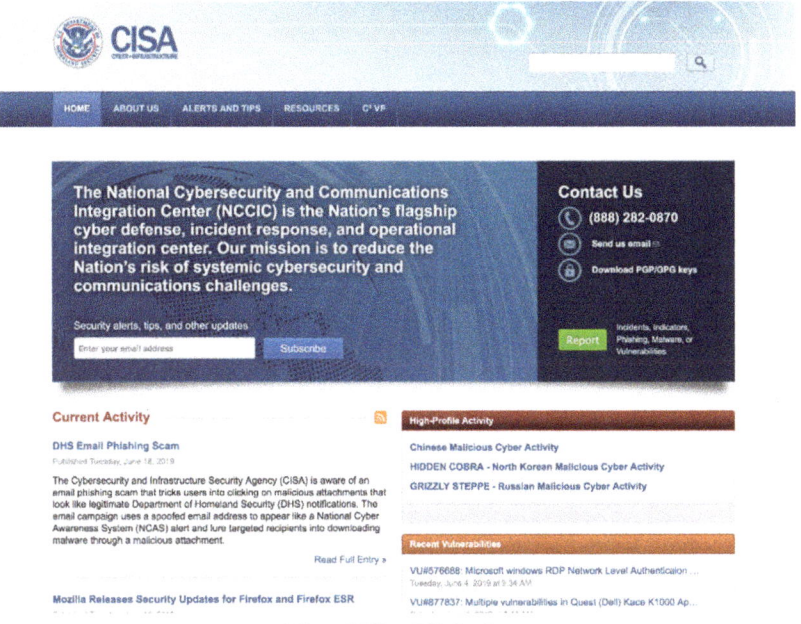

The CISA Website

CVE Details

You can search for information on these websites for CVE details:

- https://cve.mitre.org
- https://nvd.nist.gov
- https://www.us-cert.gov

You can also search for CVE information on third-party websites like the one shown in the image:

- https://www.cvedetails.com

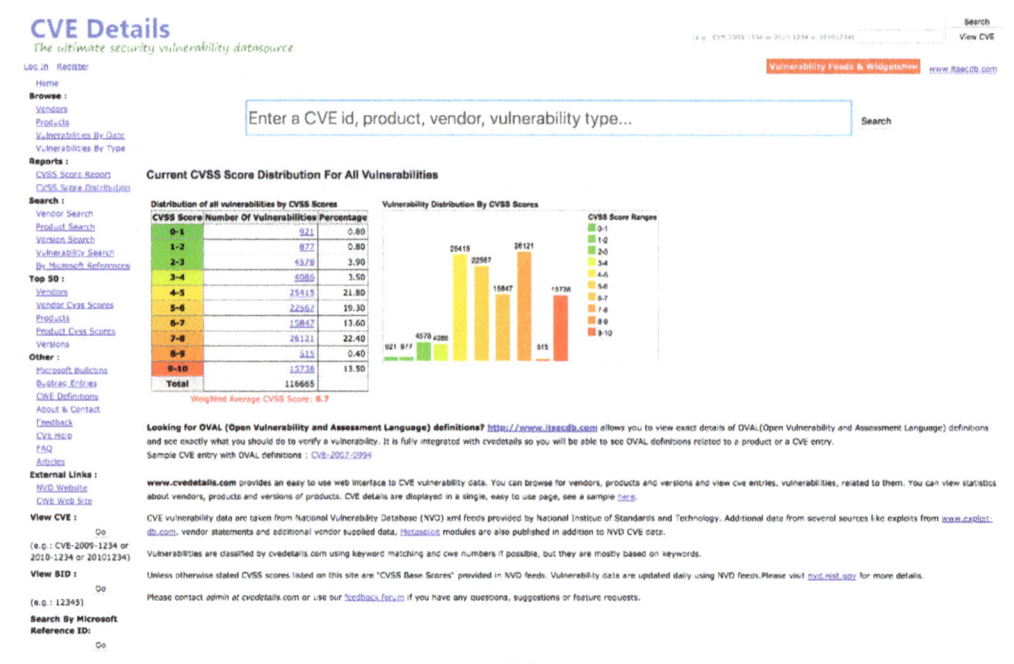

www.CVEdetails.com

Risk

Risk is the standard measurement of how you measure the possibility of an occurrence.

Risk = likelihood * impact.

The more likely something is to happen, the greater the risk. The greater the impact, the greater the risk. Once you know the risk, you can decide if you want to protect against it.

You may think that you are not at risk, but, according to the Insurance Information Institute, in just the first four months of 2019, 11.6 million records were breached because of hackers. It really is big business. Everyone wants data, and less-than-scrupulous vendors and clients are always willing to sell and pay for it. Experian suggests that one person's credit card information can be worth anywhere between $5 and $110! One victim's US passport information can sell for up to $2,000!

As an example, let's say you use a simple PSK on your network.

- Situation 1: In a hospital. Customer health details and credit card information cross the network. Here, the risk is unacceptable. The possibility of someone breaking into the network and stealing information is very likely. This kind of data breach is totally unacceptable under HIPPA and PCI rules.
- Situation 2: A coffee shop. The PSK is written on a chalkboard on the wall. Suddenly, this scenario is no longer such a big deal.

It is important to identify the risk and then calculate the worth of mitigating that risk.

Another example for consideration is determining whether you would spend $1 million dollars on protecting something worth only $10,000.

It is all about risk management.

A further example here may help to explain risk and risk mitigation.

In Florida, there are hurricanes; in California, there are not. In California there are earthquakes; in Florida, there are not. You are more likely to spend more money in California on earthquake protection and not much on hurricane protection. You are more likely to spend more money in Florida on hurricane protection, and not much on earthquake protection.

You probably do not spend much money on "alien invasion" protection, or on "large meteor hitting us" protection, probably because they are viewed as unlikely, or not worth the cost of covering. In other words, there is very little risk.

Impact

The impact is another factor to calculate. The reason you spend money to protect yourself is that the result of an earthquake or a hurricane could be high both financially and physically (loss of life). The impact of an event also needs to be planned for.

Again, if the impact is measured as $10,000, and the risk high, you still are not going to spend $1 million dollars on protecting against it. However, it is important to note that the penalty for not protecting may include fines or imprisonment for failing to adequately protect resources (e.g., if your company is subject to rules and regulations such as HIPPA, PCI, etc.)

The impact can be measured as technical impact or business impact.

The technical impact may include loss of:

- Confidentiality
- Integrity
- Availability

The business impact may include loss of:

- Money
- Reputation
- Compliance
- Privacy

To determine the risk and impact of a vulnerability, you can refer to the National Vulnerability Database (NVD), to glean information.

Let's use an example of the KRACK attack, that occurred in late 2017, and see what we can find.

1. Load your web browser and enter "nvd.nist.gov" and when the page loads select [Search].

2. Next, select the [Vulnerabilities – CVE] button.

3. When the search page loads, enter "KRACK" and select [Search].

4. The NVD finds information on your search criteria. You will see there are 10 results for "KRACK."

5. Select the top entry – [CVE-2017-13088].

6. You can see information on the vulnerability.

7. The database highlights the description of the vulnerability and also lists impact references.

8. You can follow the links to see the scoring metrics of the severity impact of the vulnerability.

9. Scrolling down the page, you can see more information on the impact, as well as references to advisories and tools.

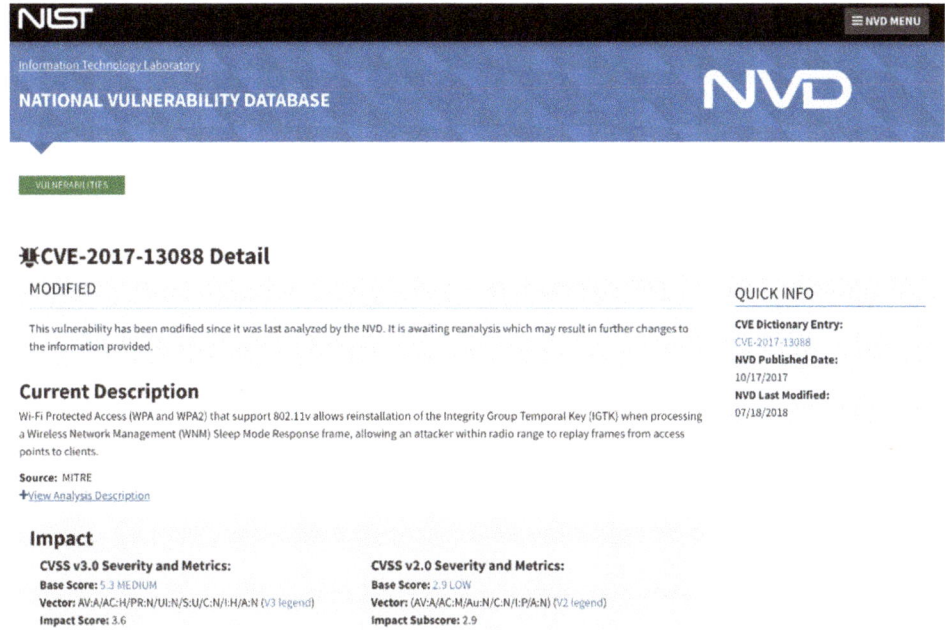

KRACK Attack CVE-2017-13088

White Box vs. Black Box

The type of pentest you may be asked to perform depends on the goals of the pentest.

White Box Testing is where you have knowledge of the systems involved and are trying to penetrate systems knowing exactly what they are and, most usually, what defenses are in place.

Black Box Testing is exactly as it sounds. You have a box, you have no knowledge about it, and try to get into it! The network in this situation is treated just like a black box. You have to discover everything, then choose what and how to attack, with no idea of defenses.

Of course, most pentests fall between these two extremes, and you will find that no two pentest jobs are alike. They all come with their own quirks and requirements, and of course, Rules of Engagement, and restrictions.

If you are conducting White Box Testing, then you will need to gather as much project documentation from the customer as possible. Network maps and layouts, design goals, equipment lists, etc. will all be useful. On the contrary, if you are performing Black Box Testing, you may need to build similar documentation as you go, because nothing is provided to you at the start.

Hardware and Software Selection

Choosing your hardware and software carefully is very important when you want to become a pentester. Even more so, when you want to become a wireless pentester.

There are not many good and reasonably priced professional platforms for wireless pentesting. Most pentesters build up their own repertoire from what is available, with Kali Linux being the most popular.

You can run Kali Linux in a Virtual Environment, or run it natively on hardware. Hardware platforms vary from Macbooks to Windows PCs, and even high-end Chromebooks. Memory and disk space form a major part of the decision process. Some pentesters are currently experimenting with the latest Raspberry Pi and Arduino platforms.

Selection of USB devices ranks very high in the decision process. You will probably need several adapters, and this impacts the number of USB ports on the platform that you want to use. Of course, you can use a USB hub, but then how are you going to power that? Not all laptops and other devices come with portability, large batteries, and multiple full-power USB ports.

Check with the platform you are using to ensure that the USB adapters are supported. It is common in the Linux world for USB drivers to take some time to come out, usually long after the adapter is released. As a result, you will frequently find that the latest and greatest adapter available, probably don't have a Linux driver available.

Jammers can come in two forms, hardware, and software. Software jammers usually work with the USB device you have available. Hardware jammers come with only one purpose: to jam! Be advised that any form of jamming of wireless frequencies is frowned upon, and most likely highly illegal, and will come with severe penalties. In fact, the mere possession of some devices in some areas of the world may end up with you having to answer some very in-depth questions from law enforcement. You may have to prove you are a pentester and may have equipment confiscated.

Scanner software on phones and tablets can be a great asset when discovering networks. Most likely you will be ignored if you are walking around or standing staring at your phone/tablet. Wi-Fi Analyzer for Android phones is a powerful tool to discover and locate networks. Apple utilities are much more scarce, but Apple's Airport utility can be put into scan mode and can reveal interesting results.

Apple restricts the operation of utilities interfacing with its Wi-Fi API. Android, to date, has not. Before the Android people get too excited here, be aware that Google has recently (2019) announced the introduction of limitations on the abilities of Wi-Fi scanners on Android platforms moving forward, so expect changes there.

Don't overlook your trusty wireless protocol analyzer here. Good old Wireshark can be a great tool once you start looking into the protocols and captured frames. If you have a more professional analyzer, such as Live Action (Savvius) Omnipeek, the visualizations provided in the software can be a great asset in your pentest.

Air Magnet Wi-Fi Analyzer Pro, is a tool that is great for troubleshooting. It has a built-in monitor, analyzer, and expert engine that help you analyze your Wi-Fi environment.

One of the advanced features of the tool is to analyze your security environment. Then, using the in-built report engine, you can create reports. The software comes with built-in analysis of common security reports, such as PCI, SOX, ISO, etc.

Discovery

Discovery is an important part of the pentest process. Discovery is where you will find the customers SSIDs, define which are your customers and which are rogue, and choose which ones to attack. If an SSID is missed in the discovery phase, it could impact the whole pentest. It is important, therefore, to ensure that all devices are powered on and functioning correctly before starting. It may be advisory to get a list of expected SSIDs, and maybe even APs, from the customer before starting the pentest.

Promiscuous and Monitor Mode

It is important to understand the different modes that a network interface can be put into, before starting pentesting.

Normal mode: this mode is where an adapter transmits and receives frames, normally. In this mode, the adapter will only accept Unicast frames addressed directly to it, Broadcast frames, and any Multicast frames it has been programmed for.

Promiscuous mode: this mode is where an adapter is configured to accept and pass all frames up to the protocol stack. Usually used on Ethernet when using a protocol analyzer.

Monitor mode: this mode is only applicable to wireless, and it is an extra mode where the wireless adapter will pass all frames it sees in the air, regardless of which BSSID they belong to, up to the protocol stack.

Microsoft limitation: Microsoft restricts monitor mode access in its APIs. Because of this, you will need dedicated hardware or a special driver written for your capture application. Usually, these applications cost money and have specific drivers written for specific chipsets. This is changing as we draw into the 2020s, with capture drivers becoming more desirable, and vendors releasing drivers that can be natively used with Windows and Wireshark (Netgear A6210, for example). Microsoft is also talking about changing this limitation in a future version of Windows.

Discovery Tools

You do not need to be a hacker to find out what is going on in the wireless world. Every wireless SSID broadcasts a beacon approximately 10 times a second advertising its functions and features. Other Management frames (Probe requests and responses, and Association requests and responses) also carry information about the networks that are running in your vicinity.

There are applications commercially available that capture these messages, and display information about the SSIDs. This information can display simple channel and name, or give a plethora of information right up to Encryption method used, power constraints, n/ac/ax features, and even proprietary Vendor specifics.

WiFi Explorer (MAC) by Adrian Granados, and Metageeks's inSSIDer (Windows and now also on MAC) are invaluable tools for researching wireless functionality.

These utilities make the job easy for you. Of course, you can power on your protocol analyzer, and manually start decoding the Information Elements yourself, should you choose to.

Once you have exhausted the features of these tools, then it is time to roll up your sleeves and start up your trusty Kali Linux.

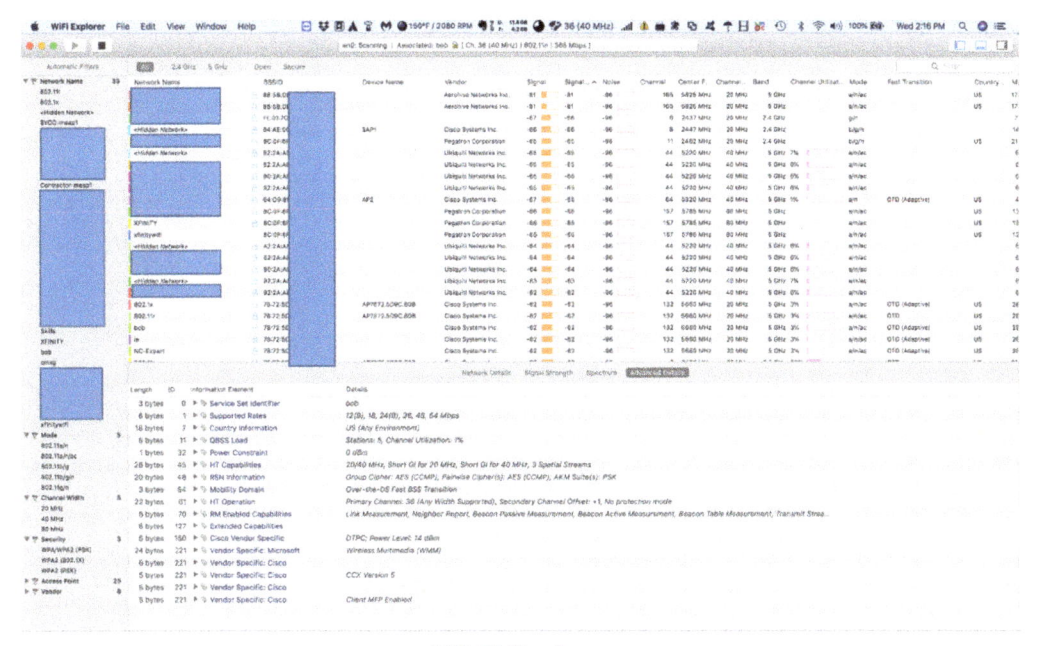

Wi-Fi Explorer

Kali Linux Discovery Tools

Kali Linux is where the great tools are, in the opinion of this author. You can run an entire pentest just from Kali Linux, and need no other software, from the planning phase through to reporting phase. You will only be exposed to some of the most commonly used pentesting discovery and attack tools here, but it is recommended to study up on other available features in Kali Linux.

You will view basic Linux wireless functionality, and then move on to the airmon-ng suite, which provides the fundamental monitor mode functionality of the Kali Linux toolset. You will be shown all-in-one tools that deal with pentesting functions in a script like manner, demonstrating to you the availability of tools that remove the need for you to have to learn the intricate details of the command line tools, so you can get up and running quickly. However, mastery of the software cannot be achieved if you shy away from learning the advanced CLI options.

> In the captures shown here, the operator was logged in as root to save time because it is easy and very convenient. Because of this, the use of the "sudo" command was not required. Some tools complain if you use them as root. Root, used to run a pentest, is frowned upon in the industry because the root user has so much power. Normally, you would create a user to connect with, then use the "sudo" command to escalate that particular user's privilege where needed.

Linux Wi-Fi Tools

The Linux command **ifconfig** is fundamental in manipulating and monitoring network interfaces.

```
root@kali:~# ifconfig
eth0: flags=4099<UP,BROADCAST,MULTICAST>  mtu 1500
        ether 00:0c:29:06:dc:7d  txqueuelen 1000  (Ethernet)
        RX packets 0  bytes 0 (0.0 B)
        RX errors 0  dropped 0  overruns 0  frame 0
        TX packets 0  bytes 0 (0.0 B)
        TX errors 0  dropped 0 overruns 0  carrier 0  collisions 0

lo: flags=73<UP,LOOPBACK,RUNNING>  mtu 65536
        inet 127.0.0.1  netmask 255.0.0.0
        inet6 ::1  prefixlen 128  scopeid 0x10<host>
        loop  txqueuelen 1000  (Local Loopback)
        RX packets 715  bytes 61292 (59.8 KiB)
        RX errors 0  dropped 0  overruns 0  frame 0
        TX packets 715  bytes 61292 (59.8 KiB)
        TX errors 0  dropped 0 overruns 0  carrier 0  collisions 0

root@kali:~#
```

Linux ifconfig Command

You can see in the image that the command was entered with no arguments from the system. This installation of Kali Linux actually has 4 wireless cards installed, but they are not shown, simply because they are not in the UP state.

If you want to see all interfaces, whatever state they are in, then use **ifconfig –a**.

The **iwconfig** command is useful to show all the connected wireless adapters.

```
root@kali:~# iwconfig
wlan3     IEEE 802.11  ESSID:off/any
          Mode:Managed  Access Point: Not-Associated   Tx-Power=20 dBm
          Retry short  long limit:2   RTS thr:off   Fragment thr:off
          Encryption key:off
          Power Management:off

eth0      no wireless extensions.

wlan2     IEEE 802.11  ESSID:off/any
          Mode:Managed  Access Point: Not-Associated   Tx-Power=20 dBm
          Retry short  long limit:2   RTS thr:off   Fragment thr:off
          Encryption key:off
          Power Management:off

wlan1     IEEE 802.11  ESSID:off/any
          Mode:Managed  Access Point: Not-Associated   Tx-Power=20 dBm
          Retry short  long limit:2   RTS thr:off   Fragment thr:off
          Encryption key:off
          Power Management:off

wlan0     IEEE 802.11  ESSID:off/any
          Mode:Managed  Access Point: Not-Associated   Tx-Power=20 dBm
          Retry short  long limit:2   RTS thr:off   Fragment thr:off
          Encryption key:off
          Power Management:off

lo        no wireless extensions.

root@kali:~#
```

Linux iwconfig Command

Here you can see there are four adapters. They are simply numbered from 0, so we have wlan0, wlan1, wlan2, and wlan3. The **iwconfig** command reports basic attributes of the interfaces.

The **airmon-ng** suite is one of the most versatile utilities ever written for wireless. It forms the foundation of most of the other tools you will use in Kali Linux.

```
File Edit View Search Terminal Help
root@kali:~# airmon-ng

PHY     Interface       Driver          Chipset

phy0    wlan0           rt2800usb       Ralink Technology, Corp. RT5370
phy1    wlan1           rt2800usb       Ralink Technology, Corp. RT5370
phy2    wlan2           rt2800usb       Ralink Technology, Corp. RT5370
phy3    wlan3           rt2800usb       Ralink Technology, Corp. RT5370

root@kali:~#
```

airmon-ng Command

The basic command **airmon-ng** displays the wireless adapters, their chipset, and the Linux driver they are using.

Note, in the image the "phy" maps to the "wlan" number. This is because this is a fresh reboot. The "phy" numbers represent the physical wireless interface and are independent of the "wlan" numbers.

In this next image, you will see that things have changed.

```
File Edit View Search Terminal Help
root@kali:~# airmon-ng

PHY     Interface       Driver          Chipset

phy5    wlan0           rt2800usb       Ralink Technology, Corp. RT5372
phy1    wlan1           rt2800usb       Ralink Technology, Corp. RT5370
phy2    wlan2           rt2800usb       Ralink Technology, Corp. RT5370
phy3    wlan3           rt2800usb       Ralink Technology, Corp. RT5370

root@kali:~#
```

airmon-ng with a New Adapter

Simply, wlan0 was removed, then a new adapter was added. This adapter, connected as phy4, but then disconnected for some reason, then reconnected as phy5. Note, however, it was given the interface name of wlan0. In Kali Linux, you have to watch out for this.

Note, also it is a slightly different chipset, but still supported by the same driver. This driver and the chipsets shown are some of the best ones to use for Kali Linux and will give you a high level of compatibility with the tools (this means fewer errors).

When putting wireless interfaces in monitor mode, and using the airmon-ng utilities, other processes on the system can interfere with the airmon-ng suite.

You can get weird errors, where the devices move channels or flip back into managed mode, and out of monitor mode. The two biggest culprits are the **NetworkManager** and the **wpa_supplicant**. These are simply client utilities on Kali that allow you to connect to wireless.

Airmon-ng has a simple argument which checks whether these are running, the command is **airmon-ng check**. If you find they are running, you can manually kill the processes, or simply use **airmon-ng check kill** to automatically stop any processes that airmon-ng doesn't like.

Now it is time to put one of our WLAN interfaces into monitor mode. Let's use wlan0.

The command is **airmon-ng start wlan0**

Note, that wlan0 is now phy7! Simply, another interface was connected, and it became phy6. Then, this new interface, and the previous one that was added (phy5), were removed, and the original one plugged back in. This means the original 4 adapters are plugged in. Linux chose to configure the adapter that was just plugged in as phy7, and allocated the first available wlan name wlan0. Note this is the exact same adapter that was originally given phy0.

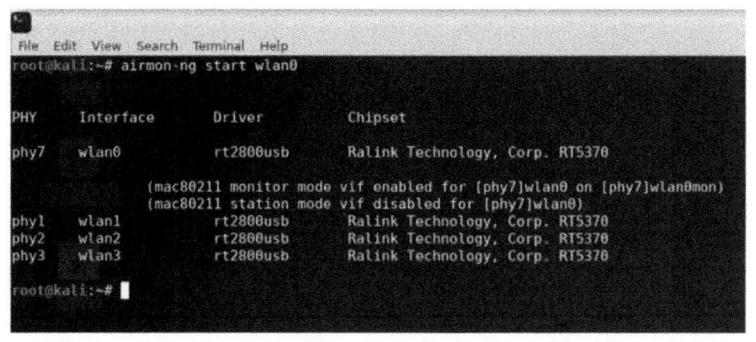

Using airmon-ng to Place an Adapter into Monitor Mode

The command shows you that wlan0 has been converted from station mode into monitor mode, and it has been renamed wlan0mon. Note that older versions of Kali Linux would create a new interface called mon0. Should you wish to turn off this monitor mode, you simply enter **airmon-ng stop wlan0mon**. This will convert the interface back to being wlan0.

We will now look at one of the airmon-ng suite of programs called **airodump-ng**. The command was **airodump-ng wlan0mon**. Notice, in the left-hand picture, it starts collecting information on SSIDs. Some SSID names and MAC addresses have been blanked to protect neighbors' networks.

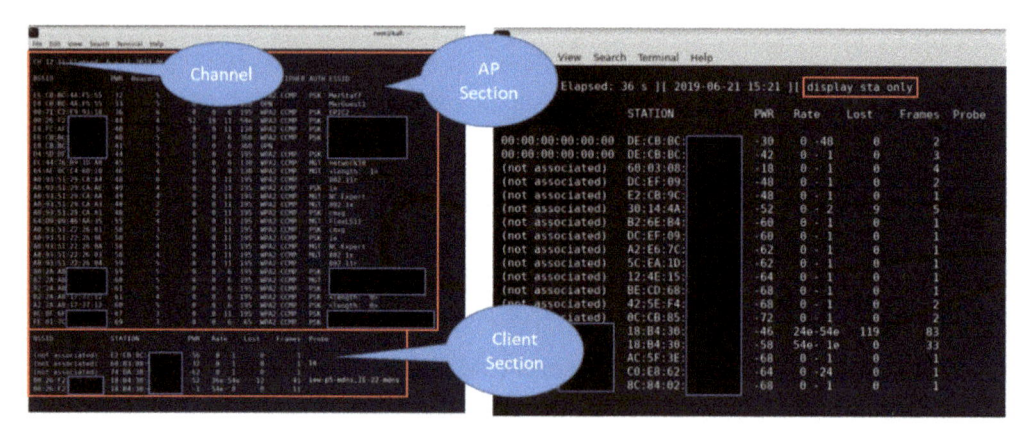

airodump-ng

The SSID name (ESSID) is shown, along with the channel (CH) it is on. It also shows the AP MAC address (BSSID) and the signal strength (PWR) the SSID can be seen at. Note, you can see hidden SSIDs this way. Their SSID name appears as <length: 0> or <length: 1>.

The current channel being scanned is shown in the top left-hand corner of the screen.

In the left-hand picture, you can see APs and clients. The clients section also shows the clients MAC address, the BSSID it is associated to (if any), the power level the client is seen at, and any SSIDs the client is probing for are shown.

Sometimes if the number of SSIDs is large, you may not see the clients section. Press the "a" key, twice, and the top right-hand text "display sta only" appears. Now you are viewing clients only. You can press "a" again to flip back to "ap+sta" view.

Kismet is a great, often overlooked tool for pentesting. Kismet has been available for some time, and is a great little text-based tool, that gives a graphical-type output using text!

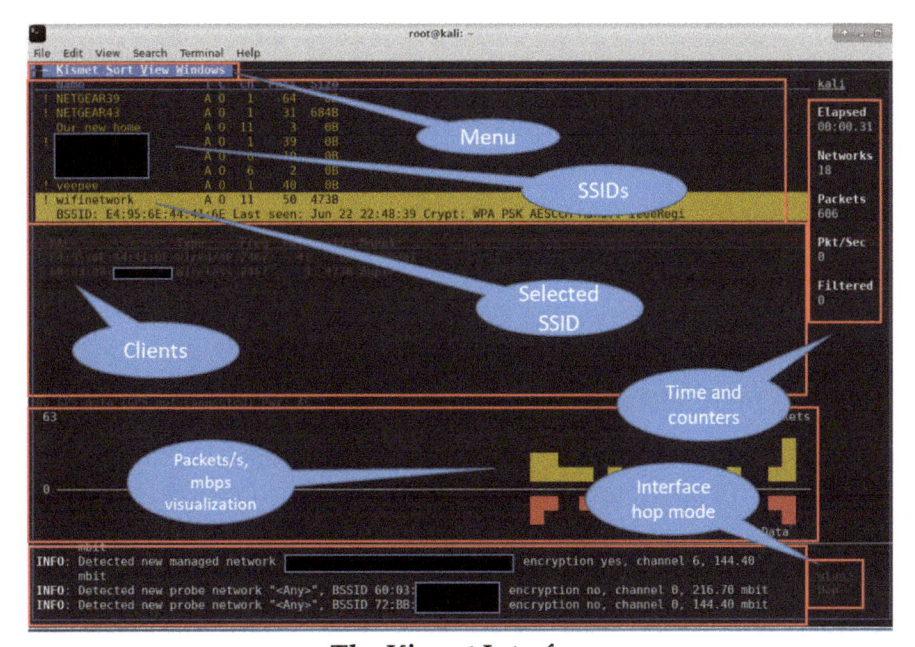

The Kismet Interface

Kismet was started with the command **kismet –c wlan3** to use interface wlan3.

The Kismet screen is split into multiple sections. Firstly, on the top is the Kismet menu options.

Below the menu options is a listing of SSIDs it is seeing. You can scroll up and down this list to move between SSIDs. The SSID selected is highlighted and shows on two lines. The lower line showing the BSSID, Security used, AP manufacturer, and time the SSID was last seen. There is a color code used that can be changed if needed. In this setup yellow shows an encrypted SSID, red would show a WEP SSID, and green denotes an unencrypted SSID.

Below the SSID section, is the client section. Here the BSSID and any client MAC addresses connected to the selected SSID are shown.

The packet/s and Mbps sections come below the client section showing a visualization of these measured flows.

Finally, at the bottom, you see the log of activity.

In the top right, you see the counters, and in the bottom right you see the interface being used, and the channel mode, in this case, Kismet is configured to hop across channels.

Kismet can be locked onto one channel by pressing the "~" key (the tilde key, usually to the left of the number 1 key, on US keyboards), then selecting "**Kismet**" menu option, then "**L**" for "**Config Channel**".

Moving through the options using the "TAB" key, select the "**Lock**" option, then select the channel or frequency you want to lock onto in the "**Chan/Freq**" section, then select "**Change**" to save the settings.

Now the visualizations will show only for the selected channel.

Attacking

Now you need to turn your attention to attacks. First, you will learn some basic attacks, then you will learn about a scripting tool that comes with menus and helps to assist you in your pentesting journey. These tools are invaluable when you start learning pentesting, as they can get you up-and-running more complex attacks very quickly.

You will first be introduced to basic attack essentials. Then you will be shown how to capture a four-way handshake and how to test it against a dictionary. Finally, you will be shown a scripting tool that automates the pentest attack.

Please note that all of the things you will be shown here must only ever be run against your own network or a network that you have clear written permission to pentest. Running any of these commands against a network without permission, can be illegal, and have severe consequences.

One of the first skills you need to learn is how to de-authenticate clients connected to a network. This skill is needed to cause clients to re-connect. This can speed up capturing of necessary communications exchanges.

Be forewarned that this is not a stealth skill. This will scream out a warning to a WIDS or WIPS system. You will most likely need to manually set the channel that the monitor interface will run on. This is a slightly more subtle attack, as it only attacks a single client on an SSID. It does mean, however, you have to track down the client you wish to attack.

In the example shown below, a locally-administered made-up MAC address was used, in a real attack, you would need to capture the MAC of the client you wish to deauth.

aireplay-ng Deauth Attack

To be able to test certain networks, you will need to be able to change your MAC address. If the client has MAC filtering configured, or if you want to test some MAC-based security, you will need to be able to emulate the client's MAC address.

The **macchanger** tool in Kali Linux is the easiest way to do this; however, the interface needs to be placed in the down state, using the **ifconfig** command.

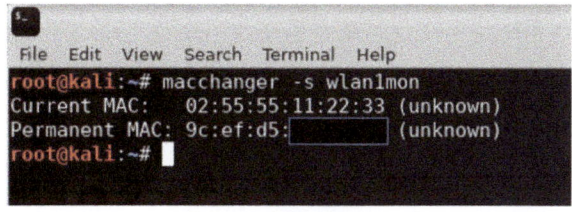

macchanger

431

A wordlist (or a dictionary) is an important tool when pentesting. A wordlist is simply a list of passwords to test for. The test analyzes every possible word that could be found in a dictionary – hence "dictionary" attack. It does not mean dictionary as in Webster, or the Oxford English. Here dictionary means "a created list." "Words" such as GoodLuck@123 are completely allowed.

The other option is a brute force attack, which goes through every possible binary combination of the password – however, this is VERY time-consuming.

Most commonly, you will pentest to see if common passwords are being used. One of the most common wordlists is provided free with Kali Linux, the "rockyou" wordlist. It stores common passwords used by people. There are many other wordlists available, as you can imagine the more comprehensive the wordlist, the larger the file.

You may want to compile your own wordlist when testing, or learning pentesting, as password cracking, even if using a wordlist, is very time-consuming.

The first thing we need to do is copy and extract the wordlist into our home directory (that is what the tilde ~ represents in the following commands).

cp /usr/share/wordlists/rockyou.txt.gz ~
gunzip rockyou.txt.gz

At this point, you have discovered the network, and you see it is using a PSK. Now it is time to learn how to attack a PSK protected network.

Assuming you have configured your interfaces to be in monitor mode, you will use the **airodump-ng** command to capture the 4-way handshake. You will use the "**--bssid**" argument to specify the bssid to attack, the "**-c**" argument to specify the channel to operate on, and the "**-w**" argument to specify a filename to write to.

airodump-ng wlan1mon
airodump-ng --bssid <mac> -c 11 -w <name> wlan1mon

In a second window, you will use the **aireplay-ng** tool to de-authenticate clients to force them to re-connect, so you can capture the 4-way handshake.

The "**-0**" argument specifies a de-authentication attack, and the "**0**" implies unlimited attempts. The "**-a**" argument specifies the BSSID to attack.

Notice, you see a variation method of using the tool, using the "**--deauth**" argument instead of "**-0**". In this example, you are attacking 10 times, and by adding the "**-c**" argument, you specific a client to attack, instead of attacking all clients.

aireplay-ng -0 0 –a <bssid> wlan2mon
aireplay-ng --deauth 10 –a <bssid> -c <client_mac> wlan2mon

You allow the attack to continue in the first window (left-hand side), while running the **aireplay-ng** tool in the second window (right-hand side).

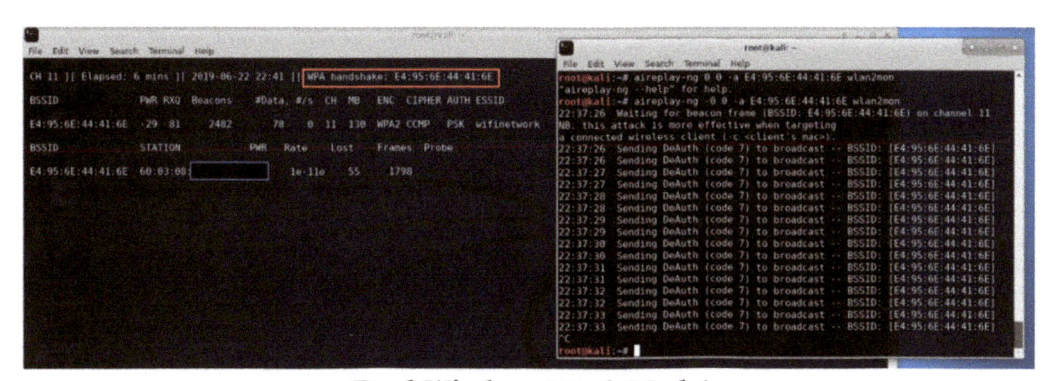

Dual-Window Attack Mode!

Continue until you see "**WPA handshake: XX:XX:XX:XX:XX:XX**" appear in the upper right of the first window (left-hand side). This signals that a 4-way handshake has been captured on the WLAN (the X's represent numbers, and they are the BSSID of the WLAN you are attacking).

You can now stop the **aireplay-ng** attack by using **CTRL+C**.

Returning to the first window, run **aircrack-ng <filename> -w rockyou.txt**

The filename will be whatever you used in step 1, but you will find the tool adds "**-#.cap**" to the end of the name you used, where "#" will represent a number, usually "1". If a file already exists with the number "1", then it will use "2" and so on.

The wordlist you will use here is rockyou.txt.

As you can see, the tool very quickly found the password used.

433

In this example, a password from very early on in the file was used, to demonstrate its use. The size of the file and the CPU/memory/disk power of the Linux platform will determine the length of time that will be taken to break the passphrase. This is a very heavy compute process, as the passphrase is mixed with the SSID name in the algorithm 4096 times to create the actual PSK. The tool has to run through every word in the wordlist, mixing the word with the SSID name 4096 times, then comparing the results with the captured 4-way handshake.

Discovered Passphrase

Remember, if the actual password isn't in the wordlist, then the passphrase will not be broken. Then you will have to try another dictionary or try a brute force attack. A brute force attack is very compute-intensive.

Wifite is a tool you can use to automate a PSK attack. You start Wifite, providing the wordlist you want to use with the "**--dict**" parameter.

First, Wifite prompts you for the interface to use, we used **wlan1**. Note, wlan1 was not in monitor mode, Wifite dealt with that for you. Wifite starts scanning for SSIDs and presents them to you. Once you see the one you want to attack, you are told to press **CTRL+C**. In the example shown, SSID number 1, "wifinetwork" was selected (by entering "**1**").

Now Wifite attempts a PMKID attack, which is not what we want. So, we again press **CTRL+C** to terminate this attack. When prompted, we press "**C**" to continue.

Wifite will now try to capture the 4-way handshake. After a short delay, if it does not see one, it will automatically start a de-auth attack. Eventually, when it captures a 4-way handshake exchange, it will automatically start the attack using the wordlist you provided.

As you can see, it found it quickly, just like the manual attack.

wifite Discovering the Passphrase

Chapter Summary

In this chapter, you learned the principles of Penetration Testing. You were introduced to the concept of hacking and penetration testing, specifically in the wireless world. You also learned about vulnerability databases and how to assess the risk and impact of a vulnerability. Now you can appreciate how vulnerable enterprise networks really are, to attack. You learned methods and practices to help you audit your networks and discover weaknesses that may be present.

Review Questions

1. What is the difference between a white hat and black hat hacker?

    a. Black hat is a pentester, white hat partakes in criminal activity.

    b. White hat is a pentester, black hat is hacker that may partake in illegal activity.

    c. There is no difference, both partake in illegal activities.

    d. Black hat is certified, white hat is not.

2. Which if the following is not a part of the hacking process?

    a. Reconnaissance.

    b. Gain access.

    c. Maintain Connection.

    d. Cover tracks.

3. Which is the most important thing to have, prior to commencing a Penetration Testing project?

    a. A report template

    b. A blueprint of the client's building

    c. A signed Statement of Work

    d. A verbal agreement of what is to be done

4. What will you not generally find in a Penetration Testing report?

    a. Executive Summary

    b. List of commands used

    c. Names of employees who caused vulnerabilities

    d. Suggestions for remedies for any vulnerabilities found

5. Which of the following is not included in the Pentesting Strategy?

    a. Mitigate

    b. Scope Out

    c. Plan

    d. Report

6. Which of the following organizations provides vulnerability information?

    a. CISSP

    b. CIA

    c. CLP

    d. CISA

7. Which is the correct formula?

    a. Risk = annulment * documentation

    b. Risk = likelihood * impact

    c. Risk = implementation * discovery

    d. Risk = mitigation * premium price

8. Which of the answers below is not a type of technical impact?

    a. Confidentiality

    b. Integrity

    c. Availability

    d. Reputation

9. What type of Impact can be accounted for, on an individual level?

    a. Emotional

    b. Mental

    c. Physical

    d. Philosophical

10. What is the name of a type of hacking attack?

    a. Karpov

    b. Kali

    c. Kasparov

    d. KRACK

Review Answers

1. **B**. A Black Hat hacker implies some kind of illegal activity. A white hat hacker uses similar techniques and methods, but only with permission

2. **C**. A hacker will maintain access, by introducing a backdoor or malware onto the system so they can easily reconnect. The do not maintain an open connection.

3. **C**. A Penetration Tester must have a signed Statement of Work, which is physical proof that they have permission from the client to test (attack) the network.

4. **C**. Penetration Testing reports contain specific details of the pentest. It will include an executive summary, the list of commands used, what was found, and suggested remedies for any vulnerabilities.

5. **A**. The Pentesting strategy we discussed is Plan, Scope Out, Attack, Report. The pentester's job is to suggest resolutions, but not to implement (mitigate) any fixes.

6. **D**. The Cybersecurity and Infrastructure Security Agency (CISA) is a government agency in the United States. It provides extensive cybersecurity and infrastructure security knowledge and practices, and shares knowledge to enable better risk management.

7. **B**. Risk is the result of the interaction between the likelihood of something happening, and its impact. The formula is expressed as risk = likelihood * impact.

8. **D**. Technical impacts can include confidentiality, integrity, and availability. Reputation is classified as a business impact.

9. **C**. Individual impacts are classified as either physical or financial. The other impacts mentioned are very difficult to quantify.

10. **D**. KRACK is a hacking attack. Kali is a pentesting platform, Karpov and Kasparov are two world class chess players.

Appendix A: Up and Running with Kali Linux

This appendix provides brief instructions for setting up a Kali Linux virtual machine. At the time of writing, all links were valid. Use the following root websites to locate the two required resources should the links cease to function:

VMware (search for VMware Workstation Player free): www.vmware.com

Offensive Security (search for Kali Linux VM): www.offensivesecurity.com

The first thing you should do is download the VMware Player and install it on your system. We assume you are installing it on a Windows system. You can download the VMware Workstation Player here (do not include the word-wrap in the URL when you enter it into your browser):

```
my.vmware.com/web/vmware/free#desktop_end_user_computing/
vmware_workstation_player/15_0
```

After downloading is complete, install the VMware Workstation Player as you would install any other Windows application. The process is straightforward.

Next, you will need to download the VMware virtual machine (VM) image from Offensive Security. This is a very big download, so be sure to have a fast Internet connection when you begin the download process. You can download the VM here (again, do not include the word-wrap):

```
www.offensive-security.com/kali-linux-vm-vmware-virtualbox-
image-download/
```

	Kali Linux VMware Images	Kali Linux VirtualBox Images			
Image Name	Torrent	Size	Version	SHA256Sum	
Kali Linux VMware 64-Bit 7z	Torrent	2.4G	2019.2	4611f3797c53ed37c89443bd8bb94ac1fd860fb807865d8933783c0f6ef21007	
Kali Linux VMware 32-Bit 7z	Torrent	2.5G	2019.2	c7f52865f5d0554ad1bc990684a0751eb46d1b8ab552d7c942d71e4fe20b7e67	

If you are running a 64-bit version of Windows, which will be required for VMware Workstation Player anyway, download the Kali Linux VMware 64-Bit 7z file. The 2+ GB files will be in the 7zip format, which can be uncompressed with most compression tools, including the free 7zip software.

After the download is complete, uncompress the files to a folder. This folder will be the resident location of your Kali Linux VM, so it should be on an internal fast drive, if possible.

Now, you can open the VMware Workstation Player and then navigate to and open the Kali Linux VM. It's that simple to get up and running with Kali Linux.

Here are a few tips to get you started:

1. The default login information is username:*root* and password:*toor*.
2. The following adapters (USB) work well when attached to a Kali Linux virtual machine:
    a. Alfa AWUS036ACH (802.11ac adapter)
        i. You can load the driver for this adapter, if required, with the following shell command:
        apt-get install realtek-rtl88xxau-dkms
    b. Alfa AWUS1900 (802.11ac adapter)
        i. Use the same driver install as AWUS036ACH
    c. TRENDnet TEW 809UB (802.11ac adapter)
        i. Use the same driver install as AWUS036ACH (Are you seeing a pattern? They all use the same chipset [RTL8814AU])
3. Remember to go to a shell periodically in your Kali Linux VM and update it by running:
    a. apt-get update
    b. apt upgrade
    c. apt dist-upgrade

With this information and configuration, you are ready to begin your Kali Linux journey. As an additional note, many different 802.11n USB adapters work very will with Kali Linux. For an exhaustive list, see: hackersgrid.com/2018/06/top-kali-wifi-adapters.html

Glossary: A CWNP Universal Glossary

40 MHz Intolerant: A bit potentially set in the 802.11 frame allowing STAs to indicate that 40 MHz channels should not be used in their BSS or in surrounding networks. The bit is processed only in the 2.4 GHz band.

4-Way Handshake: The process used to generate encryption keys for unicast frames (Pairwise Transient Key (PTK)) and transmit encryption keys for group (broadcast, multicast) (Group Temporal Key (GTK)) frames using material from the 802.1X/EAP authentication or the pre-shared key (PSK). The PTK and GTK are derived from the Pairwise Master Key (PMK) and Group Master Key (GMK) respectively.

802.11: A standard maintained by the IEEE for implementing and communicating with wireless local area networks (WLANs). Regularly amended, the standard continues to evolve to meet new demands. Several Physical Layer (PHY) methods are specified and the Medium Access Control (MAC) sublayer is also specified.

802.11a: An 802.11 amendment that operates in the 5GHz band. It uses OFDM modulation and is called the OFDM PHY. It can support data rates of up to 54 Mbps.

802.11aa: An 802.11 amendment that added support for robust audio and video streaming through MAC enhancements. It specifies a new category of station called a Stream Classification Service (SCS) station. The SCS implementation is optional for a WMM QoS station.

802.11ac: An 802.11 amendment that operates in the 5GHz band. It uses MU-MIMO, beamforming, and 256 QAM technology, up to 8 spatial streams and OFDM modulation. Support is included for data rates up to 6933.3 Mbps.

802.11ae: An 802.11 amendment that provides prioritization of management frames. It defines a new Quality of Service Management Frame (QMF). When the QMF service is used, some management frames may be transmitted using an access category other than the one used for voice (AC_VO). When communicating with stations that do not support the QMF service, the station uses access category AC_VO to transmit management frames. When QMF is supported, the beacon frame includes a QMF Policy element.

802.11ah: An 802.11 draft that specifies operations in the sub-1 GHz range. Frequencies used vary by regulatory domain. The draft supports 1, 2, 4, 8 and 16 MHz channels with OFDM modulation.

802.11ax: An 802.11 draft that will support bi-directional MU-MIMO, higher modulation rates and sub-channelization. It is too early to know the final details of this amendment at the time of writing; however, it is planned to operate in the 2.4 GHz and 5 GHz band.

802.11b: An IEEE 802.11 amendment that operates in the 2.4GHz ISM band. It uses HR/DSSS and earlier technology. It can support data rates of up to 11Mbps.

802.11e: An 802.11 amendment, now incorporated into the most recent rollup, that provided quality of service extensions to the wireless link through probabilistic prioritization based on the contention window. The Wi-Fi Multimedia (WMM) certification is based on this amendment.

802.11g: An IEEE 802.11 amendment that operates in the 2.4GHz ISM band. It uses ERP-OFDM and earlier technology. It can support data rates of up to 54Mbps.

802.11i: An 802.11 amendment, now incorporated into the most recent rollup, which provided security enhancements to the standard and resolved weaknesses in the original WEP encryption solution. It provided for TKIP/RC4 (now deprecated) and CCMP/AES cipher suites and encryption algorithms.

802.11n: An IEEE 802.11 amendment that operates in the 2.4 ISM and 5GHz UNII/ISM bands. It uses MIMO, HT-OFDM and earlier technology. It can support data rates of up to 600Mbps.

802.11k: An IEEE 802.11 amendment that specifies and defines WLAN characteristics and mechanisms.

802.11r: An IEEE 802.11 amendment that enables roaming between access points.

802.11u: An IEEE 802.11 amendment that adds features for mobile communication devices such as phones and tablets.

802.11w: An IEEE 802.11 amendment to increase security for the management frames.

802.11y: An IEEE 802.11 amendment that allows registered stations to operate at a higher power output in the 3650-3700 MHz band.

802.1X: 802.1X is an IEEE standard that uses the Extensible Authentication Protocol (EAP) framework to authenticate devices attempting to connect to the LAN or WLAN. The process involves the use of a supplicant to be authenticated, authenticator, and authentication server.

802.11 State Machine: The 802.11 state machine defines the condition of the connection of a client STA to another STA and can be in one of three states: Unauthenticated/Unassociated, Authenticated/Unassociated, or Authenticated/Associated.

802.3: A set of standards maintained by the IEEE for implementing and communicating with wired Ethernet networks and including Power over Ethernet (PoE) specifications.

AAA Framework: Authentication, Authorization, and Accounting is a framework for monitoring usage, enforcing policies, controlling access to computer resources, and providing the correct billing amount for services.

AAA Server Credential: The AAA server credential is the validation materials used for the server. When mutual authentication is required, a server certificate is typically used as the AAA server credential.

Absorption: Occurs when an obstacle absorbs some or all of a radio wave's energy.

Access Category (AC): An access category is a priority class. 802.11 specifies four different priority classes – voice (AC_VO), video (AC_VI), best effort (AC_BE), and background (AC_BK).

Access Layer Forwarding: Data forwarding that occurs at the access layer, also called *distributed data forwarding*. The data is distributed from the access layer directly to the destination without passing through a centralized controller.

Access Point: An access point (AP) is a device containing a radio that is used to create an access network, bridge network or mesh network. The AP contains the Distribution System Service.

Access Port: An AP used for mesh networks and that connects to the wired or wireless network at the edge of the mesh.

Acknowledgement Frame: A frame sent by the receiving 802.11 station confirming the received data.

Access Control List (ACL): ACLs are lists that inform a STA or user what permissions are available to access files and other resources. ACLs are also used in routers and switches to control packets allowed through to other networks.

Active Mode: A power-save mode in which the station never turns the radio off.

Active Scanning: A scanning (network location) method in which the client broadcasts probe requests and records the probe responses in order to determine the network with which it will establish an association.

Active Survey: A wireless survey conducted on location that involves measuring throughput rates, round trip time, and packet loss by connecting devices to an AP and transmitting data during the survey.

Ad-Hoc Mode: The colloquial name for an Independent Basic Service Set (IBSS). STAs connect directly with each other and an AP is not used.

Adjacent Overlapping Channels: Adjacent overlapping channels are channels whose bands interfere with their neighboring channels on the primary carrier frequencies. Non-overlapping channels are channels whose bands do not interfere with neighboring channels on the primary carrier frequencies.

Adjacent Channel Interference (ACI): ACI occurs when channels near each other (in the frequency domain) interfere with one another due to either partial frequency overlap on primary carrier frequencies or excessive output power.

AES (Advanced Encryption Standard): The encryption cipher used with CCMP and WPA2 providing improved security over WEP/RC4 or TKIP/RC4.

AID: Association ID (AID) is an identification assigned by a wireless STA (AP) to another STA (client) in order to transmit the correct data to that device in an Infrastructure Basic Service Set.

AirTime Fairness: Transmits more frames to client STAs with higher data rates than those with lower data rates so that the STAs get fair access to the air (medium) instead of having to wait for slower data rate STAs.

Aggregated MAC Protocol Data Units (A-MPDU): A-MPDU transmissions are created by transmitting multiple MPDUs as one PHY frame as opposed to A-MSDU transmissions, which are created by passing multiple MSDUs down to the PHY layer as a single MPDU.

Aggregated MAC Service Data Unit (A-MSDU): See *Aggregated MAC Protocol Data Unit*.

Amplification: The process of increase a signal's power level.

Amplifier: A device intended to increase the power level of a signal.

Amplitude: The power level of a signal.

Antenna: A device that converts electric power into radio waves and radio waves into electric power.

Association: The condition wherein a client STA is linked with an AP for frame transmission through the AP to the network.

Announcement Traffic Indication Message (ATIM): A traffic indication map (sent in a management frame) in an Ad-Hoc (IBSS) network to notify other clients of pending data transfers for power saving purposes.

Attenuation: The loss of signal strength as an RF wave passes through a medium.

Attenuator: A device that intentionally reduces the strength of an RF signal.

Authentication: The process of user or device identity validation.

Authentication and Key Management (AKM): The protocols used to authenticate a client STA on a WLAN and generate encryption key for use in frame encryption.

Authentication Server: The authentication server validates the client before allowing access to the network. In an 802.1X/EAP implementation for WLANs, the authentication server is often a RADIUS server.

Authenticator: The device that provides access to authentication services in order to allow connected devices to access network resources. In an 802.1X/EAP implementation for WLANs, the authenticator is typically the AP or controller.

Automatic Power Save Delivery (APSD): APSD is a power saving method which uses both scheduled (S-APSD) and unscheduled (U-APSD) frame delivery methods. S-APSD sends frames to a power save STA from the AP at a planned time. U-APSD sends frames to a power save STA from the AP when the STA sends a frame to the AP. The frame from the STA is considered a trigger frame.

Autonomous AP: An AP that can perform security functions, RF management, and configuration without the need for a centralized WLAN controller or any other control platform.

Azimuth Chart: A chart showing the radiation pattern of an antenna as viewed from the top of the antenna. Also called an H-Plane Chart or H-Chart.

Backoff timer: The timer used during CSMA/CA to wait for access to the medium, which is selected from the contention window.

Band Steering: A method used by vendors to encourage STAs to connect to the 5 GHz band instead of the 2.4 GHz band, which is more congested. Typically implemented by ignoring probe requests for some period of time before allowing connection to the 2.4 GHz radio by clients known to have a 5 GHz radio based on previous connections to the AP or controller.

Bandwidth: The frequencies used for transmission of data. For example, a 20 MHz wide channel has 20 MHz of bandwidth.

Basic Service Area (BSA): The coverage area provided by an AP wherein client STAs may connect to the AP to transmit data on the WLAN or through the AP to the network.

Basic Service Set (BSS): An AP and its associated STAs. Identified by the BSSID.

Basic Service Set Identification (BSSID): The ID for the BSS. Often the MAC address of the AP STA. When multiple SSIDs are used, another MAC address-like BSSID is generated.

Beacon Frame: A frame transmitted periodically from an AP that indicates the presence of a BSS network and contains capabilities and requirements of the BSS. Also colloquially called a beacon instead of the full phrase, beacon frame.

Beamforming: Directing radio waves to a specific area or device by manipulating the RF waveforms within the different radio chains.

Beamwidth: The width of the radiated signal lobe from the antenna in the intended direction of propagation. It is usually measured at the point where 3 dB of loss is experienced.

Bill of materials (BOM): A list of the materials and licenses required to assemble a system, in the case of WLANs, including APs, controllers, PoE injectors, licenses, etc.

Bit: A basic unit of information for computer systems. A bit can have a value of 1 or 0. Used in binary math.

Block Acknowledgement: An acknowledgement frame that groups together multiple ACKs instead of transmitting each individual ACK when a block transmission has been received.

Bridge: A device used to connect two networks. Wireless bridges create the connection across the wireless medium.

BSS Transition: Roaming that occurs between two BSSs that are part of the same ESS.

Byte: A basic unit of information that typically consists of 8 bits. Also called an octet.

Capacity: The number of clients and applications a network or AP can handle.

Captive Portal: Authentication technique that re-routes a user to a special webpage to verify their credentials before allowing access to the network. Commonly used in hotel and guest networks.

Guest Networks: A segregated network that is designed for use by temporary visitors.

CardBus: A PCMCIA PC Card standard interface that supports 32-bits and operates at speeds of up to 33 MHz. It is primarily used in laptops.

Carrier Frequencies: The frequency of a carrier signal or the frequencies used to modulate information.

Carrier Sense Multiple Access (CSMA): CSMA is a protocol that allows a node to detect the presence of traffic before sending data on a shared network. Used in CSMA/CA.

Carrier Sense Multiple Access with Collision Avoidance (CSMA/CA): CSMA/CA is the method in 802.11 networks in which a node only sends data if the shared network is idle in order to avoid collisions.

CCMP: Counter Cipher Mode with Block Chaining Message Authentication Code Protocol (CCMP) is an key management solution that provides for improved security over WEP.

CCMP/AES: CCMP used with AES, as it is in 802.11 networks, is a key management and encryption protocol that provides more security than WEP. It is based on the AES standard and uses a 128 bit key and 128 bit block size.

Centralized Forwarding: Every forwarding decision is made by a centralized forwarding engine, such as the WLAN controller.

Certificate Authority (CA): A server that validates the authenticity of a certificate used in authentication and encryption systems. The CA may issues certificates or it may authorize other servers to do the same.

CompactFlash (CF): Originally produced in 1994 by SanDisk, CF is a flash memory mass storage device format that can support up to 256 GB. CF devices can also function as 802.11 WLAN adapters.

Channel: A specified range of frequencies used in the 802.11 standard used by devices to communicate on the network. Channels are commonly 20, 40, 80 and 160 MHz in width in WLANs. Newer standards will support 1, 2, 4, 8 and 16 MHz channels in sub-1 GHz networks.

Channel Width: The range of frequencies a single channel encompasses.

Clear Channel Assessment (CCA): CCA is a feature defined in the IEEE 802.11 standard that allows a client to determine idle or busy state of the medium based on energy levels of a frame or raw energy levels as specified in each PHY.

Client Utilities: Software installed on devices that allows the device to connect to, authenticate with and participate in a WLAN.

Co-Channel Interference (CCI): Congestion cause by the normal operations of CSMA/CA when multiple BSSs exist on the same channel. Commonly called co-channel congestion (CCC) today as well.

Collision Avoidance (CA): A method in which devices attempt to avoid simultaneous data transmissions in order to prevent frame collisions. Used in CSMA/CA.

Coding: A process used to encode bits to be transmitted on the wireless medium such that error recovery can be achieved. Part of forward error correction (FEC) and defined in the modulation and coding schemes (MCSs) from 802.11n forward.

Containment: A process used against a detected rogue AP to prevent any connected clients from accessing the network.

Contention Window: A number range defined in the 802.11 standard and varying by QoS category from which a number is selected at random for the backoff timer in the CSMA/CA process.

Control Frame: An 802.11 frame that is used to control the communications process on the wireless medium. Control frames include, RTS frames, CTS frames, PS-Poll frames and ACK frames.

Controlled Port: In an 802.1X authentication system, the virtual port that allows all frames through to the network, but only after authentication is completed.

Controller-Based AP: An AP managed by a centralized controller device. Also called a lightweight AP or thin AP.

Coverage: 1) The colloquial term used for the BSA of an AP. 2) The requirement of available WLAN connectivity throughout a facility, campus or area. Often specified in minimum signal strength as dBm; for example, -67 dBm.

Clear-to-Send (CTS) Frame: A CTS frame sent from one STA to another to indicate that the other STA can transmit on the medium. The duration value in the CTS frame is used to silence all other STAs by setting their NAV timers.

Data Frame: An 802.11 frame specified for use in carrying data based on the general frame format. Also used for some signaling purposes as null data frames.

Data Rate: The rate at which data is sent across the wireless medium. Typically represented as megabits per second (Mbps) or gigabits per second (Gbps). The data rate should not be confused with throughput rate, which is a measurement of Layer 4 throughput or useful user data.

dBd (decibel to dipole): A relative measurement of antenna gain compared to a dipole antenna. Calculated as 2.14 dB greater than dBi as a dipole antenna already has 2.14 dBi gain.

dBi (decibel to isotropic): A relative measurement of antenna gain compared to a theoretical isotropic radiator. When necessary, calculated as 2.14 dB less than dBd.

dBm (decibel to milliwatt): An absolute measurement of the power of an RF signal based on the definition of 0 dBm = 1 milliwatt (mW).

Distributed Coordination Function (DCF): A protocol defined in 802.11 that uses carrier sensing, backoff timers, interframe spaces and frame duration values to diminish collisions on the wireless medium.

Elevation Chart: A chart showing the radiation pattern of an antenna as viewed from the side antenna. Also called an E-Plane Chart or E-Chart.

Deauthentication Frame: A notification frame sent from an 802.11 STA to another STA in order to terminate a connection between them.

Decibel (dB): A logarithmic, relative unit used when measuring antenna gain, signal attenuation, and signal-to-noise ratios. Strictly defined as 1/10 of a bel.

Delay: The time it takes for a bit of data to travel from one node to another. Also called latency.

Delivery Traffic Indication Message (DTIM): A message sent from an AP to clients in the Beacon frame indicating that it has data to transmit to the clients specified by the AIDs.

Differentiated Services Code Point (DSCP): A Layer 3 QoS marking system. IP packets can include DSCP markings in the headers. Eight precedence levels, 0-7, are defined.

Diffraction: The bending of waves around a very large object in relation to the wave.

Direct-Sequence Spread Spectrum (DSSS): A modulation technique where data is coupled with coding that spreads the data across a wide frequency range. Provides 1 or 2 Mbps data rates in 802.11 networks.

Disassociation Frame: A frame sent from one STA to another in order to terminate the association.

Distributed Forwarding: See *Access Layer Forwarding*. Also called, *distributed data forwarding*.

Distribution System (DS): The system that connects a set of BSSs and LANs such that an ESS is possible.

Distribution System Medium (DSM): The medium used to interconnect APs through the DS such that they can communicate with each other for ESS operations using either wired or wireless for the DS connection.

Domain Name System (DNS): A protocol and service that provides host name resolution (looking up the IP address of a given host name) and recursive IP address lookups (finding the host name of a known IP address). Also, colloquially used to reference the server that provides DNS lookups.

Driver: Software that allows a computer to interact with a hardware device such as a WLAN adapter.

Duty Cycle: A measure of the time a radio is transmitting or a channel is consumed by a transmitting device.

Dynamic Frequency Selection (DFS): A setting on radios that dynamically changes the channel selection based on detected interference from radar systems. Many 5 GHz channels require DFS operations.

Dynamic Rate Switching (DRS): The process of reducing a client's data rate as frame transmission failures occur or signal strength decreases. DRS results in lower data rates but fewer transmissions required to successfully transmit a frame.

Encryption: The process of converting data into a form that unauthorized users cannot understand by encoding the data with an algorithm and a key or keys.

Enhanced Distributed Channel Access (EDCA): An enhancement to DCF introduced in 802.11e that implements priority based queuing for transmissions in 802.11 networks based on access categories.

Equivalent Isotropically Radiated Power (EIRP): The output power required of an isotropic radiator to equal the measured power output from an antenna in the intended direction of propagation.

Extended Rate Physical (ERP): A physical layer technology introduced in 802.11g that uses OFDM (from 802.11a) in the 2.4 GHz band and offers data rates up to 54 Mbps.

Extended Service Set (ESS): A group of one or more BSSs that are interconnected by a DS.

Extensible Authentication Protocol (EAP): An authentication framework that defines message formats for authentication exchanges used by 802.1X WLAN authentication solutions.

Fade Margin: An amount of signal strength, in dB, added to a link budget to ensure proper operations.

Fast Fourier Transform (FFT): A mathematical algorithm that takes in a waveform as represented in the time or space domain and shows it in the frequency domain. Used in spectrum analyzers to show real-time views in the frequency domain (Real-time FFT).

Fragmentation: The process of fragmenting 802.11 frames based on the fragmentation threshold configured. Fragmented frames have a greater likelihood of successful delivery in the presence of sporadic interference.

Frame Aggregation: A feature in the IEEE 802.11n PHY and later PHYs that increases throughput by sending more than one frame in a single transmission. Aggregated MSDUs or aggregated MPDUs may be supported.

Frame: A well-defined, meaningful set of bits used to communicate management and control information on a network or transfer payloads from higher layers. Frames are defined at the MAC and PHY layer.

Free Space Path Loss: The natural loss of amplitude that occurs in an RF signal as it propagates through space and the wave front spreads.

Fresnel Zones: Ellipsoid shaped zones around the visual LoS in a wireless link. The first Freznel zone should be 60% clear and would preferably be 80% clear to allow for environmental changes.

Frequency: The speed at which a waveform cycles in a second.

Full Duplex: A communication system that allows an endpoint to send data to the network at the same time as it receives data from the network.

Gain: The increase in signal strength in a particular direction. Can be accomplished passively by directing energy into a smaller area or actively by increasing the strength of the broadcasted signal before it is sent to the antenna.

Group Key Handshake: Used to transfer the GTK among STAs in an 802.11 network if the GTK requires updating. Initiated by the AP/controller in a BSS.

Group Master Key (GMK): Used to generate the GTK for encryption of broadcast and multicast frames and is unique to each BSS.

Group Temporal Key (GTK): Used to encryption broadcast and multicast frames and is unique to each BSS.

Guard Interval (GI): A period of time between symbols within a frame used to avoid intersymbol interference.

Half Duplex: A communication system that allows only sending or receiving data by an endpoint at any given time.

Hidden Node: The problem that arises when nodes cannot receive each other's frames, which can lead to packet collisions and retransmissions.

High Density: A phrase referencing a WLAN network type that is characterized by large numbers of devices requiring access.

Highly-Directional Antenna: An antenna, such as a parabolic dish or grid antenna, that has a high gain in a specified direction and a low beamwidth measurement as compared to semi-directional and omnidirectional antennas.

High Rate Direct Sequence Spread Spectrum (HR/DSSS): An amendment-based PHY (802.11b) that increase the data rate in 2.4 GHz from the original 1 or 2 Mbps to 5.5 and 11 Mbps while maintaining backward compatibility with 1 and 2 Mbps.

High Throughput (HT): An amendment-based PHY (802.11n) that increased the data rate up to 600 Mbps and added support for transmit beamforming and MIMO.

Hotspot: A term referencing a wireless network connection point that is typically open to the public or to paid subscribers.

Independent Basic Service Set (IBSS): A set of 802.11 devices operating in ad-hoc (peer-to-peer) mode without the use of an AP.

Institute of Electrical and Electronics Engineers (IEEE): A standardization organization that develops standard for multiple industries including the networking industry with standard such as 802.3, 802.11 and 802.16.

Intentional Radiator: Any device that is purposefully sending radio waves. Signal strength of the intentional radiator is measured at the point where energy enters the radiating antennas.

Interference: In WLANs, an RF signal or incidental RF energy that is radiated in the same frequencies as the WLAN and that has sufficient amplitude and duty cycle to prevent 802.11 frames from successful delivery.

Interframe Space (IFS): A time interval that must exist between frames. Varying lengths are used in 802.11 and a references as DIFS, SIFS, EIFS and AIFS in common use.

Internet Engineering Task Force (IETF): An open group of volunteers develops Internetworking standards through request for comments (RFC) documents. Examples include RADIUS, EAP and DNS.

Isotropic Radiator: A theoretical antenna that spreads the radiaton equally in every directon as a sphere. None exist in reality, but the concept is used to measure relative antenna gain in dBi.

Jitter: The variance in delay between packets sent on a network. Excessive jitter can result in poor quality for real-time applications such as voice and video.

Jumbo Frame: An Ethernet frame that contains more than 1500 bytes of payload and up to 9000 to 9216 bytes.

Latency: The time taken data to move between places. Typically synonymous with delay in computer networking.

Layer 1: The physical layer (PHY) that is responsible for framing and transmitting bits on the medium. In 802.3 and 802.11 the entirety of Layer 1 is defined.

Layer 2: The data-link layer that deals with data frames moving within a local area network (LAN). In 802.3 and 802.11, the MAC sublayer of Layer 2 is defined.

Layer 3: The network layer where packets of data are routed between sender and receiver. Most modern networks use Internet Protocol (IP) at Layer 3.

Layer 4: The transport layer where segmentation occurs for upper layer data and TCP (connection oriented) and UDP (connectionless) are the most commonly used protocols.

Lightning Arrestor: A device that can redirect ambient energy from a lightning strike away from attached equipment.

Line of sight (LoS): When existing, the visual path between to ends. RF LoS is different from visual LoS. RF LoS does not require the same clear path for the remote receiver to hear the signal. When creating bridge links, visual LoS is often the starting point.

Link Budget: The measurement of gains and losses through an intentional radiator, antenna and over a transmission medium.

Loss: The reduction in the amplitude of a signal.

MAC filtering: A common setting that only allows specific MAC addresses onto a network. Ineffective against knowledgeable attackers because the MAC address can be spoofed to impersonate authorized devices.

Management Frame: A frame type defined in the 802.11 standard that encompasses frames used to manage access to the network including beacon, probe request, prober response, authentication, association, reassociation, deauthentication and disassociation frames.

Master Session Key (MSK): A key derived between an EAP client and EAP server and exported by the EAP method. Used to derive the PMK, which is used to derive the PTK. The MSK is used in 802.1X/EAP authentication implementations. In personal authentication implementations, the PMK is derived from the pre-shared key.

Maximal Ratio Combining (MRC): A method of increasing the signal-to-noise ratio (SNR) by combining signals received on multiple radio chains (multiple antennas and radios).

Mesh: A network that uses interconnecting devices to form a redundant set of connections offering multiple paths through the network. 802.11s defined mesh for 802.11 networks.

Mesh BSS: A basic service set that forms a self-contained network of mesh stations.

Milliwatt (mW): A unit of electrical energy used in measuring output power of RF signals in WLANs. A mW is equal to 1/1000 of a watt (W).

Mobile User: A user that physically moves while connected to the network. The opposite of a stationary user.

Modulation: The process of changing a wave by changing its amplitude, frequency, and/or phase such that the changes represent data bits.

Modulation and Coding Scheme (MCS): Term used to describe the combination of the radio modulation scheme and the coding scheme used when transmitting data, first introduced in 802.11n.

MPDU: A MAC protocol data unit (MPDU) is a portion of data to be delivered to a MAC Layer peer on a network and it is data prepared for the PHY layer by the MAC sublayer. The MAC sublayer receives the MSDU from upper layers on transmission and creates the MPDU. It receives the MPDU from the lower layer on receiving instantiation and removes the MAC header and footer to create the MSDU for the upper layers.

MSDU: A MAC service data unit is a portion of transmitted data to be handled by the MAC sublayer that has yet to be encapsulated into a MAC Layer frame.

Maximum Transmission Unit (MTU): The largest amount of data that can be sent at a particular layer of the OSI model. Typically set at layer 4 for TCP.

Multi-User MIMO (MU-MIMO): An enhancement to MIMO that allows the AP STA to transmit to multiple client STAs simultaneously.

Multipath: The phenomenon that occurs when multiple copies of the same signal reach a receiver based on RF behaviors in the environment.

Multiple Channel Architecture (MCA): A wireless network design using multiple channels strategically designed so that the implemented BSSs have minimal interference with one another.

Multiple Input/Multiple Output (MIMO): A technology used to spread a stream of data bits across multiple radio chains using spatial multiplexing at the transmitter and to recombine these streams at the receiver.

Narrowband Interference: Interference that covers a very narrow band of frequencies and typically not the full with of an 802.11 channel when used in reference to WLAN interferers.

Near-Far: A problem that occurs when a high powered device is closer to the AP in a BSS and a low powered device is farther from the AP. Most near-far problems are addressed with standard CSMA/CA operations in 802.11 networks.

Network Allocation Vector (NAV): The NAV is a virtual carrier sense mechanism used in CSMA/CA to avoid collisions and is a timer set based on the duration values in frames transmitted on the medium.

Network Segmentation: The process used to separate a larger network into smaller networks often utilizing Layer 3 routers or multi-layer switches.

Noise: RF energy in the environment that is not part of the intentional signal of your WLAN.

Noise Floor: The amount of noise that is consistently present in the environment, which is typically measured in dBm.

Network Time Protocol (NTP): A protocol used to synchronize clocks in devices using centralized time servers.

Octet: A group of eight ones and zeros. An 8-but byte. Sometimes simply called a byte.

Orthogonal Frequency Division Multiplexing (OFDM): A modulation technique and a named physical layer in 802.11 that provides data rates up to 54 Mbps and operates in the 5 GHz band. The modulation is used in all bands, but the named PHY operates only in the 5 GHz band.

Omni-Directional Antenna: An antenna that propagates in all directions horizontally. Creates a coverage area similar to a donut shape (toroidal). Also known as a dipole antenna.

Dipole Antenna: An antenna that propagates in all directions horizontally. Creates a coverage area similar to a donut (toroidal) shape. Also known as a omni-directional antenna.

Open System Authentication: A simple frame exchange, providing no real authentication, used to move through the state machine in relation to the connection between two 802.11 STAs.

Opportunistic Key Caching (OKC): A roaming solution for WLANs wherein the keys derived from the 802.1X/EAP authentication are cached on the AP or controller such that only the 4-way handshake is required at the time of roaming.

OSI (Open Systems Interconnection) Model: A theoretical model for communication systems that works by separating the communications process into seven, well-defined layers. The seven layers are Application, Presentation, Session, Transport, Network, Data Link and Physical.

Packet: Data as represented at the network layer (Layer 4) for TCP communications.

Passive Gain: An increase in strength of a signal by focusing the signal's energy rather than increasing the actual energy available, such as with an amplifier.

Passive scanning: A scanning (network location) method wherein a STA waits to receive beacon frames from an AP which contain information about the WLAN.

Passive survey: A survey conducted on location that gathers information about RF interference, signal strength and coverage areas by monitoring RF activity without active communications.

Passphrase Authentication: A type of access control that uses a phrase as the pass key. Also called personal in WPA and WPA2.

Phase: A measurement of the variance in arrival state between to copies of a wave form. Waves are said to be in phase or out of phase by some degree. The phase can be manipulated for modulation.

PHY: A shorthand notation for physical layer which is the physical means of communication on a network to transmit bits.

Physical (PHY) Layer: The physical (PHY) layer refers to the physical means by which a message is communicated. Layer one of the OSI model.

PLCP: Physical Layer Convergence Protocol (PLCP) is the name of the service within the PHY that receives data from the upper layers and sends data to the upper layers. It is the interaction point with the MAC sublayer.

PMD: Physical Medium Dependent (PMD) is the service within the PHY responsible for sending and receiving bits on the RF medium.

PMK Caching: Stores the PMK so a device only has to perform the 4 way handshake when connecting to an AP to which it has already connected.

Pairwise master Key (PMK): The key derived from the MSK, which is generated during 802.1X/EAP authentication. Used to derive the PTK. Used in unidirectional communications with a single peer.

PoE Injector: Any device that adds Power over Ethernet (PoE) to ethernet cables. Come in two variants, endpoint (such as switches) and midspan (such as inline injectors).

Point-to-Multipoint (PtMP): A connection between a single point and multiple other points for wireless bridging or WLAN access.

Point-to-Point (PtP): A connection between two points often used to connect two networks via bridging.

Polarization: The technical term used to reference the orientation of antennas related to the electric field in the electromagnetic wave.

Power over Ethernet (PoE): A method of providing power to certain hardware devices that can be powered across the Ethernet cables. Specified in 802.3 as a standard. Various classes are defined based on power requirements.

PPDU: PLCP Protocol Data Unit (PPDU) is the prepared bits for transmission on the wired or wireless medium. Sometimes also called a PHY Layer frame.

Preauthentication: Authenticating with an AP to which the STA is not intending to immediately connect so that roaming delays are reduced.

Pre-shared Key (PSK): Refers to any security protocol that uses a password or passphrase or string as the key from which encryption materials are derived.

Primary Channel: When implementing channels wider than 20 MHz in 802.11n and 802.11ac, the 20 MHz channel on which management and control frames are sent and the channel used by STAs not supporting the wider channel.

Probe Request: A type of frame sent when a client device wants information about APs in the area or is seeking a specific SSID to which it desires to connect.

Probe Response: A type frame sent in response to a probe request that contains information about the AP and the requirements of BSSs it provides.

Protected Management Frame (PMF): Frames used for managing a wireless network that are protected from spoofing using encryption. Protocol defined in the 802.11w amendment.

Protocol Analyzer: Hardware or software used to capture and analyze networking communications. WLAN protocol analyzers have the ability to capture 802.11 frames from the RF medium and decode them for display and analysis.

Protocol Decodes: The way information in captured packets or frames is interpreted for display and analysis.

PSDU: PLCP Service Data Unit (PSDU) is the name for the contents that are contained within the PPDU, the PLCP Protocol Data Unit. It is the same as the MPDU as perceived and received by the PHY.

PTK (Pairwise Transient Key): A key derived during the 4-way handshake and used for encryption only between two specific endpoints, such as an AP and a single client.

Quality of Service (QoS): Traffic prioritization and other techniques used to improve the end-user experience. IEEE 802.11e includes QoS protocols for wireless networks based on access categories.

QoS BSS: A BSS supporting 802.11e QoS features.

Radio Chains: A reference to the radio and antenna used together to transmit in a given frequency range. Multi-stream devices have multiple radio chains as one radio chain is required for each stream.

Radio Frequency (RF): The electromagnetic wave frequency range used in WLANs and many other wireless communication systems.

Radio Resource Management (RRM): Automatic management of various RF characteristics like channel selection and output power. Known by different terms among the many WLAN vendors, but referencing the same basic capabilities.

RADIUS: Remote Authentication Dial-In User Service (RADIUS) refers to a network protocol that handles AAA management which allows for authentication, authorization and accounting (auditing). Used in 802.11 WLANs as the authentication server in an 802.1X/EAP implementation.

RC4 (Rivest Cipher 4): An encryption cipher used in WEP and with TKIP. A stream cipher.

Real-Time Location Service (RTLS): A function provided by many WLAN infrastructure and overlay solutions allowing for device location based on triangulation and other algorithms.

Reassociation: The process used to associate with another AP in the same ESS. May also be used when a STA desires to reconnect to an AP to which it was formerly connected.

Received Channel Power Indicator (RCPI): Introduced in 802.11k, a power measurement calculated as INT((dBm + 110) * 2). Expected accuracy is +/- 5 dB. Ranges from 0-220 are available with 0 equaling or less than -110 dBm and 220 equaling or greater than 0 dBm. The value is calculated as an average of all received chains during the reception of the data portion of the transmission. All PHYs support RCPI and, though 802.11ac does not explicitly list its formulation, it references the 802.11n specification for calculation procedures.

Received Signal Strength Indicator (RSSI): A relative measure of signal strength for a wireless network. The method to measure RSSI is not standardized though it is constrained to a limited number of values in the 802.11 standard. Many use the term RSSI to reference dBm, and the 802.11 standard uses terms like DataFrameRSSI and BeaconRSSI and defines them as the signal strength in dBm of the specified frames, so the common vernacular is understandable. However, according to the standard, "absolute accuracy of the RSSI reading is not specified" (802.11-2012, Clause 14.3.3.3).

Reflection: An RF behavior that occurs when a wave meets a reflective obstacle large than the wavelength similar to light waves in a mirror.

Refraction: An RF behavior that occurs as an RF wave passes through material causing a bending of the wave and possible redirection of the wave front.

Regulatory Domain: A reference to geographic regions management by organizations like the FCC and ETSI that determine the allowed frequencies, output power levels and systems to be used in RF communications.

Remote AP: An AP designed to be implemented at a remote location and managed across a WAN link using special protocols.

Resolution Bandwidth (RBW): The smallest frequency that can be extracted from a received signal by a spectrum analyzer or the configuration of that frequency. Many spectrum analyzers allow for the adjustment of the RBW within the supported range of the analyzer.

Retry: That which occurs when a frame fails to be delivered successfully. A bit set in the frame to specify that it is a repeated attempt at delivery.

Return Loss: A measure of how much power is lost in delivery from a transmission line to an antenna.

RF Cables: A cable, typically coaxial, that allows for the transmission of electromagnetic waves between a transceiver and an antenna.

RF Calculator: A software application used to perform calculations related to RF signal strength values.

RF Connector: A component used to connect RF cables, antennas and transmitters. RF connectors come in many standardized forms and should match in type and resistance.

RF Coverage: Synonymous with coverage in WLAN vernacular. Reference to the BSA provided by an AP.

RF Link: An established connection between two radios.

RF Line of Sight (LoS): The existence of a path, possibly including reflections, refractions and pass-through of materials, between two RF transceivers.

RF Propagation: The process by which RF waves move throughout an area including reflection, refraction, scattering, diffraction, absorption and free space path loss.

RF Signal Splitter: An RF component that splits the RF signal with a single input and multiple outputs. Historically used with some antenna arrays, but less common today in WLAN implementations.

RF Site Survey: The process of physically measuring the RF signals within an area to determine resulting RF behavior and signal strength. Often performed as a validation procedure after implementation based on a predictive model.

Roaming: That which occurs when a wireless STA moves from one AP to another either because of end user mobility or changes in the RF coverage.

Robust Security Network (RSN): A network that supports CCMP/AES or WPA2 and optionally TKIP/RC4 or WPA. To be an RSN, the network must support only RSN Associations (RSNAs), which are only those associations that use the 4-way handshake. WEP is not supported in an RSN.

Robust Security Network Association (RSNA): An association between a client STA and an AP that was established through authentication resulting in a 4-way handshake to derive unicast keys and transfer group keys. WEP is not supported in an RSNA.

Rogue Access Point: An access point that is connected to a network without permission from a network administrator or other official.

Rogue Containment: Procedures used to prevent clients from associating with a rogue AP or to prevent the rogue AP from communicating with the wired network.

Rogue Detection: Procedures used to identify rogue devices. May include simple identification of unclassified APs or algorithmic processes that identify likely rogues.

Role-Based Access Control (RBAC): An authorization system that assigns permissions and rights based on user roles. Similar to group management of authorization policies.

RSN Information Element: A portion of the beacon frame that specifies the security used on the WLAN.

Request to Send/Clear to Send (RTS/CTS): A frame exchange used to clear the channel before transmitting a frame in order to assist in the reduction of collisions on the medium. Also used as a backward compatible protection mechanism.

RTS Threshold: The minimum size of a frame required to use RTS/CTS exchanges before transmission of the frame.

S-APSD: See *Automatic Power Save Delivery*.

Scattering: An RF behavior that occurs when an RF wave encounters reflective obstacles that are smaller than the wavelength. The result is multiple reflections or scattering of the wave front.

Secondary Channel: When implementing channels wider than 20 MHz in 802.11n and 802.11ac, the second channel used to form a 40 MHz channel for data frame transmissions to and from supporting client STAs.

Semi-Directional Antenna: An antenna such as a yagi or a patch that has a propagation pattern which maximizes gain in a given direction rather than an omni-directional pattern, having a larger beamwidth than highly directional antennas.

Service Set Identifier (SSID): The BSS and ESS name used to identify WLAN. Conventionally made to be readable by humans. Maximum of 32 bytes long.

Signal Strength: A measure of the amount of RF energy being received by a radio. Often specified as the RSSI, but referenced in dBm, which is not the proper definition of RSSI from the 802.11 standard.

Single Channel Architecture (SCA): A WLAN architecture that places all APs on the same channel and uses a centralized controller to determine when each AP can transmit a frame. No control of client transmissions to the network is provided.

Single Input Single Output (SISO): A radio transmitter that supports one radio chain and can send and receive only a single stream of bits.

Signal to Noise Ratio (SNR): A comparison between the received signal strength and the noise floor. Typically presented in dB. For example, given a noise floor of -95 dBm and a signal strength of -70 dBm, the SNR is 25 dB.

Space-Time Block Coding (STBC): The use of multiple streams of the same data across multiple radio chains to improve reliability of data transfer through redundancy.

Spatial Multiplexing (SM): Used with MIMO technology to send multiple spatial streams of data across the channel using multiple radio chains (radios coupled with antennas).

Spatial Multiplexing Power Save (SMPS): A power saving feature from 802.11n that allows a station to use only one radio (or spatial stream).

Spatial Streams: The partitioning of a stream of data bits into multiple streams transmitted simultaneously by multiple radio chains in an AP or client STA.

Spectrum Analysis: The inspection of raw RF energy to determine activity in an area on monitored frequencies. Useful in troubleshooting and design planning.

Spectrum Analyzer: A hardware and software solution that allows the inspection of raw RF energy.

Station (STA): Any device that can use IEEE 802.11 protocol. Includes both APs and clients.

Supplicant: In 802.1X, the device attempting to be authenticated. Also the term used for the client software on a device that is capable of connecting to a WLAN.

Sweep Cycle: The time it takes a spectrum analyzer to sweep across the frequencies monitored. Often a factor of the number of frequencies scanned and the RBW.

System Operating Margin (SOM): The actual positive difference in the required link budget for a bridge link to operate properly and the received signal strength in the link.

Temporal Key Integrity Protocol (TKIP): The authentication and key management protocol supported by WPA systems and implemented as an interim solution between WEP and CCMP.

Transition Security Network (TSN): A network that allows WEP connections during the transition period over to more secure protocols and an eventual RSN. An RSN does not allow WEP connections.

Transmit Beamforming (TxBF): The use of multiple antennas to transmit a signal strategically with varying phases so that the communication arrives at the receiver such that the signal strength is increased.

Transmit Power Control (TPC): A process implemented in WLAN devices allowing for the output power to be adjusted according to local regulations or by an automated management system.

Uncontrolled Port: In an 802.1X authentication system, the virtual port that allows only authentication frames/packets through to the network and, when authentication is successfully completed, provides the 802.1X service with the needed information to open the controlled port.

User Priority (UP): A value (from 0-7) assigned to prioritize traffic that correspond to different access categories for WMM QoS.

Virtual Carrier Sense: The 802.11 standard currently defines the Network Allocation Vector (NAV) for use in virtual carrier sensing. The NAV is set based on the duration value in perceived frames within the channel.

Voltage Standing Wave Ratio (VSWR): The Voltage Standing Wave Ratio is the ratio between the voltage at the maximum and minimum points of a sanding wave.

Watt: A unit of power. Strictly defined as the energy consumption rate of one joule per second such that 1 W is equal to 1 joule per 1 second.

Wavelength: The distance between two repeating points on a wave. Wavelength is a factor of the frequency and the constant of the speed of light.

Wired Equivalent Privacy (WEP): A legacy method of security defined in the original IEEE 802.11 standard in 1997. Used the RC4 cipher like TKIP (WPA), but implemented it poorly. WEP is deprecated and should no longer be used.

Wi-Fi Alliance: An association that certifies WLAN equipment to interoperate based on selected portions of the 802.11 standard and other standards. Certifications include those based on each PHY as well as QoS and security.

Wi-Fi Multimedia (WMM): A QoS certification created and tested by the Wi-Fi Alliance using traffic prioritizing methods defined in the IEEE 802.11e.

Wi-Fi Multimedia Power Save (WMM-PS): A power saving certification designed by the Wi-Fi Alliance and optimized for mobile devices and implementing methods designated in the IEEE 802.11e amendment.

Wireless Intrusion Prevention System (WIPS): A system used to detect and prevent unwanted intrusions in a WLAN by detecting and preventing rogue APs and other WLAN threats.

Wireless Local Area Network (WLAN): A local area network that connects devices using wireless signals based on the 802.11 protocol rather than wires and the common 802.3 protocol.

WPA-Enterprise: A security protocol designed by the Wi-Fi Alliance. Requires an 802.1X authentication server. Uses the TKIP encryption protocol with the RC4 cipher. Implements a portion of 802.11i and the older, no deprecated TKIP/RC4 solution.

WPA-Personal: A security protocol designed by the Wi-Fi Alliance. Does not require an authentication server. Uses the TKIP encryption protocol with the RC4 cipher. Also known as WPA-PSK (Pre-Shared Key).

WPA2-Enterprise: A security protocol designed by the Wi-Fi Alliance. Requires an 802.1X authentication server. Uses the CCMP key management protocol with the AES cipher. Also known as WPA2-802.1X. Implements the non-deprecated portion of 802.11i.

WPA2-Personal: A security protocol designed by the Wi-Fi Alliance. Does not require an authentication server. Uses the CCMP key management protocol with the AES cipher. Also known as WPA2-PSK (Pre-Shared Key).

Wi-Fi Protected Setup (WPS): A standard designed by the Wi-Fi Alliance to secure a network without requiring much user knowledge. Users connect either by entering a PIN associated with the device or by Push-Button which allows users to connect when a real or virtual button is pushed.

CPSIA information can be obtained
at www.ICGtesting.com
Printed in the USA
BVHW020437120121
597606BV00001B/2